Michael M. Abbott
20 October 1987

Benchmark Papers on Energy

Series Editors:
R. Bruce Lindsay, Brown University
Mones E. Hawley, Professional Services International

Benchmark Papers on Energy/7

A BENCHMARK ® Books Series

EARLY CONCEPTS OF ENERGY IN ATOMIC PHYSICS

Edited by

ROBERT LINDSAY

Trinity College

Dowden, Hutchinson & Ross, Inc.

STROUDSBURG, PENNSYLVANIA

Benchmark Papers on Energy, Volume 7
Library of Congress Catalog Card Number: 78-23311
ISBN: 0-87933-340-5

81 80 79 1 2 3 4 5
Manufactured in the United States of America.

LIBRARY OF CONGRESS CATALOGING IN PUBLICATION DATA
Main entry under title:
Early concepts of energy in atomic physics.
(Benchmark papers on energy ; v. 7)
Includes indexes.
1. Atomic theory—History. 2. Force and energy —History. I. Lindsay, Robert, 1924–
QC173.E524 530.1 78-23311
ISBN 0-87933-340-5

Distributed world wide by Academic Press,
a subsidiary of Harcourt Brace Jovanovich,
Publishers.

SERIES EDITOR'S FOREWORD

The Benchmark Papers on Energy constitute a series of volumes that makes available to the reader in carefully organized form important and seminal articles on the concept of energy, including its historical development, its applications in all fields of science and technology, and its role in civilization in general. This concept is generally admitted to be the most far-reaching idea that the human mind has developed to date, and its fundamental significance for human life and society is everywhere evident.

One group of volumes in the series contains papers bearing primarily on the evolution of the energy concept and its current applications in the various branches of science. Another group of volumes concentrates on the technological and industrial applications of the concept and its socio-economic implications.

Each volume has been organized and edited by an authority in the area to which it pertains and offers the editor's careful selection of the appropriate seminal papers, that is, those articles which have significantly influenced further development of that phase of the whole subject. In this way every aspect of the concept of energy is placed in proper perspective, and each volume represents an introduction and guide to further work.

Each volume includes an editorial introduction by the volume editor, summarizing the significance of the field being covered. Every article or group of articles is accompanied by editorial commentary, with explanatory notes where necessary. Both an author index and subject index are provided for ready reference. Articles in languages other than English are either translated or summarized in English. It is the hope of the publisher and editor that these volumes will serve as a working library of the most important scientific, technological, and social literature connected with the idea of energy.

The present volume, *Early Concepts of Energy in Atomic Physics*, has been prepared by Professor Robert Lindsay of the Department of Physics of Trinity College, Hartford, Connecticut. In 33 carefully selected papers the editor portrays the development of the atomic theory in physics with reference to the incorporation of the concept of energy. Though a few very early papers are included, main emphasis is laid on the work of the nineteenth and early twentieth centuries after the concept of energy had become well established in macroscopic physics. Ample editorial commentaries discuss the significance of each group of papers. The scope of the volume carries the story up to the middle 1920s and does not include quantum mechanics and modern nuclear physics, which will be the subject of a subsequent volume.

R. BRUCE LINDSAY

PREFACE

The present volume traces the development of the atomic idea in physics as expressed in published papers and seeks to show specifically the stages at which the concept of energy became more and more integral to the formulation of a precise description of atomic scale phenomena. The volume contains 33 papers, beginning with Greek atomism and reviewing briefly seventeenth- and eighteenth-century contributions. The major emphasis is on articles from the nineteenth and early twentieth centuries. No attempt has been made to cover material after 1925, and hence there is no reference to the role of energy in quantum mechanics and modern nuclear physics.

The volume editor is deeply indebted to the staffs of the Sciences Library of Brown University (and especially Patricia Galkowski and Caroline Helie) and the Library of Trinity College (including the Watkinson Collection) for their efficient and helpful aid in the location of source material. The opportunity to use the South Kensington Science Library in London, England, in 1974 has also been much appreciated.

The volume editor expresses his gratitude to the series editor for his helpful advice and for his translation of numerous articles.

Finally, both volume and series editors are deeply indebted to Denise Perreault and Nancy Smelzer of the Department of Physics in Brown University for their typing of translated material and editorial commentaries.

ROBERT LINDSAY

CONTENTS

PART IX: BROWNIAN MOTION AND MOLECULAR REALITY

PART X: ENERGY AND THE PROPERTIES OF SOLIDS

CONTENTS BY AUTHOR

INTRODUCTION

This collection of papers and associated commentary traces the development of the concept of energy as it evolved into an integral and important part of atomic physics. Although the Benchmark volume *Energy: Early Historical Development of the Concept* (edited by R. Bruce Lindsay, Dowden, Hutchinson and Ross, Stroudsburg, Pennsylvania, 1975) emphasizes that the basic idea of energy has roots going back at least as far as the introduction of the atomic concept, it was not until comparatively recent times (i.e., in the middle of the nineteenth century) that serious and successful efforts were made to connect the two concepts. The main barriers to the development of this connection were the long continued adherence to the caloric or substantial theory of heat and a lack of precise experimental observations that clearly demonstrated the existence of atomic behavior.

The following brief sketch of the evolution of the atomic point of view, which stresses the places in its development where the energy concept began to play a role, will provide a framework for the articles that follow.

THE EARLY ATOMISTS

The initiation of an atomic theory of matter (at any rate, in Europe) is commonly attributed to the Greek philosopher Democritus (465–400 B.C.), though his views are believed to have been strongly influenced by another pre-Socratic philosopher Leucippus (fl. ca 400 B.C.). The atomic notion set forth by them was characterized by the following features.

1. The universe is composed of *atoms* and *void,* with the atoms being infinite in number and the void infinite in extent.
2. All atoms are very small solids, composed of the same uniform substance but differing from each other in size and shape. They are incapable of being either created or destroyed or experiencing internal change or subdivision. (The Greek word *atomos* means "that which cannot be *cut*.")
3. Atoms can be arranged and combined in many different ways, thereby causing the differences in appearance of the familiar objects of experience.
4. The atoms in the void are in a state of ceaseless motion, though this motion is not apparent to the eye of the observer.

Two fundamental ideas are thus involved in the atomic theory of Democritus and Leucippus. The first is that of an immutable, indestructible unit as the fundamental constituent of all matter. The second is that of ceaseless motion, an interesting hypothesis in light of the theories on heat and energy that followed. We may also think of the notion of immutability as containing some germ of the idea of conservation, so significant for the future concept of energy.

The atomic concept did not fare well at the hands of Aristotle 384–322 B.C.). He by no means took kindly to atoms, partly because he could not believe in the existence of a void for them to move in and partly because he felt that our only really believable conception of the nature of a substance is acquired by direct observation through the senses, an argument against atomism also used, though much later, by the Austrian scientist Ernst Mach (1838–1916). Aristotle's influence in the two millenia after his death is usually viewed as one of the chief deterrents to the acceptance of the atomic point of view. Of course, the concept was by no means completely neglected even in antiquity. An example is provided by the Roman poet Titus Lucretius Carus (ca. 99–54 B.C.). Lucretius, as he is commonly known, drew his interpretation of the atomic nature of matter from the materialistic philosopher Epicurus (347–271 B.C.), who found the atomism of Democritus congenial to his assurance that no life persists after death. Lucretius enshrined his conception of atomism in elegant fashion in his famous poem "De Rerum Natura" (On the Nature of Things). Paper 1 reproduces several extracts from his poem in English.

Lucretius' poem undoubtedly helped to keep the atomic idea alive during the early years of the Christian Era and paved the way for the revival of atomism in the Renaissance. However, during the first 1500 years of the Christian Era, relatively little decisive conjecture about atomism has come to the attention of science historians. Of course, the idea never died out completely, and a number of medieval philosophers, among them the Arab Rhazes (ca. A.D. 925) and Giles of Rome (ca. A.D. 1300), used it in some form to develop their own ideas about the constitution of matter. For reference to this period see the work of A. C. Crombie, Volume II, pp. 33 ff. in the general bibliography for this volume. In the scientific revolution of the sixteenth and seventeenth centuries, the attention of those scientists, whose work has long been recognized as decisive in the advancement of science, was focused mainly on the behavior of large-scale objects, as well as on the structure of the universe and the position of the earth therein. The Copernican model of the solar system was followed by Kepler's laws and the formulation of the fundamental theory of classical mechanics by Galileo (1564–1642) and Newton (1642–1727). The great scientific revolution was really the beginning of a new scientific methodology in which observation and theory worked in tandem to describe and explain all mechanical phenomena in nature. The sophisticated concept of energy as we use it today had its origin in the idea that certain physical quantities remain constant in the midst of change, a viewpoint that first attracted the attention of Simon Stevin (1548–1620) and Galileo. The eighteenth century saw the ardent debate between the adherents of René Descartes (1596–1650) and Gottfried Wilhem Leibniz (1646–1716) over the differing views of these savants concerning the true nature of force or what we should now call "energy" and of what God intended to be conserved in nature. Descartes felt that momentum (mass times velocity) is the divinely ordained property whereas Leibniz was equally convinced that *vis viva* (mass times the square of the velocity) plays the predominant role. Leibniz's view signalled the emergence of energy as a quantifiable quantity since *vis viva* is twice what in time came to be called *kinetic energy.* As a matter of fact, both momentum and *vis viva* are conserved under certain circumstances, but it has since turned out that the energy concept can be extended to a host of natural circumstances on which momentum casts little or no light. This fact was made clear by the epoch-making contribution to mechanics of the great applied mathematicians Lagrange and Laplace in the late eighteenth and early nineteenth centuries and the equally important contributions of J. P. Joule (1818–1889), J. R. Mayer (1814–1878), and H. von Helmholtz (1821–1894) in the mid-nineteenth century. The overriding importance of energy as a physical concept could not be brought to bear on the atomic hypothesis until the work of Joule, Mayer, and von Helmholtz had been consolidated.

The first fresh insights that eventually led to a successful particulate and dynamic description of the nature of matter emerged from the experimental studies of the behavior of air. Galileo, E Torricelli (1608–1647), B. Pascal (1623–1662), and O. von Guericke (1602–1686) showed by observation that the earth's atmosphere is a fluid capable of exerting pressure. Robert Boyle (1627–1691) conducted an extensive series of experiments on the expansion and compression of air, and his results led to the famous law associated with his name. Actually, the commonly accepted statement of the law was deduced by Richard Towneley (1628–1707), a contemporary of Boyle, from an examination of Boyle's data. Boyle advanced two hypotheses to explain the apparent elastic behavior of air exemplified by his results. The first hypothesis postulated that air is composed of many tiny hair-like springs that become coiled up under compression. The second proposed that air is made up of many small particles of various sizes that execute whirling vortex motions and exert repulsive forces on each other when compressed.

A further refinement of Boyle's second model of a gas was introduced by Robert Hooke (1635–1703). In his work *De Potentia Restitutiva* (1678), Hooke outlined the first kinetic theory of matter with analytical possibilities by associating a vibratory motion with each particle of a substance, the vibrational frequency being supposed to be dependent on the size of the particle. These vibrations were thought to be due to impulses received from particle collisions. On this view the pressure of a gas in a confined vessel is a consequence of the impact of the vibrating particles on the walls. This kinetic behavior of the gas particles was different from the essentially static behavior of atoms in a gas that Newton appears to have accepted. In his *Principia* (1687), he dealt briefly with a gas composed of particles at rest subject to repulsive forces varying as the first power of the distance of separation. From this postulate it is possible to derive Boyle's law. However, this derivation implies that the pressure depends on the shape of the containing vessel, which is contrary to fact, though this was not well established in Newton's time. It is possible that Newton's prestigious position effectively denigrated Hooke's idea and prevented its exploitation in the latter part of the seventeenth century. By that time, indeed, the value of the atomic concept in physics was admitted by many active scientists. The trouble came with the detailed application. The fact that Newton's static model of a gas as well as Hooke's dynamic model were both able to account for the existence of Boyle's law obviously did not tend to arouse particular interest in Hooke's hypothesis. What was lacking was more precise experimental data on the behavior of fluids and a more sophisticated mathematical approach through which the laws of mechanics could be applied to a system containing a very large number of particles. A further complicating factor was the solid entrenchment of the caloric theory of heat. The fixation on this point of view for such a long time did much to discourage any approach to the behavior of a gas which sought to associate atomic motion with heat and temperature.

The Swiss Daniel Bernoulli (1700–1792) was the first person to follow the essence of Hooke's atomism in a sophisticated attempt to provide a genuine kinetic theory of gases. In his famous treatise of 1738 on hydrodynamics, he applied the principles of mechanics to the particles of a gas confined in a vessel and was able to derive Boyle's law. It does not appear that this result attracted much attention. So far as can be determined from the easily available scientific literature, it was ignored for about three quarters of a century. The essential method was revived by the English amateur scientist John Herapath (1790–1868) in 1814. Apparently without any knowledge of Bernoulli's work, he derived the mathematical relation connecting the product of pressure and volume of a gas with the product of the mean square velocity of the constituent molecules and the molecular mass. This, of course, at once implied a definite connection with the *vis viva* or kinetic energy, but Herepath either did not notice or he ignored this association. In fact, somewhat curiously, in his theory he laid principal emphasis on the concept of momentum and its conservation, probably influenced in this respect by the work of Descartes. A more elaborate theory of the behavior of gases based on the molecular point of view was proposed in 1845 by the Scottish engineer John James Waterston (1811–1883). His fundamental paper on the subject was refused publication in the *Philosophical Transactions* of the Royal Society of London, and though he wrote further on the subject and in particular on its connection with the theory of sound propagation, he received little or no recognition until Lord Rayleigh's influence brought about the posthumous publication of his original paper in 1892). Ironically, his ideas were essentially like those published by Clausius and Maxwell in the late 1850s and early 1860s, which the scientific community accepted as valid approximations.

Though somewhat off the track of the role of energy in atomic physics, the point atom theory of the Croatian priest Roger Joseph Boscovich (1711–1787) deserves mention, and an extract from his work is included in this volume (Paper 5).

The middle of the nineteenth century saw the establishment of the concept of energy as an idea of overriding importance in natural phenomena of all kinds. This development was accomplished through the pioneer work of J. R. Mayer, J. P. Joule, and H. von Helmholtz. Their studies led rather quickly to the incorporation of the concept in a decisive way in the further elaboration of the atomic or molecular theory of matter through the investigations of Joule himself as well as those of Krönig, Clausius, Maxwell, and Boltzmann. With the exception of Boltzmann, these studies are all represented by articles in this volume. The first significant relevance of their researches to the theme of this volume was the association of the average energy of the molecule with the temperature of the gas.

As introduced by Democritus and utilized in the early kinetic theory of gases, the atom or molecule was conceived of as the fundamental irreducible unit of matter. By the beginning of the twentieth century, it was clear that the atom has a structure of its own, and increasing interest was directed to the problem of describing the energy of its constituent parts. This interest is represented in this volume by the reprinting of important pioneer papers by J. J. Thomson, N. Bohr, J. Franck, and G. Hertz, A. H. Compton, and A. Einstein. These papers have been selected because of the significant role that energy plays in the results set forth in each.

One of the greatest achievements of the past hundred years in physics was what may appropriately be called "the atomization of radiant energy" through the agency of the quantum theory of Max Planck. It is altogether proper that this great achievement should find a place in a review of papers on the role of energy in atomic physics. Planck's theory opened up a vast realm for further investigation in which the concept of energy became paramount. We point particularly to the theory of the photoelectric effect that Einstein devised as well as the papers on atomic structure cited previously.

No collection of papers on energy in atomic physics would be complete without a reference to Einstein's famous mass-energy relation based on the special theory of relativity.

Radioactive emission as indicative of energy conversions at the nuclear level is represented by the fundamental papers of Becquerel, Curie and Laborde, and Rutherford and Barnes. Brownian motion, as further convincing proof of molecular reality and its relation to molecular energy, starts with Robert Brown's original paper and was subsequently extended by the fundamental theoretical papers of Langevin and Einstein.

Energy flow involved in the motion of electrons in metals is represented by P. Drude's pioneer paper of 1900. Major contributions to the theories of the specific heat and magnetic behavior of solids, which invoke the distribution of molecules over internal energy states, are represented by the papers of J. P. Debye and P. Langevin, respectively.

The commentary on the individual groups of papers will provide more details about individual entries and in each case will seek to emphasize the key role of energy in atomic physics.

As has already been made clear in the preface, the scope of this volume stops with the introduction of wave and quantum mechanics in 1925 and does not discuss the development in the last half-century of nuclear atomic and solid-state physics and their various ramifications.

BIBLIOGRAPHY

Bernoulli, D., and J. Bernoulli, 1968. *Hydrodynamics and Hydraulics.* New York: Dover.

Black, J., 1807. *Lectures on the Elements of Chemistry, Given at the University of Edinburgh, 1766–1797.* Philadelphia: Matthew Carey.

Boorse, H. A., and Motz, eds., 1966. *The World of the Atom,* 2 vols. New York: Basic Books.

Born, M., and K. Huang, 1954. *Dynamical Theory of Crystal Lattices.* London: Oxford.

Boscovich, R. J., 1966. *A Theory of Natural Philosophy.* Cambridge, Massachusetts: MIT Press.

Broglie, Louis de, 1953. *The Revolution in Physics.* New York: The Noonday Press.

Brush, S. G., 1965. *Kinetic Theory, Vol. 1, The Nature of Gases and of Heat.* Oxford: Pergamen Press.

Centore, F. W., 1970. *Robert Hooke's Contributions to Mechanics.* Hague: Martinus Nijhoff.

Chalmers, T. W., 1949. *Historic Researches.* London: Morgan Brothers.

Crombie, A. C., 1959. *Medieval and Early Modern Science,* Vol. I and Vol. II. New York: Doubleday/Anchor Books.

Dalton, J., 1912. *New System of Chemical Philosophy.* London: William Dawsons and Sons.

Dampier, W. C., and M. Dampier, 1959. *Readings in the Literature of Science.* New York: Harper and Row (Torch books).

Daub, E. E., 1971. "Waterston's Influence on Krönig Kinetic Theory of Gases," *ISIS,* **62,** 512.

Davy, H., 1812. *Elements of Chemical Philosophy.* London: J. Johnson.

Debus, Allen G., ed., 1968. *World Who's Who in Science.* Chicago: A. N. Marquis.

Farrington, B., 1944. *Greek Science and Its Meaning to Us.* London: Penguin Books.

Gillispie, C. C., ed., 1970–1976. *Dictionary of Scientific Biography.* New York: Charles Scribner's Sons.

Haldane, J. S., ed., 1928. *The Collected Papers of John James Waterston.* Edinburgh: Oliver and Boyd.

Hoffman, Banesh, with Helen Dukas, 1971. *Albert Einstein: Creator and Rebel.* New York: Viking.

Hull, L. W. H., 1959. *History and Philosophy of Science.* London: Longmans, Green and Co.

Jammer, M., 1966. *The Conceptual Development of Quantum Mechanics.* New York: McGraw-Hill.

Kemble, E. C., 1966. *Physical Science, Its Structure and Development,* Vol. I. Cambridge, Massachusetts: MIT Press.

Lindsay, R. B., ed.: *Acoustics: Historical and Philosophical Development.* Stroudsburg, Pennsylvania: Dowden, Hutchinson & Ross, 1973. *Energy: Historical Development of the Concept.* Stroudsburg, Pennsylvania: Dowden, Hutchinson & Ross, 1975; and *Applications of Energy: Nineteenth Century.* Stroudsburg, Pennsylvania: Dowden, Hutchinson & Ross, 1976.

Magie, W. F., ed., 1935. *A Source Book in Physics.* New York: McGraw-Hill.

Koenigsberger, Leo, 1965. *Hermann von Helmholtz* (trans. F. A. Welby). New York: Dover.

Maxwell, J. C., 1965. *Scientific Papers.* New York: Dover Publications.

McCormmack, R., ed., 1969. *Historical Studies in the Physical Sciences, I.* University of Pennsylvania Press, Philadelphia, Pennsylvania, 1969 (article by Martin J. Klein, "Gibbs on Clausius," pp. 127–149).

Millar, W. J., ed., 1881. *Miscellaneous Scientific Papers of W. J. M. Rankine.* London: G. Griffith.

Pledge, H. T., 1959. *Science Since 1500.* New York: Harper and Row.

Preston, T., 1904. *Theory of Heat.* London: Macmillan.

Richtmyer, F. K., E. H. Kennard, and T. Lauritsen, 1955. *Introduction to Modern Physics,* 5th ed. New York: McGraw-Hill.
Rinard, P. M., 1976. "Quarks and Boscovich," *Am. J. Phys.,* **44,** 704–705.
Schlipp, Paul A., ed., 1959. *Albert Einstein: Philosopher-Scientist,* Vols. I and II, Harper Torchbooks/Science Library. New York: Harper and Brothers.
Slater, John C., 1960. *Quantum Theory of Matter,* Vol. I. New York: McGraw-Hill.
Ter Haar, D., 1954. *Elements of Statistical Mechanics.* New York: Rinehart.
Van Meken, A. G., 1952. *From Atomos to Atom.* Pittsburgh, Pennsylvania: Duquesne University Press.
Weisskopf, V. F., 1970. "Three Steps in the Structure of Matter," *Phys. Today,* **23,** 17–24.
Whittaker, Edmund, 1951. *A History of the Theories of Aether and Electricity, The Classical Theories.* London: Thomas Nelson.
——, 1958. *From Euclid to Eddington.* New York: Dover Publications.
Whyte, L. L., 1961. *Atomism—From Democritus to 1960.* Wesleyan University Press.

Part I

BEGINNINGS OF ATOMISM

Editor's Comments on Papers 1, 2 and 3

1 **LUCRETIUS**
Excerpts from *De Rerum Natura*

2 **BOYLE**
Excerpts from *New Experiments Physico-Mechanical—Touching the Spring of the Air*

3 **HOOKE**
Excerpt from *Lectures de Potentia Restitutiva, or of Spring: Explaining the Power of Springing Bodies*

The Introduction pointed out that the atomic idea in western civilization had its earliest roots in the Greek philosophic schools of Leucippus and Democritus. Although their ideas were for the most part rejected by contemporary and later Greek and Roman philosophers, one group did maintain a firm belief in the atomic point of view: the followers of Epicurus of Samos (341–270 B.C.). The main thrust of Epicurus' teachings was moral philosophy rather than natural science; yet he taught that in order for an individual to achieve peace of mind, he must have a scientific understanding of the nature of his surroundings. Scientific knowledge could banish superstition and fears of the unknown, thereby making for a happier existence on earth. For example, Epicurus explained violent environmental happenings like earthquakes, thunder storms, and volcanic eruptions as having natural causes and in no way signifying retributive action by the gods. Moreover, Epicurus adopted the atomic hypothesis in his explanations. In the Mediterranean and Asia Minor cultures, his philosophy found sufficient support to flourish in the face of Aristotelian orthodoxy, albeit as a minority view, well into the Christian Era. It had its disciples in Rome, and among its most articulate and eloquent was Titus Lucretius Carus (99–55 B.C.), a well-educated Roman who apparently moved in the highest social circles of that city. He was evidently well acquainted with the literature of both Greece and Rome. Short extracts from his famous but unfinished poem "De Rerum Natura" are reprinted here as Paper 1. This poem is essentially a didactic presentation of the physical picture of the world according to the Epicurean atomic theory. (For a thorough analysis of the entire poem, the reader is referred to "Lucretius, De Rerum Natura," 1975, introduction by Martin F. Smith). The excerpt produced here emphasizes some of the atomic ideas the poet weaves into his discourse. It should be observed that he uses the term "seeds" (Latin, *semen*) in several places to refer to the elementary bodies conceived of as being the constituents of all things in the natural world. It is tempting to think that Lucretius came close to anticipating the genetic code and DNA when he asserts that "because every kind is produced from fixed seeds, the source of everything that is born and comes forth into the borders of light is that in which resides the material of it and its first bodies." Further, "it is impossible that all things be born from all things." Lucretius also suggests that the observed motion of the dust motes in a sunbeam is analogous to the invisible motion of the primordial atoms. Thus he not only theorized that atoms exist and are the basic units in the structure of matter but also by ascribing motion to the atoms, he connected them with the physical property that was later to become known as "energy."

The atomism of antiquity was essentially subjective and qualitative in nature. The first major steps to provide an experimental basis for a particulate theory of matter were not taken until the seventeenth century. The basic theme of this volume is the gradual convergence of the ideas of atomism and energy and the powerful advances in knowledge that this merger made possible. From the point of view of our theme, the modern era can be said to have begun with the work of Robert Boyle (1627-1691) and Robert Hooke (1635–1702). Their investigation of the physical properties of air as well as of solid matter in which they made careful physical measurements served as the beginning of what was to become known as the "kinetic" or "molecular theory" of matter two hundred years later.

The Honorable Robert Boyle was born in Ireland, the seventh son of the first Earl of Cork. After completing his formal education at Eton, he spent several years on the continent of Europe. There he was introduced to and became interested in the post-Renaissance revival of science including the work of Galileo. Most of his detailed knowledge of science was derived during a period of intensive independent study while residing on a family estate in Dorset after his return. By 1654 his abilities and knowledge had become recognized, and he took up residence at Oxford at the invitation of John Wilkins, warden of Wadham College. Boyle's interests ranged from the emerging experimental sciences of physics and chemistry to philosophy and theology, and he wrote extensively in all these fields. However, the work for which he is most remembered in physics was the investigation into the mechanical properties of air. He carried out his research during the first years of his stay at Oxford. In his experiments Boyle was assisted by Hooke, who constructed the air pump that made possible the variation in pressure in the vessel within which Boyle studied the relation between pressure and volume. Paper 2 consists of extracts from his publication, "New Experiments Physico-Mechanicall, Touching the Spring of Air and its Effects," which appeared originally in 1660 and in modified form in 1662. It should be noted that Boyle himself did not claim priority for the actual inference of the law that the product of the air's pressure and volume is a constant at a constant temperature. He gave credit for this insight to Richard Towneley, a landowner in Lancashire, who was an amateur scientist and a correspondent of Boyle. Towneley drew attention to the now-famous relationship when he was shown Boyle's data.

Boyle also made theoretical speculations about the nature of air in order to account for the behavior he had observed. Though never an atomist in the Epicurean sense, he was influenced by Francis Bacon (1561-1626) and others to adopt the view that air and indeed all other forms of matter have a particulate constitution. He did not call the constituent particles "atoms" but instead referred to them as "corpuscles." He imagined various ways in which the corpuscles of air, through their interaction (e.g., as if they were springs) and through motions of special kinds, might account for the observed pressure. Boyle never devised a thoroughgoing kinetic or molecular theory like that of Daniel Bernoulli (1700-1782) a century later, but with his emphasis on careful experiment and what he called the "corpuscular philosophy," he provided strong support for the later developments. This corpuscular view is perhaps even more obvious in his writings on heat. He definitely ascribed the heat of bodies to the motion of the constituent corpuscles and hence gave impetus to what later proved to be the important connection between the atomic idea and energy. Boyle's ideas on this subject are presented in extracts reproduced in the Benchmark volume *Energy: Historical Development of the Concept,* edited by R. Bruce Lindsay, 1975.

During his early days in Oxford, Robert Hooke materially assisted Boyle in his experimental work. Hooke then went on to become an important scientific figure through his own experimental and theoretical investigations into which the prevalent mechanical and corpuscular ideas entered decisively. In 1662 Hooke became apparatus curator of the newly established Royal Society of London and shortly thereafter was appointed professor of geometry in Gresham College in London. Of his important contributions, many resulted from experiments performed while he was curator. Hooke's commitment to science, like Boyle's, extended beyond the mere performance of experiments to speculation on the meaning of many of the phenomena his observations revealed. His experiments with springs led to the formulation of the familiar law known by his name, expressing the linear relation between the stress and strain of an elastic material. Some of his thoughts on the nature of matter were presented in the Cutlerian lectures, delivered as part of his duties at Gresham College. Hooke may be regarded as more of an atomist than Boyle. He assumed that the atoms constituting matter execute motions of a vibratory character. His lecture "De Potentia Restitutiva" (Concerning Restoring Force), which appeared in 1678 and is reproduced in part here as Paper 3, went considerably beyond the discussion of the behavior of springs and applied the atomic hypothesis to matter spread throughout space. Even though Hooke's model did not endow atoms with unrestricted translational motion, it did represent a significant forward step in introducing a form of rectilinear motion as characteristic of the behavior of atoms. By emphasizing such motion, Hooke's speculation foreshadowed the important future role of the concept of energy in atomic physics.

The pioneer significance of the early work of Boyle and Hooke and their contemporaries becomes clearer in the next chapter when we pay attention to the beginnings of quantitative kinetic theory.

REFERENCES

Boyle, Robert, 1772. "The Nature of Heat," *Works of the Honorable Robert Boyle,* Vol. 4. London, pp. 244–251. In R. Bruce Lindsay, ed., 1975. *Energy: Historical Development of the Concept. Benchmark Papers in Energy, Vol. 1.* Stroudsburg, Pennsylvania: Dowden, Hutchinson & Ross, pp. 182–189.

"Lucretius, De Rerum Natura," with an introduction by Martin F. Smith, Harvard University Press, Cambridge, Massachusetts, 1975.

1

Reprinted from pages 15, 17, 21, 23, 49, 51, 53, 101, 103, and 105 of *De Rerum Natura,* M. F. Smith, ed., W. H. D. Rouse, trans. Cambridge Mass.: Harvard University Press, 1972, 602pp.

DE RERUM NATURA

Lucretius

[*Editor's Note:* In the original, material precedes these excerpts. Also, the Latin text has been completely omitted.]

Knowledge of Nature's laws will dispel this fear. I. Nothing can arise out of nothing.

146 This terror of mind therefore and this gloom must be dispelled, not by the sun's rays or the bright shafts of day, but by the aspect and law of nature.[e] The first principle of our study we will derive from this, that no thing is ever by divine power produced from nothing.[f] For assuredly a dread holds all mortals thus in bond, because they behold many things happening in heaven and earth whose causes they can by no means see, and they think them to be done by divine power. For which reasons, when we shall perceive that nothing can be created from nothing, then we shall at once more correctly understand from that principle what we are seeking, both

[e] 146-148 = 2.59-61, 3.91-93, 6.39-41.

[f] *Cf.* Epicurus, *Ep. ad Hdt.* 38: πρῶτον μὲν ὅτι οὐδὲν γίνεται ἐκ τοῦ μὴ ὄντος.

the source from which each thing can be made and the manner in which everything is done without the working of gods.

159 For if things came out of nothing, all kinds of things could be produced from all things, nothing would want a seed.[a] Firstly, men could arise from the sea, from the earth scaly tribes, and birds could hatch from the sky ; cattle and other farm animals and every kind of wild creature would fill desert and cultivated land alike, with no certainty as to birth.[b] Nor would trees be constant in bearing the same fruit, but they would interchange : all would be able to bear all. Seeing that there would be no bodies apt to generate each kind, how could there be a constant unchanging mother for things ? But as it is, because every kind is produced from fixed seeds, the source of everything that is born and comes forth into the borders of light is that in which is the material of it and its first bodies ; and therefore it is impossible that all things be born from all things, because in particular things resides a distinct power.

or else (1) anything could arise out of anything,

174 Besides, why do we see the rose put forth in spring, corn in the heat, grapes under persuasion of autumn, unless because each created thing discloses itself when at their own time the fixed seeds of things have streamed together, while the due seasons are present and the lively earth safely brings out things young and tender into the borders of light ? But if they came from nothing, suddenly they would arise at uncertain intervals and at unsuitable times of the year ; naturally, for there would be no first-beginnings to be restrained from generative union by the unfavourable season.

(2) things could be created at any season,

[a] Lucr. is translating Epicurus, *Ep. ad Hdt.* 38.

[b] That is, there would be no certainty that a wild animal would be born in the wilds, a farm animal on the farm.

[*Editor's Note:* Material has been omitted at this point.]

II. Nothing can be reduced to nothing, or else (1) things would be annihilated suddenly,

215 Add to this that nature resolves everything again into its elements, and does not reduce things to nothing.[a] For if anything were perishable in all its parts, each thing would then perish in a moment snatched away from our sight. For there would be no need of any force, to cause disruption of its parts and dissolve their connexions. But as it is, because the seed of all things is everlasting, nature allows no destruction of anything to be seen, until a force has met it, sufficient to shatter it with a blow, or to penetrate within through the void places and break it up.

(2) there would be no supply of material for the creation and replenishment of things,

225 Besides, if time consuming all the material utterly destroys whatever by lapse of years it removes, whence does Venus restore living creatures to the light of life each after its kind, or, when they are restored, whence does the wonder-working earth nourish them and make them grow, providing food for each after its kind? Whence is the sea supplied by the springs within it, and by the rivers without, flowing from afar? Whence does the ether nourish the stars? For all things that are of perishable body must have been consumed by infinite time and ages past. But if through that space of time past there have been bodies from which this sum of things subsists being made again, imperishable indeed must their nature be; therefore things cannot severally return to nothing.

(3) all things could be destroyed by the same force,

238 Again, the same force and cause would destroy all things without distinction, unless everlasting matter held them together entangled more or less closely in their interlacing bonds; for just a touch would be cause enough for destruction, inasmuch as there would be no particles of everlasting body, whose contexture a special force would be needed to dissolve. But as it is, since the bonds which combine the elements are different, and their matter is everlasting, things abide with body intact until a force meet them that is found vigorous enough to affect the texture of each. Therefore no single thing returns to nothing, but all by disruption return to the elements of matter.

[a] *Cf.* Epicurus, *Ep. ad Hdt.* 39.

[*Editor's Note:* Material has been omitted at this point.]

There is also Void

329 Yet everything is not held close and packed everywhere in one solid mass, for there is void in things : which knowledge will be useful to you in many matters, and will not allow you to wander in doubt and always to be at a loss as regards the universe and to distrust my words. Therefore there is intangible space, void, emptiness.[a] But if there were none, things could not in any way move ; for that which is the province of body, to prevent [b] and to obstruct, would at all times be present to all things ; therefore nothing would be able to move forward, since nothing could begin to give place. But as it is, we discern before our eyes, throughout seas and lands and the heights of heaven, many things moving in many ways and various manners, which, if there were no void, would not so much lack altogether their restless motion, as never would have been in any way produced at all, since matter would have been everywhere quiescent packed in one solid mass.

or (1) things could not move,

[*Editor's Note:* Material has been omitted at this point.]

(9) the atom consists of smallest parts inseparably cohering ;

599 Then further, since there is always an extreme point on that body which our senses are no longer able to perceive, that point undoubtedly is without parts, and is the smallest possible existence, and it

[a] *Cf.* Epicurus, *Ep. ad Hdt.* 40.

[b] " Province . . . prevent " (M. F. Smith) is an attempt to reproduce the verbal play *officium . . . officere.*

has never existed apart by itself nor will ever have force to do so, since it is essentially a part of something else, a first part with unity of its own, and then other and other like parts, each in its own place, in close formation fill up the nature of the atom; and since these cannot exist separately, they must necessarily so adhere to the whole that they cannot by any means be torn away.[a] The first-beginnings, therefore, are of solid singleness, made of these smallest parts closely packed and cohering together, not compounded by the gathering of these parts, but strong rather by their eternal singleness, and from these nature allows nothing to be torn away or diminished any longer, but keeps them as seeds for things.

(10) If there were infinite division, the smallest thing would be equal to the sum of things;

Besides, unless there is to be a smallest something, each littlest[b] body will consist of infinite parts, since of course a half of the half of anything will always have a half of its own, and there will be no limit to the division. Then what difference will there be between the sum of things and the least of things? There will be no difference; for although the whole sum of things be absolutely infinite, yet the bodies which are littlest will equally consist of infinite parts.[c] But since true reasoning protests against this, and denies that the mind can believe it, you must yield and confess that there are things which no longer consist of any parts and are of the smallest possible nature. And since these exist, you must also confess that the first-beginnings are solid and everlasting.

[a] For the subtle and difficult doctrine of minimal parts (*minimae partes*, *minima* = ἐλάχιστα), *cf.* Epicurus, *Ep. ad Hdt.* 56-59. Epicurus could not accept that matter is infinitely divisible and so postulated the existence of minute, physically indivisible particles, *i.e.* atoms, but at the same time believed that each atom, since it has magnitude, must also have parts which, though they are physically inseparable from the atom, can be distinguished in thought. Lucr. returns to the doctrine in 2.478-499 in connexion with atomic shapes.

[b] *parvissima* is used here to avoid confusion with *minimum*.

[c] The fallacious assumption that all infinities are equal is refuted by Newton in a passage quoted by Munro. Lucr.'s argument is probably aimed chiefly at Anaxagoras, perhaps also at the Stoics.

[*Editor's Note:* Material has been omitted at this point.]

Some move free through the void;

80 If you think the first-beginnings of things can stand still, and by standing still can beget new motions amongst things, you are astray and wander far from true reasoning.[b] For since the first-beginnings of things wander through the void, they must all be carried on either by their own weight or by a chance blow from another atom. For when in quick motion they have often met and collided, it follows that they leap apart suddenly in different directions; and no wonder, since they are perfectly hard in their solid weight and nothing obstructs them from behind. And to show you more clearly that all the bodies of matter are constantly being tossed about, remember that there is no bottom in the sum of things and the first bodies have nowhere to rest, since space is without end or limit, and I have shown at large and proved by irrefragable reasoning that it extends immeasurable from all sides in all directions.[a] Since this stands firm, beyond doubt no rest is granted to the first bodies throughout the profound void, but rather driven by incessant and varied motions, some after being pressed together then leap back with wide intervals, some again after the blow are tossed about within a narrow compass. And all those which being held in combination more closely condensed collide and leap back through tiny intervals, caught fast in the complexity of their own shapes, these constitute the strong roots of stone and the bulk of fierce iron and the others of their kind. Of the rest which go on wandering through the great void, a very few leap far apart and pass far back with long intervals between: these supply thin air for us and the gleaming light of the sun. And many besides wander through the great void which have been rejected from combination with things, and have nowhere been able to obtain admittance and also harmonize their motions.[b]

those which are combined into groups move also.

The motion of

112 Of this fact there is, I recall, an image and similitude always moving and present before our eyes. Do but apply your scrutiny whenever the

[b] Epicurus deals briefly with atomic motion in *Ep. ad Hdt.* 43-44, 61-62.

[a] See 1.958-1007.

[b] The point is that an atom cannot join a compound body, unless (as well as being of suitable size and shape) it can move in harmony with the other component atoms of the object.

sun's rays are let in and pour their light through a dark room: you will see many minute particles mingling in many ways throughout the void [a] in the light itself of the rays, and as it were in everlasting conflict struggling, fighting, battling in troops without any pause, driven about with frequent meetings and partings; so that you may conjecture from this what it is for the first-beginnings of things to be ever tossed about in the great void. So far as it goes, a small thing may give an analogy of great things, and show the tracks of knowledge.

motes in a sunbeam illustrates atomic motion,

125 Even more for another reason it is proper that you give attention to these bodies which are seen to be in turmoil within the sun's rays, because such turmoil indicates that there are secret and unseen motions also hidden in matter. For there you will see how many things set in motion by unseen blows change their course and beaten back return back again, now this way, now that way, in all directions. You may be sure that all take their restlessness from the first-beginnings. For first the first-beginnings of things move of themselves; then the bodies that form a small combination [b] and, as one may say, are nearest to the powers of the first-beginnings, are set moving, driven by the unseen blows of these, while they in their turn attack those that are a little larger. Thus the movement ascends from the first-beginnings and by successive degrees emerges upon our senses,[c] so that those bodies also are moved which we are able to perceive in the sun's light, yet it does not openly appear by what blows they are made to do so.

and indeed the motes derive their motion from the atoms.

[*Editor's Note:* Material has been omitted at this point.]

[a] *inane* (116) refers to the air (*cf.* Virgil, *Aen.* 12.906) through which the motes move, not to void in the strict scientific sense. But, like *corpora* in 117, the word is carefully chosen in order to emphasize the parallel with the behaviour of the atoms.

[b] Small atomic nuclei.

[c] For the imperceptibility of the motions of the atoms, see 308-332.

2

Reprinted from pages 11–12 and 156–163 of *Works of the Honorable Robert Boyle,* T. Birch, ed., London, 1772.

NEW EXPERIMENTS PHYSICO-MECHANICAL—TOUCHING THE SPRING OF THE AIR

Robert Boyle

[*Editor's Note:* In the original, material precedes these excerpts.]

For the more eaſy underſtanding of the experiments triable by our engine, I thought it not ſuperfluous nor unſeaſonable in the recital of this firſt of them, to inſinuate that notion, by which it ſeems likely, that moſt, if not all of them, will prove explicable. Your Lordſhip will eaſily ſuppoſe, that the notion I ſpeak of is, that there is a ſpring, or elaſtical power in the air we live in. By which ἐλατὴρ or ſpring of the air, that which I mean is this; that our air either conſiſts of, or at leaſt abounds with, parts of ſuch a nature, that in caſe they be bent or compreſſed by the weight of the incumbent part of the atmoſphere, or by any other body, they do endeavour, as much as in them lieth, to free themſelves from that preſſure, by bearing againſt the contiguous bodies that keep them bent; and, as ſoon as thoſe bodies are removed, or reduced to give them way, by preſently unbending and ſtretching out themſelves, either quite, or ſo far forth as the contiguous bodies that reſiſt them will permit, and thereby expanding the whole parcel of air, theſe elaſtical bodies compoſe.

This notion may perhaps be ſomewhat further explained, by conceiving the air near the earth to be ſuch a heap of little bodies, lying one upon another, as may be reſembled to a fleece of wool. For this (to omit other likeneſſes betwixt them) conſiſts of many ſlender and flexible hairs; each of which may indeed, like a little ſpring, be eaſily bent or rolled up; but will alſo, like a ſpring, be ſtill endeavouring to ſtretch itſelf out again. For though both theſe hairs, and the aëreal corpuſcles to which we liken them, do eaſily yield to external preſſures; yet, each of them (by virtue of its ſtructure) is endowed with a power or principle of ſelf-dilatation; by virtue whereof, though the hairs may by a man's hand be bent and crouded cloſer together, and into a narrower room than ſuits beſt with the nature of the body; yet, whilſt the compreſſion laſts, there is in the fleece they compoſe an endeavour outwards, whereby it continually thruſts againſt the hand that oppoſes its expanſion. And upon the removal of the external preſſure, by opening the hand more or leſs, the compreſſed wool doth, as it were, ſpontaneouſly expand or diſplay itſelf towards the recovery of its former more looſe and free condition, till the fleece hath either regained its former dimensions, or at least approached them as near as the compressing hand (perchance not quite opened) will permit. This power of self-dilatation is somewhat more conspicuous in a dry spunge compressed, than in a fleece of wool. But yet we rather chose to employ the latter on this occasion, because it is not, like a spunge, an entire body, but a number of slender and flexible bodies, loosely complicated, as the air itself seems to be.

THERE is yet another way to explicate the spring of the air; namely, by supposing with that most ingenious gentleman, Monsieur *Des Cartes*, that the air is nothing but a congeries or heap of small and (for the most part) of flexible particles, of several sizes, and of all kind of figures, which are raised by heat (especially that of the sun) into that fluid and subtle ethereal body that surrounds the earth, and by the restless agitation of that celestial matter, wherein those particles swim, are so whirled round, that each corpuscle endeavours to beat off all others from coming within the little sphere requisite to its motion about its own centre; and in case any, by intruding into that sphere, shall oppose its free rotation, to expel or drive it away: so that, according to this doctrine, it imports very little, whether the particles of the air have the structure requisite to springs, or be of any other form (how irregular soever) since their elastical power is not made to depend upon their shape or structure, but upon the vehement agitation, and (as it were) brandishing motion, which they receive from the fluid æther, that swiftly flows between them, and whirling about each of them (independently from the rest) not only keeps those slender aëreal bodies separated and stretched out (at least, as far as the neighbouring ones will permit) which otherwise, by reason of their flexibleness and weight, would flag or curl; but also makes them hit against, and knock away each other, and consequently require more room than that, which, if they were compressed, they would take up.

BY these two differing ways, my Lord, may the springs of the air be explicated. But though the former of them be that, which by reason of its seeming somewhat more easy, I shall for the most part make use of in the following discourse; yet am I not willing to declare peremptorily for either of them against the other. And indeed, though I have in another treatise endeavoured to make it probable, that the returning of elastical bodies (if I may so call them) forcibly bent, to their former position, may be mechanically explicated; yet I must confess, that to determine whether the motion of restitution in bodies proceed from this, that the parts of a body of a peculiar structure are put into motion by the bending of the spring, or from the endeavour of some subtle ambient body, whose passage may be opposed or obstructed, or else its pressure unequally resisted by reason of the new shape or magnitude, which the bending of a spring may give the pores of it: to determine this, I say, seems to me a matter of more difficulty, than at first sight one would easily imagine it. Wherefore I shall decline meddling with a subject, which is much more hard to be explicated than necessary to be so by him, whose business it is not, in this letter, to assign the adequate cause of the spring of the air, but only to manifest, that the air hath a spring, and to relate some of its effects.

[*Editor's Note:* Material has been omitted at this point.]

CHAP. V.

Two new Experiments touching the meaſure of the force of the ſpring of air compreſſed and dilated.

THE other thing, that I would have conſidered touching our adverſary's hypotheſis is, that it is needleſs. For whereas he denies not, that the air has ſome weight and ſpring, but affirms, that it is very inſufficient to perform ſuch great matters as the counterpoiſing of a mercurial cylinder of 29 inches, as we teach that it may; we ſhall now endeavour to manifeſt by experiments purpoſely made, that the ſpring of the air is capable of doing far more than it is neceſſary for us to aſcribe to it, to ſolve the phænomena of the Torricellian experiment.

WE took then a long glaſs-tube, which, by a dexterous hand and the help of a lamp, was in ſuch a manner crooked at the bottom, that the part turned up was almoſt parallel to the reſt of the tube, and the orifice of this ſhorter leg of the ſiphon (if I may ſo call the whole inſtrument) being hermetically ſealed, the length of it was divided into inches (each of which was ſubdivided into eight parts) by a ſtreight liſt of paper, which containing thoſe diviſions, was carefully paſted all along it. Then putting in as much quickſilver as ſerved to fill the arch or bended part of the ſiphon, that the mercury ſtanding in a level might reach in the one leg to the bottom of the divided paper, and juſt to the ſame height or horizontal line in the other; we took care, by frequently inclining the tube, ſo that the air might freely paſs from one leg into the other by the ſides of the mercury (we took, I ſay, care) that the air at laſt included in the ſhorter cylinder ſhould be of the ſame laxity with the reſt of the air about it. This done, we began to pour quickſilver into the longer leg of the ſiphon, which by its weight preſſing up that in the ſhorter leg, did by degrees ſtreighten the included air: and continuing this pouring in of quickſilver till the air in the ſhorter leg was by condenſation reduced to take up but half the ſpace it poſſeſſed (I ſay, poſſeſſed, not filled) before; we caſt our eyes upon the longer leg of the glaſs, on which was likewiſe paſted a liſt of paper carefully divided into inches and parts, and we obſerved, not without delight and ſatisfaction, that the quickſilver in that longer part of the tube was 29 inches higher than the other. Now that this obſervation does both very well agree with and confirm our hypotheſis, will be eaſily diſcerned by him, that takes notice what we teach; and Monſieur *Paſchal* and our Engliſh friend's experiments prove, that the greater the weight is that leans upon the air, the more forcible is its endeavour of dilatation, and conſequently its power of reſiſtance (as other ſprings are ſtronger when bent by greater weights). For this being conſidered, it will appear to agree rarely-well with the hypotheſis, that as according to it the air in that degree of denſity and correſpondent meaſure of reſiſtance, to which the weight of the incumbent atmoſphere had brought it, was able to counterbalance and reſiſt the preſſure of a mercurial cylinder of about 29 inches, as we are taught by the Torricellian experiment; ſo here the ſame air being brought to a degree of denſity about twice as great as that it had before, obtains a ſpring twice as ſtrong as formerly. As may appear by its being able to ſuſtain or reſiſt a cylinder of 29 inches in the longer tube, together with the weight of the atmoſpherical cylinder, that leaned upon thoſe 29 inches of mercury; and, as we juſt now inferred from the Torricellian experiment, was equivalent to them.

WE were hindered from proſecuting the trial at that time by the caſual breaking of the tube. But becauſe an accurate experiment of this nature would be of great importance to the doctrine of the ſpring of the air, and has not yet been made (that I know) by any man; and becauſe alſo it is more uneaſy to be made than one would think, in regard of the difficulty as well of procuring crooked tubes fit for the purpoſe, as of making a juſt eſtimate of the true place of the protuberant mercury's ſurface; I ſuppoſe it will not be unwelcome to the reader, to be informed, that after ſome other trials, one of which we made in a tube whoſe longer leg was perpendicular, and the other, that contained the air, parallel to the horizon, we at laſt procured a tube of the figure expreſt in the ſcheme; which tube, though of a pretty bigneſs, was ſo long, that the cylinder, whereof the ſhorter leg of it conſiſted, admitted a liſt of paper, which had before been divided into 12 inches and their quarters, and the longer leg admitted another liſt of paper of divers feet in length, and divided after the ſame manner. Then quickſilver being poured in to fill up the bended part of the glaſs, that the ſurface of it in either leg might reſt in the ſame horizontal line, as we lately taught, there was more and more quickſilver poured into the longer tube; and notice being watchfully taken how far the mercury was riſen in that longer tube, when it appeared to have aſcended to any of the diviſions in the ſhorter tube, the ſeveral obſervations, that were thus ſucceſſively made, and as they were made ſet down, afforded us the enſuing table:

A table of the condenſation of the air.

A	*A*	*B*	*C*	*D*	*E*
48	12	00	Added to $22\frac{1}{8}$ makes	$29\frac{2}{16}$	$29\frac{2}{16}$
46	$11\frac{1}{2}$	$01\frac{7}{16}$		$30\frac{9}{16}$	$33\frac{6}{16}$
44	11	$02\frac{13}{16}$		$31\frac{15}{16}$	$31\frac{12}{16}$
42	$10\frac{1}{2}$	$04\frac{6}{16}$		$33\frac{8}{16}$	$33\frac{1}{7}$
40	10	$06\frac{3}{16}$		$35\frac{5}{16}$	35 - -
38	$9\frac{1}{2}$	$07\frac{14}{16}$		37	$36\frac{15}{19}$
36	9	$10\frac{2}{16}$		$39\frac{5}{16}$	$38\frac{7}{8}$
34	$8\frac{1}{2}$	$12\frac{8}{16}$		$41\frac{10}{16}$	$41\frac{2}{17}$
32	8	$15\frac{1}{16}$		$44\frac{3}{16}$	$43\frac{11}{16}$
30	$7\frac{1}{2}$	$17\frac{15}{16}$		$47\frac{1}{16}$	$46\frac{3}{5}$
28	7	$21\frac{3}{16}$		$50\frac{5}{16}$	50 - -
26	$6\frac{1}{2}$	$25\frac{3}{16}$		$54\frac{5}{16}$	$53\frac{10}{13}$
24	6	$29\frac{11}{16}$		$58\frac{13}{16}$	$58\frac{2}{8}$
23	$5\frac{3}{4}$	$32\frac{3}{16}$		$61\frac{5}{16}$	$60\frac{18}{23}$
22	$5\frac{1}{2}$	$34\frac{15}{16}$		$64\frac{1}{16}$	$63\frac{6}{11}$
21	$5\frac{1}{4}$	$37\frac{15}{16}$		$67\frac{1}{16}$	$66\frac{4}{7}$
20	5	$41\frac{9}{16}$		$70\frac{11}{16}$	70 - -
19	$4\frac{3}{4}$	45 - -		$74\frac{2}{16}$	$73\frac{11}{19}$
18	$4\frac{1}{2}$	$48\frac{12}{16}$		$77\frac{14}{16}$	$77\frac{2}{3}$
17	$4\frac{1}{4}$	$53\frac{11}{16}$		$82\frac{12}{16}$	$82\frac{4}{17}$
16	4	$58\frac{2}{16}$		$87\frac{14}{16}$	$87\frac{3}{8}$
15	$3\frac{3}{4}$	$63\frac{15}{16}$		$93\frac{1}{16}$	$93\frac{1}{5}$
14	$3\frac{1}{2}$	$71\frac{5}{16}$		$100\frac{7}{16}$	$99\frac{6}{7}$
13	$3\frac{1}{4}$	$78\frac{11}{16}$		$107\frac{13}{16}$	$107\frac{7}{13}$
12	3	$88\frac{7}{16}$		$117\frac{9}{16}$	$116\frac{4}{8}$

AA. The number of equal ſpaces in the ſhorter leg, that contained the ſame parcel of air diverſly extended.

B. The height of the mercurial cylinder in the longer leg, that compreſſed the air into thoſe dimenſions.

C. The height of the mercurial cylinder, that counterbalanced the preſſure of the atmoſphere.

D. The aggregate of the two laſt columns *B* and *C*, exhibiting the preſſure ſuſtained by the included air.

E. What that preſſure ſhould be according to the hypotheſis, that ſuppoſes the preſſures and expanſions to be in reciprocal proportion.

FOR the better underſtanding of this experiment, it may not be amiſs to take notice of the following particulars:

1. THAT the tube being ſo tall, that we could not conveniently make uſe of it in a chamber, we were fain to uſe it on a pair of ſtairs, which yet were very lightſome, the tube being for preſervation's ſake by ſtrings ſo ſuſpended, that it did ſcarce touch the box preſently to be mentioned.

2. THE lower and crooked part of the pipe was placed in a ſquare wooden box, of a good largeneſs and depth, to prevent the loſs of the quickſilver, that might fall aſide in the transfuſion from the veſſel into the pipe, and to receive the whole quickſilver in caſe the tube ſhould break.

3. THAT we were two to make the obſervation together, the one to take notice at the bottom, how the quickſilver roſe in the ſhorter cylinder, and the other to pour in at the top of the longer; it being very hard and troubleſome for one man alone to do both accurately.

4. That the quickſilver was poured in but by little and little, according to the direction of him that obſerved below; it being far eaſier to pour in more, than to take out any, in caſe too much at once had been poured in.

5. THAT at the beginning of the operation, that we might the more truly difcern where the quickfilver refted from time to time, we made ufe of a fmall looking-glafs, held in a convenient pofture to reflect to the eye what we defired to difcern.

6. THAT when the air was fo compreffed, as to be crouded into lefs than a quarter of the fpace it poffeffed before, we tried whether the cold of a linen cloth dipped in water would then condenfe it. And it fometimes feemed a little to fhrink, but not fo manifeftly as that we dare build any thing upon it. We then tried likewife, whether heat would, notwithftanding fo forcible a compreffure, dilate it; and approaching the flame of a candle to that part where the air was pent up, the heat had a more fenfible operation than the cold had before; fo that we fcarce doubted, but that the expanfion of the air would, notwithftanding the weight that oppreft it, have been made confpicuous, if the fear of unfeafonably breaking the glafs had not kept us from increafing the heat.

Now although we deny not, but that in our table fome particulars do not fo exactly anfwer to what our formerly mentioned hypothefis might perchance invite the reader to expect; yet the variations are not fo confiderable, but that they may probably enough be afcribed to fome fuch want of exactnefs as in fuch nice experiments is fcarce avoidable. But for all that, till further trial hath more clearly informed me, I fhall not venture to determine, whether or no the intimated theory will hold univerfally and precifely, either in condenfation of air, or rarefaction: all that I fhall now urge being, that however the trial already made fufficiently proves the main thing, for which I here allege it; fince by it, it is evident, that as common air, when reduced to half its wonted extent, obtained near about twice as forcible a fpring as it had before; fo this thus compreft air being further thruft into half this narrow room, obtained thereby a fpring about as ftrong again as that it laft had, and confequently four times as ftrong as that of the common air. And there is no caufe to doubt, that if we had been here furnifhed with a greater quantity of quickfilver and a very ftrong tube, we might, by a further compreffion of the included air, have made it counterbalance the preffure of a far taller and heavier cylinder of mercury. For no man perhaps yet knows, how near to an infinite compreffion the air may be capable of, if the compreffing force be competently increafed. So that here our adverfary may plainly fee, that the fpring of the air, which he makes fo light of, may not only be able to refift the weight of 29 inches, but in fome cafes of above a hundred inches of quickfilver, and that without the affiftance of his Funiculus, which in our prefent cafe has nothing to do. And to let you fee, that we did not (as a little above) inconfiderately mention the weight of the incumbent atmofpherical cylinder as a part of the weight refifted by the imprifoned air, we will here annex, that we took care, when the mercurial cylinder in the longer leg of the pipe was about an hundred inches high, to caufe one to fuck at the open orifice; whereupon (as we expected) the mercury in the tube did notably afcend. Which confiderable phænomenon cannot be afcribed to our examiner's Funiculus, fince by his own confeffion that cannot pull up the mercury, if the mercurial cylinder be above 29 or 30 inches of mercury And therefore we fhall render this reafon of it, that the preffure of the incumbent air being in part taken off by its expanding itfelf into the fucker's dilated cheft; the imprifoned air was thereby enabled to dilate itfelf manifeftly, and repel the mercury, that compreft it, till there was an equality of force betwixt the ftrong fpring of that compreft air on the one part, and the tall mercurial cylinder, together with the contiguous dilated air, on the other part.

Now, if to what we have thus delivered concerning the compreffion of the air, we add fome obfervations concerning its fpontaneous expanfion, it will the better appear, how much the phænomena of thefe mercurial experiments depend upon the differing meafures of ftrength to be met with in the air's fpring, according to its various degrees of compreffion and laxity. But, before I enter upon this fubject, I fhall readily acknowledge, that I had not reduced the trials I had made about meafuring the expanfion of the air to any certain hypothefis, when that ingenious gentleman Mr. *Richard Townley* was pleafed to inform me, that having by the perufal of my phyfico-mechanical experiments been fatisfied that the fpring of the air was the caufe of it, he endeavoured (and I wifh in fuch attempts other ingenious men would follow his example) to fupply what I had omitted concerning the reducing to a precife eftimate, how much air dilated of itfelf lofes of its elaftical force, according to the meafures of its dilatation. He added, that he had begun to fet down what occurred to him to this purpofe in a fhort difcourfe, whereof he afterwards did me the favour to fhew me the beginning, which gives me a juft curiofity to fee it perfected. But, becaufe I neither know, nor (by reafon of the great diftance betwixt our places of refidence) have at prefent the opportunity to inquire, whether he will think fit to annex his difcourfe to our appendix, or to publifh it by itfelf, or at all; and becaufe he hath not yet, for aught I know, met with fit glaffes to make an any-thing-accurate table of the decrement of the force of the dilated air; our prefent defign invites us to prefent the reader with that which follows, wherein I had the affiftance of the fame perfon, that I took notice of in the former chapter, as having written fomething about rarefaction: whom I the rather make mention of on this occafion, becaufe when he firft heard me fpeak of Mr. *Townley*'s fuppofitions about the proportion, wherein air lofes of its fpring by dilatation, he told me he had the year before (and not long after the publication of my pneumatical treatife) made obfervations to the fame purpofe, which he acknowledged to agree well enough with Mr. *Townley*'s theory: and fo did (as their author was pleafed to tell me) fome trials made about the fame time by that noble virtuofo and eminent mathematician the Lord *Brouncker*, from whofe further enquiries into this mattter, if his occafions will allow him to make them, the curious may well hope for fomething very accurate.

A table of the rarefaction of the air.

A. The number of equal fpaces at the top of the tube, that contained the fame parcel of air.
B. The height of the mercurial cylinder, that together with the fpring of the included, air counterbalanced the preffure of the atmofphere.
C. The preffure of the atmofphere.
D. The complement of *B* to *C*, exhibiting the preffure fuftained by the included air.
E. What that preffure fhould be, according to the hypothefis.

A	*B*	*C*	*D*	*E*
1	$00\frac{0}{0}$	Subftracted from $29\frac{3}{4}$ leaves	$29\frac{3}{4}$	$29\frac{3}{4}$
$1\frac{1}{2}$	$10\frac{5}{8}$		$19\frac{1}{8}$	$19\frac{5}{6}$
2	$15\frac{3}{8}$		$14\frac{3}{8}$	$14\frac{7}{8}$
3	$20\frac{2}{8}$		$9\frac{4}{8}$	$9\frac{15}{12}$
4	$22\frac{5}{8}$		$7\frac{1}{8}$	$7\frac{7}{16}$
5	$24\frac{1}{8}$		$5\frac{5}{8}$	$5\frac{19}{20}$
6	$24\frac{7}{8}$		$4\frac{7}{8}$	$4\frac{27}{26}$
7	$25\frac{4}{8}$		$4\frac{2}{8}$	$4\frac{1}{4}$
8	$26\frac{0}{0}$		$3\frac{6}{8}$	$3\frac{23}{32}$
9	$26\frac{3}{8}$		$3\frac{3}{8}$	$3\frac{11}{36}$
10	$26\frac{6}{8}$		$3\frac{0}{0}$	$2\frac{39}{40}$
12	$27\frac{1}{8}$		$2\frac{5}{8}$	$2\frac{23}{48}$
14	$27\frac{4}{8}$		$2\frac{2}{8}$	$2\frac{1}{8}$
16	$27\frac{6}{8}$		$2\frac{0}{0}$	$1\frac{55}{64}$
18	$27\frac{7}{8}$		$1\frac{7}{8}$	$1\frac{47}{72}$
20	$28\frac{0}{0}$		$1\frac{6}{8}$	$1\frac{9}{80}$
24	$28\frac{2}{8}$		$1\frac{4}{8}$	$1\frac{23}{96}$
28	$28\frac{3}{8}$		$1\frac{3}{8}$	$1\frac{1}{16}$
32	$28\frac{4}{8}$		$1\frac{2}{8}$	$0\frac{117}{128}$

To make the experiment of the debilitated force of expanded air the plainer, it will not be amiſs to note ſome particulars, eſpecially touching the manner of making the trial; which (for the reaſons lately mentioned) we made on a lightſome pair of ſtairs, and with a box alſo lined with paper to receive the mercury that might be ſpilt. And in regard it would require a vaſt, and in few places procurable quantity of quick-ſilver, to imploy veſſels of ſuch kind as are ordinary in the Torricellian experiment, we made uſe of a glaſs-tube of about ſix feet long; for that being hermetically ſealed at one end, ſerved our turn as well as if we could have made the experiment in a tub or pond of ſeventy inches deep.

SECONDLY, We alſo provided a ſlender glaſs-pipe of about the bigneſs of a ſwan's quill, and open at both ends; all along which was paſted a narrow liſt of paper, divided into inches and half quarters.

THIRDLY, This ſlender pipe being thruſt down into the greater tube almoſt filled with quickſilver, the glaſs helped to make it ſwell to the top of the tube; and the quickſilver getting in at the lower orifice of the pipe, filled it up till the mercury included in that was near about a level with the ſurface of the ſurrounding mercury in the tube.

FOURTHLY, There being, as near as we could gueſs, little more than an inch of the ſlender pipe left above the ſurface of the reſtagnant mercury, and conſequently unfilled therewith, the prominent orifice was carefully cloſed with ſealing-wax melted; after which the pipe was let alone for a while, that the air dilated a little by the heat of the wax, might, upon refrigeration, be reduced to its wonted denſity. And then we obſerved by the help of the above-mentioned liſt of paper, whether we had not included ſomewhat more or ſomewhat leſs than an inch of air; and in either caſe we were fain to rectify the error by a ſmall hole made (with a heated pin) in the wax, and afterwards cloſed up again.

FIFTHLY, Having thus included a juſt inch of air, we lifted up the ſlender pipe by degrees, till the air was dilated to an inch, an inch and an half, two inches, *&c.* and obſerved in inches and eighths the length of the mercurial cylinder, which at each degree of the air's expanſion was impelled above the ſurface of the reſtagnant mercury in the tube.

SIXTHLY, The obſervations being ended, we preſently made the Torricellian experiment with the above-mentioned great tube of ſix feet long, that we might know the height of the mercurial cylinder, for that particular day and hour; which height we found to be $29\frac{3}{4}$ inches.

SEVENTHLY, Our obſervations made after this manner furniſhed us with the preceding table, in which there would not probably have been found the difference here ſet down betwixt the force of the air, when expanded to double its former dimenſions, and what that force ſhould have been preciſely according to the theory, but that the included inch of air received ſome little acceſſion during the trial; which this newly mentioned difference making us ſuſpect, we found by replunging the pipe into the quickſilver, that the included air had gained about half an eighth, which we gueſſed to have come from ſome little aërial bubbles in the quickſilver, contained in the pipe (ſo eaſy is it in ſuch nice experiments to miſs of exactneſs). We tried alſo with 12 inches of air ſhut up to be dilated; but being then hindered by ſome unwelcome avocations to proſecute thoſe experiments, we ſhall elſewhere, out of other notes and trials (God permitting) ſet down ſome other accurate tables concerning this matter. By which poſſibly we may be aſſiſted to reſolve, whether the atmoſphere ſhould be looked upon (as it uſually is) as a limited and bounded portion of the air; or whether we ſhould, in a ſtricter ſenſe than we did before, uſe the atmoſphere and aërial part of the world for almoſt equivalent terms; or elſe whether we ſhould allow the word atmoſphere ſome other notion in relation to its extent and limits; (for as to its ſpring and weight, theſe experiments do not queſtion, but evince them.) But we are willing, as we ſaid, to refer the matters to our Appendix, and till then to retain our wonted manner of ſpeaking of the air and atmoſphere. In the mean time (to return to our laſt-mentioned experiments) beſides that ſo little a variation may be in great part imputed to the difficulty of making experiments of this nature exactly, and perhaps a good part of it to ſomething of inequality in the cavity of the pipe, or even in the thickneſs of the glaſs; beſides this, I ſay, the proportion betwixt the ſeveral preſſures of the included air undilated and expanded, eſpecially when the dilatation was great (for when the air ſwelled but to four times its firſt extent, the mercurial cylinder, though of near 23 inches, differed not a quarter of an inch from what it ſhould have been according to mathematical exactneſs) the proportion, I ſay, was ſuitable enough to what might be expected, to allow us to make this reflection upon the whole; that whether or no the intimated theory will hold exactly (for about that, as I ſaid above, I dare determine nothing reſolutely till I have further conſidered the matter) yet ſince the inch of air, when it was firſt included, was ſhut up with no other preſſure than that, which it had from the weight of the incumbent air, and was no more compreſſed than the reſt of the air we breathed and moved in; and ſince alſo this inch of air, when expanded to twice its former dimenſions, was able with the help of a mercurial cylinder of about 15 inches to counterpoiſe the weight of the atmoſphere, which the weight of the external air gravitating upon the reſtagnant mercury was able to impel up into the pipe, and ſuſtain above twenty-eight inches of mercury, when the internal air, by its great expanſion, had its ſpring too far debilitated to make any conſiderable (I ſay conſiderable, for it was not yet ſo dilated as not to make ſome) reſiſtance: ſince, I ſay, theſe things are ſo, the free air here below appears to be almoſt as ſtrongly compreſſed by the weight of the incumbent air, as it would be by the weight of a mercurial cylinder of twenty eight or thirty inches; and conſequently is not in ſuch a ſtate of laxity and freedom as men are wont to imagine; and acts like ſome mechanical agent, the decrement of whoſe force holds a ſtricter proportion to its increaſe of dimenſion, than has been hitherto taken notice of.

I MUST not now ſtand to propoſe the ſeveral reflections, that may be made upon the foregoing obſervations touching the compreſſion and expanſion of air; partly becauſe we could ſcarce avoid making the hiſtorical part ſomewhat prolix; and partly becauſe I ſuppoſe we have already ſaid enough to ſhew what was intended: namely, that to ſolve the phænomena there is not of our adverſary's hypotheſis any need: the evincing of which will appear to be of no ſmall moment in our preſent controverſy to him that conſiders, that the two main things, that induced the learned examiner to reject our hypotheſis, are, that nature abhors a vacuum; and that though the air have ſome weight and ſpring, yet, theſe are inſufficient to make out the known phænomena; for which we muſt therefore have recourſe to his Funiculus. Now as we have formerly ſeen, that he has not ſo ſatisfactorily diſproved as reſolutely rejected a vacuum ſo we have now manifeſted, that the ſpring of the air may ſuffice to perform greater things than what our explication of the Torricellian experiments and thoſe of our engine obliges us to aſcribe to it. Wherefore ſince beſides the ſeveral difficulties, that incumber the hypotheſis we oppoſe, and eſpecially its being ſcarce, if at all, intelligible, we can add that it is unneceſſary; we dare expect, that ſuch readers as are not biaſſed by their reverence for *Ariſtotle*, or the Peripatetick ſchools, will hardly reject an hypotheſis, which, beſides that it is very intelligible, is now proved to be ſufficient, only to imbrace a doctrine, that ſuppoſes ſuch a rarefaction and condenſation, as many famous Naturaliſts rejected for its not being comprehenſible, even when they knew of no other way (that was probable) of ſolving the phænomena wont to be explicated by it.

[*Editor's Note:* Material has been omitted at this point.]

3

Reprinted by permission from pages 331 and 338–348 of *Early Science in Oxford,* R. T. Gunther. The Cutler Lectures of Robert Hooke, Vol. VIII. Oxford: Oxford University Press, 1931.

LECTURES
De Potentia Reſtitutiva,
OR OF
SPRING

Explaining the Power of Springing Bodies.

To which are added ſome

COLLECTIONS

Viz.

*A Deſcription of Dr.*Pappins *Wind-Fountain and Force-Pump.*
*Mr.*Young's *Obſervation concerning natural Fountains.*
Some other Conſiderations concerning that Subject.
Captain Sturmy's *remarks of a Subterraneous Cave and Ciſtern.*
Mr. G. T. *Obſervations made on the* Pike *of* Teneriff, 1674.
Some Reflections and Conjectures occaſioned thereupon.
A Relation of a late Eruption in the Iſle of Palma.

By *ROBERT HOOKE.* S.R.S.

LONDON,

Printed for *John Martyn* Printer to the *Royal Society,* at the Bell in St. *Pauls* Church-Yard, 1678.

[*Editor's Note:* In the original, material precedes and follows this excerpt.]

In the year 1660. I printed a little Tract, which I called, *An Attempt for the explication of the Phenomena, &c.* of the riſing of water in the pores of very ſmall Pipes, Filtres, *&c.* And being unwilling then to publiſh this Theory, as ſuppoſing it might be prejudicial to my deſign of Watches, which I was then procuring a Patent for, I only hinted the principle which I ſuppoſed to be the cauſe of theſe Phænomena of ſprings in the 31 page thereof in the Engliſh Edition, and in the 38 page of the Latine Edition, tranſlated by *M. Behem*, and printed at *Amſterdam*, 1662. But referred the further explication thereof till ſome other opportunity.

The Principles I then mentioned I called by the names of *Congruity* and *Incongruity* of bodies. And promiſed a further explanation of what I thereby meant on ſome other occaſion. I ſhall here only explain ſo much of it as concerns the explication of this preſent Phænomenon.

By *Congruity* and *Incongruity* then I underſtand nothing elſe but an agreement or diſagreement of Bodys as to their Magnitudes and motions.

Thoſe Bodies then I ſuppoſe congruous whoſe particles have the ſame Magnitude, and the ſame degree of Velocity, or elſe an harmonical proportion of Magnitude, and harmonical degree of Velocity. And thoſe I ſuppoſe incongruous which have neither the ſame Magnitude, nor the ſame degree of Velocity, nor an harmonical proportion of Magnitude nor of Velocity.

I ſuppoſe then the ſenſible Univerſe to conſiſt of body and motion.

By Body I mean ſomewhat receptive and communicative of motion or progreſſion. Nor can I have any other Idea thereof, for neither Extention nor Quantity, hardneſs nor ſoftneſs, fluidity nor fixedneſs, Rarefaction nor Denſation are the proprieties of Body, but of Motion or ſomewhat moved.

By Motion I underſtand nothing but a power or tendency progreſſive of Body according to ſeveral degrees of Velocity.

Theſe two do always counterballance each other in all the effects, appearances, and operations of Nature, and therefore it is not impoſſible but that they may be one and the ſame; for a little body with great motion is equivalent to a great body with little motion as to all its ſenſible effects in Nature.

I do further ſuppoſe then that all things in the Univerſe that become the objects of our ſenſes are compounded of theſe two (which we will for the preſent ſuppoſe diſtinct eſſences, though poſſibly they may be found hereafter to be only differing conceptions of one and the ſame eſſence) namely, *Body*, and *Motion*. And that there is no one ſenſible Particle of matter but owes the greateſt part of its ſenſible Extenſion to Motion whatever part thereof it ows to Body according to the common notion thereof: Which is, that

Body is ſomewhat that doth perfectly fill a determinate quantity of ſpace or extenſion ſo as neceſſarily to exclude all other bodies from being comprehended within the ſame Dimenſions.

I do therefore define a ſenſible Body to be a determinate Space or Extenſion defended from being penetrated by another, by a power from within.

To make this the more intelligible, Imagine a very thin plate of Iron, or the like, a foot ſquare, to be moved with a Vibrative motion forwards and backwards the flat ways the length of a foot with ſo ſwift a motion as not to permit any other body to enter into that ſpace within which it Vibrates, this will compoſe ſuch an eſſence as I call in my ſenſe a Cubick foot of ſenſible Body, which differs from the common notion of Body as this ſpace of a Cubick foot thus defended by this Vibrating plate doth from a Cubick foot of Iron, or the like, throughout ſolid. The Particles therefore that compoſe all bodies I do ſuppoſe to owe the greateſt part of their ſenſible or potential Extenſion to a Vibrative motion.

This Vibrative motion I do not ſuppoſe inherent or inſeparable from the Particles of body, but communicated by Impulſes given from other bodies in the Univerſe. This only I ſuppoſe, that the Magnitude or bulk of the body doth make it receptive of this or that peculiar motion that is communicated, and not of any other. That is, every Particle of matter according to its determinate or preſent Magnitude is receptive of this or that peculiar motion and no other, ſo that Magnitude and receptivity of motion ſeems the ſame thing: To explain this by a ſimilitude or example. Suppoſe a number of muſical ſtrings, as A B C D E, *&c.* tuned to certain tones, and a like number of other ſtrings, as *a*, *b*, *c*, *d*, *e*, &c. tuned to the ſame ſounds reſpectively, A ſhall be receptive of the motion of *a*, but not of that of *b*, *c*, nor *d*; in like manner B ſhall be receptive of the motion of *b*, but not of the motion of *a*, *c* or *d*. And ſo of the reſt. This is that which I call *Congruity* and *Incongruity*.

Now as we find that muſical ſtrings will be moved by Uniſons and Eighths, and other harmonious chords, though not in the ſame degree; ſo do I ſuppoſe that *the particles of matter* will be moved principally by ſuch motions as are Uniſons, as I may call them, or of equal Velocity with their motions, and by other harmonious motions in a leſs degree.

I do further ſuppoſe, A ſubtil matter that incompaſſeth and pervades all other bodies, which is the Menſtruum in which they ſwim which maintains and continues all ſuch bodies in their motion, and which is the medium that conveys all Homogenious or Harmonical motions from body to body.

Further I ſuppoſe, that all ſuch particles of matter as are of a like nature, when not ſeparated by others of a differing nature will remain together, and ſtrengthen the common Vibration of them all againſt the differing Vibrations of the ambient bodies.

According to this Notion I ſuppoſe the whole Univerſe and all the particles thereof to be in a continued motion, and every one to take its ſhare of ſpace or room in the ſame, according to the bulk of its body, or according to the particular power it hath to receive, and continue this or that peculiar motion.

Two or more of theſe particles joyned immediately together, and coaleſcing into one become of another nature, and receptive of another degree of motion and Vibration, and make a compounded particle differing in nature from each of the other particles.

All bulky and ſenſible bodies whatſoever I ſuppoſe to be made up or compoſed of ſuch particles which have their peculiar and appropriate motions which are kept together by the differing or diſſonant Vibrations of the ambient bodies or fluid.

According to the difference of theſe Vibrative motions of the Incompaſſing bulks. All bodies are more or leſs powerful in preſerving their peculiar ſhapes.

All bodies neer the Earth are incompaſſed with a fluid ſubtil matter by the differing Velocity of whoſe parts all ſolid bodies are kept together in the peculiar ſhapes, they were left in when they were laſt fluid. And all fluid bodies whatſoever are mixed with this fluid, and which is not extruded from them till they become ſolid.

Fluid bulks differ from ſolids only in this, that all fluids conſiſt of two ſorts of particles, the one this common Menſtruum near the Earth, which is interſperſed between the Vibrating particles appropriated to that bulk, and ſo participating of the motions and Vibrations thereof: And the other, by excluding wholly, or not participating of that motion.

Though the particles of ſolid bodies do by their Vibrative motions exclude this fluid from coming between them where their motions do immediately touch, yet are there certain ſpaces between them which are not defended by the motion of the particles from being pervaded by the Heterogeneous fluid menſtruum.

Theſe ſpaces ſo undefended by the bodies and Vibrative motion of the particles, and conſequently pervaded by the ſubtil incompaſſing Heterogeneous fluid are thoſe we call the inſenſible pores of bodies.

According to the bigneſs of the bodies the motions are, but in reciprocal proportion: That is, the bigger or more powerful the body is, the ſlower is its motion with which it compounds the particles; and the leſs the body is, the ſwifter is its motion.

The ſmaller the particles of bodies are, the nearer do they approach to the nature of the general fluid, and the more eafily do they mix and participate of its motion.

The Particles of all folid bodies do immediately touch each other; that is, the Vibrative motions of the bodies do every one touch each other at every Vibration. For explication, Let A B C reprefent three bodies, each of thefe bodies I fuppofe to have a Vibrative motion on either fide of it, A between D and E, B between E and F, and C between F and G.

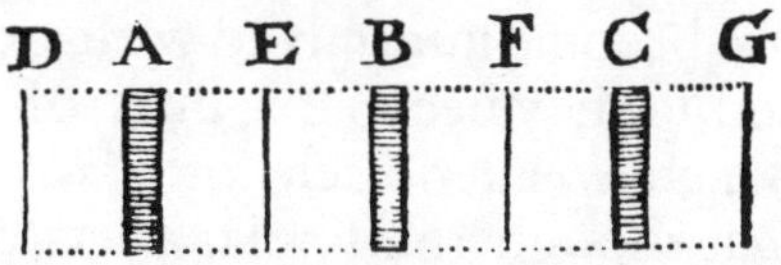

I fuppofe then that B in every one of its Vibrations doth meet A at E, and C at F, and fo the motions are continually interchanged: That is, B communicates its motion to A at E, and A at the fame time and place communicates its motion to B, which returning to F meets there with C, and communicates its received motion to C, which at the fame inftant and place communicates its own motion to B, which returns it back to E: So that the Velocity of thefe bodies is always the fame, and each body impreffeth on the contiguous bodies fuch a determinate number of pulfes within a certain fpace of time. Suppofe for inftance, in every fecond of time B communicates to A and to C one million of pulfes, and hath received as many from each of them, by which means each of them doth preferve its own fpace of Vibration, according to the power of its Vibration, that neither of the contiguous bodies can enter into it. The extreme particles A and C are repercuffed by the motion of the ambient Heterogeneous fluid, whereof though the bodies are of differing magnitudes, yet the body and motion of the one are equivalent to the body and motion of the other, fo that whatever the body be lefs, the motion is quicker; and where the body is bigger, the motion is lefs. But the Particles of fluid bodies do not immediately touch each other, but permit the mixture of the other Heterogeneous fluid near the Earth, which ſerves to communicate the motion from particle to particle without the immediate contact of the Vibrations of the Particles.

All ſolid Bodies retain their ſolidity till by other extraordinary motions their natural or proper motions become intermixed with other differing motions, and ſo they become a bulk of compounded motions, which weaken each others Vibrative motions. So that though the ſimilar parts do participate of each others motions, whereby they indeavour to joyn or keep together, yet do they alſo participate of an Heterogeneous motion which endeavours to ſeparate or keep them aſunder. And according to the prevalency of the one or the other is the body more or leſs fluid or ſolid.

All bodies whatſoever would be fluid were it not for the external Heterogeneous motion of the Ambient.

And all fluid bodies whatſoever would be unbounded, and have their parts fly from each other were it not for ſome prevailing Heterogeneous motion from without them that drives them more powerfully together.

Heterogeneous motions from without are propagated within the ſolid in a direct line if they hit perpendicular to the ſuperficies or bounds, but if obliquely in ways not direct, but different and deflected, according to the particular inclination of the body ſtriking, and according to the proportion of the Particles ſtriking and being ſtruck.

All ſpringy bodies whatſoever conſiſt of parts thus qualified, that is, of ſmall bodies indued with appropriate and peculiar motions, whence every one of theſe particles hath a particular Bulk, Extenſion, or Sphere of activity which it defends from the ingreſs of any other incompaſſing Heterogeneous body whilſt in its natural eſtate and balance in the Univerſe. Which particles being all of the ſame nature, that is, of equal bodies, and equal motions, they readily coaleſce and joyn together, and make up one ſolid body, not perfectly every where contiguous, and wholly excluding the above mentioned ambient fluid, but permitting it in many places to pervade the ſame in a regular order, yet not ſo much but that they do wholly exclude the ſame from paſſing between all the ſides of the compounding particles.

The parts of all ſpringy bodies would recede and fly from each other were they not kept together by the Heterogeneous compreſſing motions of the ambient whether fluid or ſolid.

Theſe principles thus hinted, I ſhall in the next place come to the particular explication of the manner how they ſerve to explain the Phænomena of ſpringing bodies whether ſolid or fluid.

Firſt for ſolid bodies, as Steel, Glaſs, Wood, *&c.* which have a Spring both inwards and outwards, according as they are either compreſſed or dilated beyond their natural ſtate.

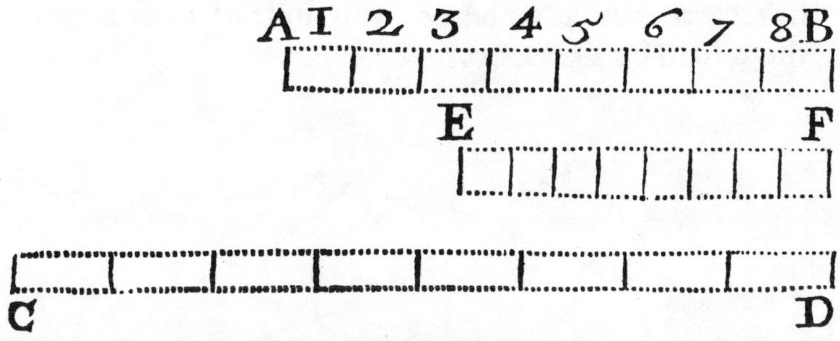

Let A B repreſent a line of ſuch a body compounded of eight Vibrating particles, as 1, 2, 3, 4, 5. 6, 7, 8, and ſuppoſe each of thoſe Particles to perform a million of ſingle Vibrations, and conſequently of occurſions with each other in a ſecond minute of time, their motion being of ſuch a Velocity impreſſed from the Ambient on the two extreme Particles 1 and 8. Firſt, if by any external power on the two extremes 1 and 8, they be removed further aſunder, as to C D, then ſhall all the Vibrative Particles be proportionably extended, and the number of Vibrations, and conſequently of occurſions be reciprocally diminiſhed, and conſequently their endeavour of receding from each other be reciprocally diminiſhed alſo. For ſuppoſing this ſecond Dimenſion of Length be to the firſt as 3 to 2, the length of the Vibrations, and conſequently of occurſions, be reciprocally diminiſhed. For whereas I ſuppoſed 1000000 in a ſecond of the former, here can be but *666666* in this, and conſequently the Spring inward muſt be in proportion to the Extenſion beyond its natural length.

Secondly, if by any external force the extreme particles be removed a third part nearer together than (the external natural force being alway the ſame both in this and the former inſtance, which is the ballance to it in its natural ſtate) the length of the Vibrations ſhall be proportionably diminiſhed, and the number of them, and conſequently of the occurſions be reciprocally augmented, and inſtead of 1000000, there ſhall be 1500000.

Having thus explained the moſt ſimple way of ſpringing in ſolid bodies, it will be very eaſie to explain the compound way of ſpringing, that is, by flexure, ſuppoſing only two of theſe lines joyned

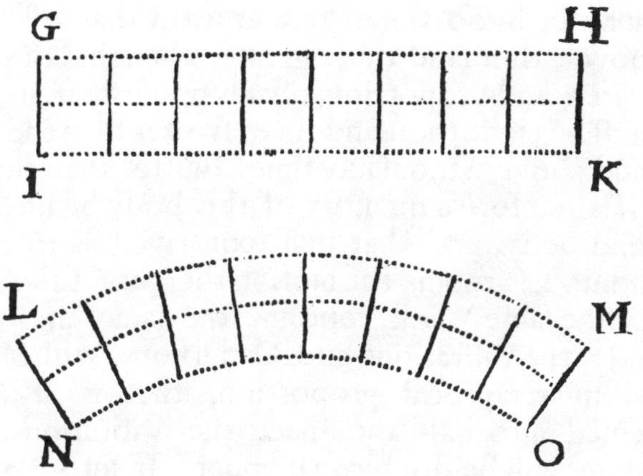

together as at G H I K, which being by any external power bended into the form L N N O, L M will be extended, and N O will be diminiſhed in proportion to the flexure, and conſequently the ſame proportions and Rules for its endeavour of reſtoring it ſelf will hold.

In the next place for fluid bodies, amongſt which the greateſt inſtance we have is air,though the ſame be in ſome proportion in all other fluid bodies.

The Air then is a body conſiſting of particles ſo ſmall as to be almoſt equal to the particles of the Heterogeneous fluid medium incompaſſing the earth. It is bounded but on one ſide, namely, towards the earth, and is indefinitely extended upward being only hindred from flying away that way by its own gravity, (the cauſe of which I ſhall ſome other time explain.) It conſiſts of the ſame particles ſingle and ſeparated, of which water and other fluids do, conjoyned and compounded, and being made of particles exceeding ſmall, its motion(to make its ballance with the reſt of the earthy bodies) is exceeding ſwift, and its Vibrative Spaces exceeding large, comparative to the Vibrative Spaces of other terreſtrial bodies. I ſuppoſe that of the Air next the Earth in its natural ſtate may be 8000 times greater than that of Steel, and above a thouſand times greater than that of common water, and proportionably I ſuppoſe that its motion muſt be eight thouſand times ſwifter than the former, and above a thouſand times ſwifter than the later. If therefore a quantity of this body be incloſed by a ſolid body, and that be ſo contrived as to compreſs it into leſs room, the motion thereof (ſuppoſing the heat the ſame) will continue the ſame, and conſequently the Vibrations and Occurſions will be increaſed in reciprocal proportion, that is, if it be Condenſed into half the ſpace the Vibrations and Occurſions will be double in number : If into a quarter the Vibrations and Occurſions will be quadruple, *&c.*

Again, If the conteining Veſſel be ſo contrived as to leave it more ſpace, the length of the Vibrations will be proportionably inlarged, and the number of Vibrations and Occurſions will be reciprocally diminiſhed, that is, if it be ſuffered to extend to twice its former dimenſions, its Vibrations will be twice as long, and the number of its Vibrations and Occurſions will be fewer by half, and conſequently its indeavours outward will be alſo weaker by half.

Theſe Explanations will ſerve *mutatis mutandis* for explaining the Spring of any other Body whatſoever.

Part II
BEGINNINGS OF KINETIC THEORY

Editor's Comments on Papers 4 Through 9

4 BERNOUILLI
Excerpt from *Concerning Properties and Motions of Elastic Fluids, but Especially of Air*

5 BOSCOVICH
Synopsis from *A Theory of Natural Philosophy*

6 BROWN
A Brief Account of Microscopical Observations Made in the Months of June, July, and August, 1827 on the Particles Contained in the Pollen of Plants; and on the General Existence of Active Molecules in Organic and Inorganic Bodies

7 DAVY
An Essay on Heat, Light, and the Combination of Light

8 HERAPATH
Excerpts from *Mathematical Inquiry into the Causes, Laws, and Principal Phenomena of Heat, Gases, Gravitation, etc.*

9 WATERSTON
Excerpts from *On the Physics of Media That Are Composed of Free and Perfectly Elastic Molecules in a State of Motion*

It was Daniel Bernoulli (1700–1782) who developed analytically Hooke's ideas on the possible association between the pressure of a gas and the motion of the constituent atoms. He was able to show that Boyle's law can be derived theoretically by calculating the average force exerted by atoms colliding with the walls of the containing vessel. Daniel Bernoulli was the son of Johann Bernoulli, an eminent Swiss mathematician. The Bernoulli family formed a veritable dynasty of scholars; Daniel's two brothers Nickolaus and Johann and their uncle Jacob all achieved widespread recognition in European scientific circles during the eighteenth century. There were even others of the same family whose biographies appear in the *Dictionary of Scientific Biography* (Gillespie, 1970). Daniel was born in Groningen, the Netherlands, but he received his formal education in Basel where his father held the chair in mathematics at the university. After spending several of his early professional years in Saint Petersburg, he returned to Basel in 1733 where in 1750 he became professor of physics. While in Saint Petersburg, Bernoulli worked with Leonhard Euler and began writing his famous treatise *Hydrodynamics.* This volume was published in Germany in 1738 and contains Bernoulli's most significant contributions to atomism and kinetic theory, including the derivation of Boyle's law. Extracts from it are reproduced here as Paper 4.

Bernoulli was influenced by his father in the belief that *vis viva* (the product of the mass and the square of the velocity or twice the quantity now known as "kinetic energy") is ultimately the more important dynamical quantity than momentum, the vector product of mass and velocity (see Lindsay, 1975). Although in his work on the properties of air he correctly relates an increase in elasticity (pressure) to an increase in heat and reasons the pressure is proportional to the square of the molecular velocity, he does not specifically connect *vis viva* with the concept of heat.

A somewhat isolated figure in the early history of kinetic theory was Roger Boscovich (in the Jugoslav language Rudjer J. Bośkovic) (1711–1787). Although his work had no systematic experimental foundation and did not by itself influence to any great extent the subsequent eighteenth- and nineteenth- century development of kinetic theory, it contains enough allusions to what are now considered atomic characteristics, and indeed from the standpoint of energy, to justify some attention to him in this volume. He was born in Dubrovnik, Jugoslavia, and received his early education in Jesuit schools there. After an extended period of study and teaching at the Collegium Romanum in Rome, he was in 1744 ordained a priest in the Society of Jesus. In 1764 he was appointed professor of mathematics in Pavia, then part of Austria. From Pavia he went to Milan where he played an important part in setting up the observatory at Brera near Milan. In 1773 Boscovich went to Paris where he served several years as director of naval optics for King Louis XV. He returned to Italy in 1782 and spent the remaining five years of his life editing his collected works for publication.

Although primarily a mathematician, Boscovich had much interest in the practical applications of physics and astronomy. He was an ardent traveler, and his wide-ranging intellectual activities assured his acceptance in leading scientific circles throughout Europe. He was elected a foreign member of the Royal Society of London in 1764.

Boscovich's best-known work *Theoria Philosophiae Naturalis* was first published in 1758. It was the culmination of efforts to attribute all natural phenomena to an atomic constitution of matter. He rejected the hard sphere atomic model of Newton and others. His atoms are represented by identical and unchangeable mathematical points subject to alternating forces of either attraction or repulsion depending on their distance of separation. These forces had to be chosen heuristically in order to explain such phenomena as liquefaction, gas pressure, surface tension, and crystalization. Boscovich's theory did not attract much attention at the time since it could not be tested effectively with the experimental techniques available. However, in retrospect, the attention he focused on the atom as the source of all forces in the physical world was significant. The emphasis on more abstract forms of mathematical representation of atoms indeed anticipated developments in twentieth-century atomic and nuclear physics. In fact, Boscovich's atoms have recently been cited as having rather striking resemblances to the quarks, now being proposed as constituents of the atomic nucleus (see Rinard 1976). Boscovich's emphasis on the importance of the interaction of his point atoms in the explanation of the properties of matter contained a germ of the idea of potential energy as used in modern atomic physics.

Paper 5 consists of extracts from the English translation of Boscovich's *Theoria Philosophiae Naturalis.*

Some visual observations suggesting the physical plausibility of atomic theory were described and interpreted by the celebrated Scottish botanist Robert Brown (1773–1858). The name "Brownian motion" persists to this day to designate the random motion of microscopic particles suspended in a fluid. Born in Montrose, Scotland, Brown studied medicine at the University of Edinburgh but did not stay long enough to receive a degree there. After a brief period of service in the British army, he devoted the rest of his life to the study of botany, being appointed in 1817 keeper of the botanical department of the British Museum, a position he held until his death.

Brown's observations of the motion of plant pollen particles suspended in a liquid led to his famous 1828 paper, which also indicated that inorganic particles exhibited the same type of random motion as well. Brown's work, as subject to more detailed study and speculation later in the century, was a pioneer effort in the establishment of the molecular theory. We reproduce Brown's celebrated paper here as Paper 6. We return to a consideration of the Brownian motion and the later developments concerning it in Papers 29 and 30.

The caloric theory of heat, which treated heat as a substance, was the prevailing theory at the end of the eighteenth century and for nearly all of the first half of the nineteenth century. The acceptance of the theory had much to do with the difficulty of establishing the idea of energy as a general concept, applying it to all physical phenomena; yet it did not go unchallenged during that time. Two people stand out prominently as critics of the theory: Benjamin Thompson (Count Rumford) (1753–1814) and Humphry Davy (1778–1829).

Rumford experimented on the heat produced by friction in the boring of cannon and through his researches concluded the caloric hypothesis was not tenable. His fundamental paper is reproduced in the Benchmark volume *Energy: Historical Development of the Concept* (Lindsay, 1975).

Closely associated with Rumford in his opposition to the caloric theory was Humphry Davy, a native of Penzance in Cornwall who had a distinguished career in the Royal Institution of Great Britain. Largely self-taught in his chosen field of chemistry, he became interested in the nature of heat early in his studies. In 1798 he became an assistant to Dr. Thomas Beddoes of Bristol, a physician interested in the use of gases for therapeutic purposes. With laboratory facilities thus made available, Davy began a wide variety of experimental investigations in chemistry. In 1799, when he was barely twenty years old, he devised an experiment that he felt would decisively test the caloric theory: melting ice by friction in a vacuum. His results were presented in "An Essay on Heat, Light, and the Combinations of Light," printed in a book brought out by Dr. Beddoes. Paper 7 contains Davy's announced results and his interpretation of them. Davy's experiment has been viewed in modern times with grave doubt. Even in his own time it produced considerable controversy and can hardly be taken as decisive proof of the invalidity of the caloric theory. Nevertheless, the idea was ingenious and did serve to focus Davy's attention on the nature of heat and strengthened his adherence to the mechanical theory. The experiment also helped to bring Davy to the attention of Rumford and hence undoubtedly assisted in promoting the former's scientific career. For Davy's continued opposition to the caloric theory of heat, the reader may examine Paper 27 in the Benchmark volume *Energy: Historical Development of the Concept* (Lindsay, 1975).

The inability of Rumford and Davy to win acceptance of a molecular theory of heat was due in considerable measure to their omission of any mention of how the heat transferred across empty space by radiation could be the same kind of heat allegedly transferred by motion of material particles. After all, the experiments of both could be explained readily by the caloric theory. There was always enough caloric available for any purpose, and as a substance, it could travel both through matter and empty space. A successful molecular theory would have to be one that would yield unequivocally a quantitative connection between the postulated mechanical properties and observed physical phenomena, including heat.

John Herapath (1790–1868), an Englishman, and John Waterston (1811–1883), a Scot, used mathematical lines of reasoning in atomic physics, which others following in their wake were able to exploit successfully. Neither name figures prominently in modern textbooks, but their basic contributions called attention to the possibilities of a mathematically-based molecular theory of matter.

John Herapath, born in Bristol, England, was a self-educated amateur mathematician and physicist who spent a considerable part of his early years trying to establish a mathematical basis for the explanation of heat, gases, and gravitation. His efforts led to a paper that appeared in the *Annals of Philosophy* in 1821, after having first been rejected by the Royal Society. Apparently he had no knowledge of Bernoulli's earlier work along similar lines. Herapath postulated a model of colliding elastic molecules, but he made the unfortunate assumption that heat and temperature are proportional to the individual momenta of the molecule. However, because he made the additional error of setting the pressure proportional to the square of the temperature, he still arrived at the implicit result that the product of the pressure and volume is proportional to the square of the mean molecular velocity and the molecular mass. This at once implied a definite connection with *vis viva,* although Herapath ignored it and chose to emphasize the fundamental importance of momentum, possibly because of Descartes' influence. Nevertheless, Herapath should be credited with reviving Bernoulli's approach and recognizing the dependence of temperature on molecular velocity even if not to the correct power. Herapath never achieved acceptance as a member of the scientific community and spent most of his life associated with private tutoring and the proprietorship of a railway magazine. Herapath's paper from the *Annals of Philosophy* is reprinted here as Paper 8.

James Waterston was a more tragic figure because his analysis turned out to have been correct from the beginning, while his critics were wrong. Waterston was a graduate of the University of Edinburgh where he studied physics and mathematics. After doing some railway engineering work in England, he went to Bombay in 1839 to be an instructor at the East India Company's cadet school. While there, he had the time and facilities to indulge his interests in fundamental science. In 1845 he submitted a paper to the Royal Society in which he developed kinetic theory in a fashion that was substantially correct. Starting from an elastic particle model, he was able to confirm Boyle's law by showing that at constant molecular velocity, the pressure is proportional to the density and inversely proportional to the volume. He deduced the law that the pressure is directly proportional to the mean square molecular velocity. By using the gas laws of Gay-Lussac and Charles, which state the direct dependence of volume on absolute temperature, Waterston inferred that the temperature of a gas depends directly on its *vis viva,* or kinetic energy. At the time he did this, Waterston was not aware of Herapath's work; later when he read Herapath's paper, he pointed out the error in the assumed dependence of temperature on momentum and graciously recognized the importance of Herapath's general line of reasoning. Waterston further established as an important corollary the property that for gases at the same temperature, the root-mean-square molecular velocity is proportional to the inverse of the square root of the molecular mass. The Royal Society rejected Waterston's paper, but in 1892 Lord Rayleigh reactivated it from the society's archives and arranged for its publication in the transactions of the society. This paper is reproduced here as Paper 9. Included among Rayleigh's introductory comments is the following: "The character of the advance to be dated from this paper will be at once understood when it is realized that Waterston was the first to introduce into the theory that heat and temperature are to be measured by *vis viva.* This enabled him at a stroke to complete Bernoulli's explanation of pressure by showing the accordance of the hypothetical medium with the law of Dalton and Gay-Lussac." [From Rayleigh's memoir at the time of publication (Lindsay 1973).]

Herapath's and Waterston's works concluded the beginning stage of the formulation of a kinetic theory of atomic particles. The transition had been made from levels of description that were either empirically based on the gas laws or highly idealized theoretical models to a systematic deductive approach in which the first major consequences were a plausible association between the kinetic energy of atoms and temperature and the impending unification of theories of mechanics and heat.

REFERENCES

Gillespie, C. C., ed., 1970–1976. *Dictionary of Scientific Biography.* New York: Charles Scribner's Sons.

Lindsay, R. Bruce, ed., 1973. *Acoustics: Historical and Philosophical Development. Benchmark Papers in Acoustics,* Vol. 2. Stroudsburg, Pennsylvania: Dowden, Hutchinson & Ross, pp. 407–410.

Lindsay, R. Bruce, ed., 1975. *Energy: Historical Development of the Concept.* Stroudsburg, Pennsylvania: Dowden, Hutchinson & Ross, pp. 108–165, 206–227, 246–251.

Rinard, Phillip M., 1976. "Quarks and Boscovich," *Am. J. Phys.,* **44,** 704.

4

Reprinted from pages 226–230 of *Hydrodynamics and Hydraulics.* New York: Dover Publications, 1968.

CONCERNING PROPERTIES AND MOTIONS OF ELASTIC FLUIDS, BUT ESPECIALLY OF AIR

D. Bernouilli

1. Now being about to consider elastic fluids, we may ascribe to them a constitution that coincides with all properties known so far, in order that thus also a path be provided to the remaining, still insufficiently explored properties. But the outstanding properties of elastic fluids are stated as follows: (1) they are heavy, (2) they extend in all directions, unless they are confined, and (3) they allow themselves to be compressed continuously more and more as the compressing forces increase. Air, to which our present considerations pertain mostly, is composed in this way.

§2. And so consider a cylindrical vessel *ACDB*, placed vertically (Fig. 56), and in it a movable lid *EF*, on top of which lies the weight *P*. Let the cavity *ECDF* contain extremely small bodies agitated in a

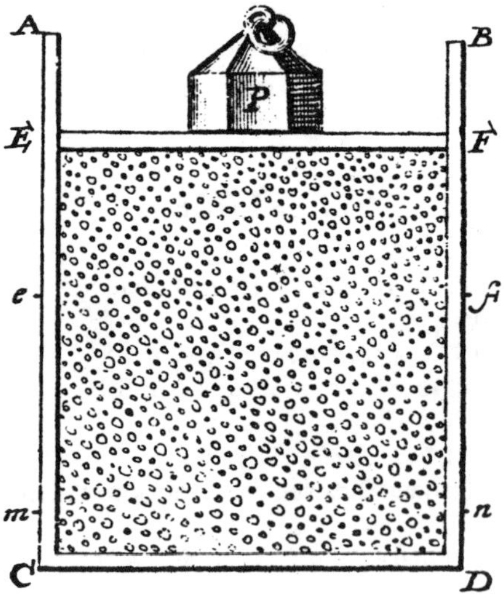

FIGURE 56

very rapid motion; thus the small bodies, while they impinge on the lid *EF* and also support the same by their continually repeated impacts, compose an elastic fluid which expands if the weight *P* is removed or diminished; this is compressed if the same is increased, and it gravitates on the horizontal base *CD* not at all differently than if it existed with no elastic property. Indeed, whether the small bodies are at rest or agitated, they do not change gravity, so that the base sustains either the weight or the elasticity of the fluid. Therefore, let us substitute for air a fluid that is consistent with the primary properties of elastic fluids, and thus we will explain some properties which have been already detected in air, and we will illustrate further some others [that are] not yet sufficiently investigated.

§3. We shall consider the small bodies enclosed by the cavity of the cylinder as infinite in number, and since they occupy the space *ECDF*, let us say that the latter forms the natural air, to the measures of which all [other measures] are to be referred; and thus the weight *P* holding the lid in the position *EF* does not differ from the pressure of the Atmosphere lying above it, which we therefore shall designate henceforth by *P*.

But let it be noted that this pressure is not at all equal to the absolute weight of the vertical cylinder of air lying above the lid *EF* in the atmosphere, which authors so far have affirmed inconsiderately; but that pressure is equal to the fourth proportional to the surface of the earth, the size of the lid *EF*, and the weight of the entire atmosphere on the surface of the earth.

§4. Now the weight π is sought which can compress the air *ECDF* into the space *eCDf*, the velocities of the particles in either air (the natural and the compressed), of course, having been assumed the same; moreover, let $EC = 1$, and $eC = s$; but since the lid *EF* is transferred to *ef*, it suffers a greater pressure from the fluid in two ways: *firstly*, because the number of particles is now greater in proportion to the space in which they are contained, and *secondly*, because any particle repeats the impetus more often. In order to perform correctly the calculation of the increment which depends on the *first cause*, we shall consider the particles as resting, and we shall make *n* the number of those which are adjacent to the lid in the position *EF*, and the equivalent number for the location of the lid at *ef* will be $n\Big/\left(\frac{eC}{EC}\right)^{2/3}$, or $n/s^{2/3}$.

But let it be noted that the fluid is considered by us not more compressed in the lower part than in the upper part, which is so because the weight *P* is just as infinitely much larger than the very weight of the fluid. Hence it is clear that with this designation the force of the fluid varies in proportion to the numbers n and $n/s^{2/3}$, that is, as $s^{2/3}$ is to 1. But what pertains to the other increment arising from the *second cause* is found by observing the motion of the particles; and thus it is apparent that the impulse occurs the more often, the closer the particles are located to each other; of course, the number of impulses will be reciprocal to the median distance between the surfaces of the particles, and these median distances will be determined as follows:

We assume the particles to be spherical, and we shall call D the median distance between the centers of the small spheres for the position EF of the lid, and the diameter of a small sphere we shall designate by d; so the median distance between the surfaces of the small spheres will be $D - d$; but it is evident that at the position ef of the lid the median distance between the centers of the small spheres will be $D\sqrt[3]{s}$, and therefore the median distance between the surfaces of the small spheres is $D\sqrt[3]{s} - d$. Therefore, with respect to the second cause the force of the natural air $ECDF$ will be to the force of the compressed air $eCDf$ as $\frac{1}{D-d}$ is to $\frac{1}{D\sqrt[3]{s}-d}$, or as $D\sqrt[3]{s} - d$ is to $D - d$; for both causes together, however, the aforementioned forces will be as $s^{2/3} \cdot (D\sqrt[3]{s} - d)$ is to $D - d$.

For the ratio D to d we can substitute another, more intelligible one: namely, if we consider that the lid EF, [when] depressed by an infinite weight, descends to the position mn at which all particles touch each other, and if we designate the line mC by m, D will be to d as 1 is to $\sqrt[3]{m}$, which ratio being substituted, finally the forces of the natural air $ECDF$ and of the compressed, $eCDf$, will be as $s^{2/3} \cdot (\sqrt[3]{s} - \sqrt[3]{m})$ is to $1 - \sqrt[3]{m}$, or as $s - \sqrt[3]{mss}$ to $1 - \sqrt[3]{m}$. Therefore,

$$\pi = \frac{1 - \sqrt[3]{m}}{s - \sqrt[3]{mss}} \cdot P.$$

§5. From all phenomena we can judge that natural air can be compressed into an almost infinitely small space; therefore, with $m = 0$ having been assumed, π becomes $\frac{P}{s}$, so that the compressing weights are almost in inverse proportion to the space which the air occupies when compressed differently; manifold experience has confirmed this. This rule can also certainly be accepted safely for air rarer than natural; but I have not explored sufficiently whether it can be also [accepted] for very much denser air; and indeed experiments have not yet been performed with that accuracy which is required here; for defining the value of the letter m, there is need of only one, but of one to be formed most accurately and certainly with violently compressed air; however, let the degree of heat in the air, while it is being compressed, be carefully kept unchanged.

§6. Meanwhile, the elasticity of air is increased not only by compression but also by increased heat, and since it is established that heat is spread out everywhere by increasing internal motion of the particles, it follows that an increased elasticity of air not changing volume discloses a more intensive motion in the air particles, which agrees correctly with our hypothesis; it is indeed evident that the greater the weight P required for keeping the air in the position $ECDF$, the greater the velocity at which the air particles are agitated. By all means it is not difficult to see that the weight P will follow the ratio of the square of that velocity, for the reason that the number of impacts as well as their intensity increases equally with an increased velocity; but [each one] separately is proportional to the weight P.

If, therefore, the velocity of the air particles is called v, the weight which it can sustain at the position EF of the lid will be vvP and in the position ef it will be $\frac{1 - \sqrt[3]{m}}{s - \sqrt[3]{mss}} \cdot vvP$, or approximately $\frac{vvP}{s}$, because, as we have seen the number m is extremely small with respect to unity and the number s.

§7. That theorem which I presented in the preceding paragraph, by which, namely, it is indicated that *in all air of any density whatever but of the same prescribed degree of heat, the elasticities are as the densities, and therefore also the increments of the elasticities, which are created by equally increased heat, are proportional to the densities*, that theorem, I say, Mr. Amontons was taught by experience, and he recorded it in *the memoirs of the Royal Academy of Science of Paris for the year 1702*. The sense of this theorem is that if, for example, natural air of moderate heat sustains a weight of 100 pounds imposed on a given surface, and then its heat is increased until it can carry 120 pounds on the same surface and at the same volume, then it will occur that the same air, compressed to half the space and possessing the same degrees of heat, can carry, respectively, 200 pounds and 240 pounds, so that increments of 20 pounds and 40 pounds, proportional to the densities, are generated in either case by the increased heat. He further affirms that the expansion of air, which he calls tempered, is to the expansion of the same air with the heat of boiling water approximately as 3 to 4 or, more accurately as 55 to 73. However, I have learned from performed experiments that very hot air, such as it is in the hottest summer in this country, is not yet of such an expansion as Mr. Amontons attributes to tempered air; I believe that not at the equator itself is the air ever of this heat. But I believe that my experiments are to be trusted more than the Amontonian ones for the reason that in the latter the air will not preserve its volume, and to this variation no consideration was given by the Author in the calculation. I learned that the expansion of air which was very cold here in Petersburg on 25 December 1731 (old statutes) was to the expansion of similar air attributed with the same heat as boiling water as 523 is to 1000.

But in the year 1733, on the 21st day of January the cold was much more intense, and for this I observed that the elasticity of the air corresponds to within half of that which the same air has when heated to boiling water. But when the heat of air in a shadowed place was greatest in the year 1731, it had an elasticity of approximately $\frac{4}{3}$ and more accurately $\frac{100}{76}$ of that which the coldest air had and $\frac{2}{3}$ of that which air of the same heat as boiling water has; therefore, the greatest variations in the air in these places are contained within the limits 3 and 4; I have read that in England they do not exceed the limits 7 and 8. But I believe that the heat of air, the elasticity of which equals three quarters of the elasticity of air as hot as boiling water, is almost intolerable for an animal body.

[*Editor's Note:* Material has been omitted at this point.]

5

Reprinted from pages 10–18 of *A Theory of Natural Philosophy.* Cambridge, Mass.: The M.I.T. Press, 1966.

SYNOPSIS OF THE WHOLE WORK (FROM THE VIENNA EDITION)

Roger Joseph Boscovich, S. J.

PART I

IN the first six articles, I state the time at which I evolved my Theory, what led me to it, & where I have discussed it hitherto in essays already published: also what it has in common with the theories of Leibniz and Newton; in what it differs from either of these, & in what it is really superior to them both. In addition I state what I have published elsewhere about equilibrium & the centre of oscillation; & how, having found out that these matters followed quite easily from a single theorem of the most simple & elegant kind, I proposed to write a short essay thereon; but when I set to work to deduce the matter from this principle, the discussion, quite unexpectedly to me, developed into a whole work of considerable magnitude.

From this until Art. 11, I explain the Theory itself: that matter is unchangeable, and consists of points that are perfectly simple, indivisible, of no extent, & separated from one another; that each of these points has a property of inertia, & in addition a mutual active force depending on the distance in such a way that, if the distance is given, both the magnitude & the direction of this force are given; but if the distance is altered, so also is the force altered; & if the distance is diminished indefinitely, the force is repulsive, & in fact also increases indefinitely; whilst if the distance is increased, the force will be diminished, vanish, be changed to an attractive force that first of all increases, then decreases, vanishes, is again turned into a repulsive force, & so on many times over; until at greater distances it finally becomes an attractive force that decreases approximately in the inverse ratio of the squares of the distances. This connection between the forces & the distances, & their passing from positive to negative, or from repulsive to attractive, & conversely, I illustrate by the force with which the two ends of a spring strive to approach towards, or recede from, one another, according as they are pulled apart, or drawn together, by more than the natural amount.

From here on to Art. 16 I show that it is not merely an aggregate of forces combined haphazard, but that it is represented by a single continuous curve, by means of abscissæ representing the distances & ordinates representing the forces. I expound the construction & nature of this curve; & I show how it differs from the hyperbola of the third degree which represents Newtonian gravitation. Finally, here too I set forth the scope of the whole work & the nature of the parts into which it is divided.

These statements having been made, I start to expound the whole of the analysis, by which I came upon a Theory of this kind, & from which I believe I have deduced the whole of it by a straightforward & perfectly rigorous chain of reasoning. I contend indeed, from here on until Art. 19, that, in the collision of solid bodies, either there must be compenetration, or the Law of Continuity must be violated by a sudden change of velocity, if the bodies come into immediate contact with unequal velocities. Now since the Law of Continuity must (as I prove that it must) be observed in every case, I infer that, before the bodies reach the point of actual contact, their velocities must be altered by some force which is capable of destroying the velocity, or the difference of the velocities, no matter how great that may be.

From Art. 19 to Art. 28 I consider the artifice, adopted for the purpose of evading the strength of my argument by those who deny the existence of hard bodies; as a matter of fact this cannot be used as an argument against me by the Newtonians, or the Corpuscularians in general, for they assume that the elementary particles of solids are perfectly hard. Moreover, those who admit that all the particles of solids, however small they may be, are soft or elastic, yet do not escape the difficulty, but transfer it to prime surfaces, or points; & here a sudden change would be made & the Law of Continuity violated. In the same connection I consider a certain verbal quibble, used in a vain attempt to foil the force of my reasoning.

In the next articles, 28 & 29, I refute a further pair of arguments advanced by others; in the first of these, in order to evade my reasoning, someone states that there is compenetration of the primary elements of matter; in the second, the points of matter are said to be moved with regard to one another, even when they are absolutely at rest as regards position. In reply to the first artifice, I prove the principle of impenetrability by induction; & in reply to the second, I expose an equivocation in the meaning of the term *motion*, an equivocation upon which the whole thing depends.

Then, in Art. 30, 31, I show in what respect I differ from Maclaurin, who, having considered the same point as myself, came to the conclusion that in the collision of bodies the Law of Continuity was violated; whereas I obtained the whole of my Theory from the assumption that this law must be unassailable.

At this point therefore, in order that the strength of my deductive reasoning might be shown, I investigate the Law of Continuity; and from Art. 32 to Art. 38, I set forth its nature, & what is meant by a continuous change through all intermediate stages, such as to exclude any sudden change from any one magnitude to another except by a passage through intermediate stages; & I call in geometry as well to help my explanation of the matter. Then I investigate its truth first of all by induction; &, investigating the principle of induction itself, as far as Art. 44, I show whence the force of this principle is derived, & where it can be used. I give by way of illustration an example in which impenetrability is derived entirely by induction; & lastly I apply the force of the principle to demonstrate the Law of Continuity. In the articles that follow I consider certain cases of two kinds, in which the Law of Continuity appears to be violated, but is not however really violated.

After this proof of the principle of continuity procured through induction, in Art. 48, I undertake another proof of a metaphysical kind, depending upon the necessity of a limit on either side for either real quantities or for a finite series of real quantities; & indeed it is impossible that these limits should be lacking, either at the beginning or the end. I demonstrate the force of this reasoning in the case of local motion, & also in geometry, in the next two articles. Then in Art. 52 I explain a certain difficulty, which is derived from the fact that, at the instant at which there is a passage from *non-existence* to *existence*, it appears according to a theory of this kind that we must have at the same time both *existence* and *non-existence*. For one of these belongs to the end of the antecedent series of states, & the other to the beginning of the consequent series. I consider fairly fully the solution of this problem; and I call in geometry as well to assist in giving a visual representation of the matter.

In Art. 63, after summing up all that has been said about the Law of Continuity, I apply the principle to exclude the possibility of any sudden change from one velocity to another, except by passing through intermediate velocities; this would be contrary to the very full proof that I give for continuity, as it would lead to our having two velocities at the instant at which the change occurred. That is to say, there would be the final velocity of the antecedent series, & the initial velocity of the consequent series; in spite of the fact that it is quite impossible for a moving body to have two different velocities at the same time. Moreover, in order to illustrate & prove the point, from here on to Art. 72, I consider velocity itself; and I distinguish between a potential velocity, as I call it, & an actual velocity; I also investigate carefully many matters that relate to the nature of these velocities & to their changes. Further, I settle several difficulties that can be brought up in opposition to the proof of my Theory, in consequence.

This done, I then conclude from the principle of continuity that, when one body with a greater velocity follows after another body having a less velocity, it is impossible that there should ever be absolute contact with such an inequality of velocities; that is to say, a case of the velocity of each, or of one or the other, of them being changed suddenly at the instant of contact. I assert on the other hand that the change in the velocities must begin before contact. Hence, in Art. 73, I infer that there must be a cause for this change: which is to be called "force." Then, in Art. 74, I prove that this force is a mutual one, & that it acts in opposite directions; the proof is by induction. From this, in Art. 75, I infer that such a mutual force may be said to be repulsive; & I undertake the investigation of the law that governs it. Carrying on this investigation as far as Art. 80, I find that this force must increase indefinitely as the distance is diminished, in order that it may be capable of destroying any velocity, however great that velocity may be. Moreover, I find that, whilst the force must be indefinitely increased as the distance is indefinitely decreased, it must be on the contrary attractive at very great distances, as is the case for gravitation. Hence I infer that there must be a limit-point forming a boundary between attraction & repulsion; & then by degrees I find more, indeed very many more, of such limit-points, or points of transition from attraction to repulsion, & from repulsion to attraction; & I determine the form of the entire curve, that expresses by its ordinates the law of these forces.

So far I have been occupied in deducing and settling the law of these forces. Next, in Art. 81, I derive from this law the constitution of the elements of matter. These must be quite simple, on account of the repulsion at very small distances being immensely great; for if by chance those elements were made up of parts, the repulsion would destroy all connections between them. Then, as far as Art. 88, I consider the point, as to whether these elements, as they must be simple, must therefore be also of no extent; &, having explained what is called "virtual extension," I reject it by the principle of induction. I then consider the difficulty which may be brought forward from an example of this kind of extension; such as is generally admitted in the case of the indivisible and one-fold soul pervading a divisible & extended portion of the body, or in the case of the omnipresence of God. Next I consider the difficulty that may be brought forward from an analogy with rest; for here in truth one point of space must be connected with a continuous series of instants of time, just as in virtual extension a single instant of time would be connected with a continuous series of points of space. I show that there can neither be perfect rest anywhere in Nature, nor can there be at all times a perfect analogy between time and space. In this connection, I also gather a large harvest from such a conclusion as this; showing, as far as Art. 91, the great advantage of simplicity, indivisibility, & non-extension in the elements of matter. For they do away with the idea of a passage from a continuous vacuum to continuous matter through a sudden change. Also they render unnecessary any limit to density: this, in a Theory like mine, can be just as well increased to an indefinite extent, as it can be indefinitely decreased: whilst in the ordinary theory, as soon as contact takes place, the density cannot in any way be further increased. But, most especially, they do away with the idea of everything continuous coexisting; & when this is done away with, the majority of the greatest difficulties vanish. Further, nothing infinite is found actually existing; the only thing possible that remains is a series of finite things produced indefinitely.

These things being settled, I investigate, as far as Art. 99, the point as to whether elements of this kind are to be considered as being homogeneous or heterogeneous. I find my first evidence in favour of homogeneity—at least as far as the complete law of forces is concerned—in the equally great homogeneity of the first repulsive branch of my curve of forces for very small distances, upon which depends impenetrability, & of the last attractive branch, by which gravity is represented. Moreover I show that there is nothing that can be proved in opposition to homogeneity such as this, that can be derived from either the Leibnizian principle of "indiscernibles," or by induction. I also show whence arise those differences, that are so great amongst small composite bodies, such as we see in boughs & leaves; & I prove, by induction & analogy, that the very nature of things leads us to homogeneity, & not to heterogeneity, for the elements of matter.

These matters are all connected with the proof of my Theory. Having accomplished this, before I start to gather the manifold fruits to be derived from it, I proceed to consider the objections to my theory, such as either have been already raised or seem to me capable of being raised; first against forces in general, secondly against the law of forces that I have enunciated & proved, & finally against those indivisible, non-extended points that are deduced from a law of forces of this kind.

First of all then, in order that I may satisfy even those who are confused over the empty sound of certain terms, I show, in Art. 101 to 104, that these forces are not some sort of mysterious qualities; but that they form a readily intelligible mechanism, since both the idea of them is perfectly distinct, as well as their existence, & in addition the law that governs them is demonstrated in a direct manner. To Mechanics belongs every discussion concerning motions that arise from given forces without any direct impulse. In Art. 104 to 106, I show that no sudden change takes place in passing from repulsions to attractions or *vice versa*; for this transition is made through every intermediate quantity. Then I pass on to consider the objections that are made against the whole form of my curve. I show indeed, from here on to Art. 116, that all repulsions cannot be taken to come from a decreased attraction; that repulsions belong to the self-same series as attractions, differing from them only as less does from more, or negative from positive. From the very nature of the curves (for which, the higher the degree, the more points there are in which they can intersect a right line, & vastly more such curves there are), I deduce that there is more reason for assuming a curve of the nature of mine (so that it may cut a right line in a large number of points, & thus give a large number of transitions of the forces from repulsions to attractions), than for assuming a curve that, since it does not cut the axis anywhere, will represent attractions alone, or repulsions alone, at all distances. Further, I point out that repulsive forces, and a multiplicity of transitions are directly demonstrated, & the whole form of the curve is a matter of deduction; & I also show that it is not formed of a number of arcs differing in nature connected together haphazard;

but that it is absolutely one-fold. This one-fold character I demonstrate in the Supplements in a very evident manner, giving a method by which a simple and uniform equation may be obtained for a curve of this kind. Although, as I there point out, this law of forces may be mentally resolved into several, and these may be represented by several corresponding curves, yet that law, actually unique, may be compounded from all of these together by means of the unique, continuous & one-fold curve that I give.

In Art. 121, I start to give a refutation of those objections that may be raised from a consideration of the fact that the law of gravitation, decreasing in the inverse duplicate ratio of the distances, demands that there should be an attraction at very small distances, & that it should increase indefinitely. However, I show that the law is nowhere exactly in conformity with a ratio of this sort, unless we add explanations that are merely imaginative; nor, I assert, can a law of this kind be deduced from astronomy, that is followed with perfect accuracy even at the distances of the planets & the comets, but one merely that is at most so very nearly correct, that the difference from the law of inverse squares is very slight. From Art. 124 onwards, I examine the value of the argument that can be drawn in favour of a law of this sort from the view that, as some have thought, it is the best of all, & that on that account it was selected by the Founder of Nature. In connection with this I examine the principle of Optimism, & I reject it; moreover I prove conclusively that there is no reason why this sort of law should be supposed to be the best of all. Further in the Supplements, I show to what absurdities a law of this sort is more likely to lead; & the same thing for other laws of an attraction that increases indefinitely as the distance is diminished indefinitely.

In Art. 131 I pass from forces to elements. I first of all show the reason why we may not appreciate the idea of non-extended points; it is because we are unable to perceive them by means of the senses, which are only affected by masses, & these too must be of considerable size. Consequently we have to build up the idea by a process of reasoning; & this we can do without any difficulty. In addition, I point out that I am not the first to introduce indivisible & non-extended points into physical science; for the "monads" of Leibniz practically come to the same thing. But I show that, by rejecting the idea of continuous extension, I remove the whole of the difficulty, which was raised against the disciples of Zeno in years gone by, & has never been answered satisfactorily; namely, the difficulty arising from the fact that by no possible means can continuous extension be made up from things of no extent.

In Art. 140 I show that the principle of induction yields no argument against these indivisibles; rather their existence is demonstrated by that principle, for continuity is self-contradictory. On this assumption it may be proved, by arguments originated by myself, that the primary elements are indivisible & non-extended, & that there does not exist anything possessing the property of continuous extension. From Art. 143 onwards, I point out the only connection in which I shall admit continuity, & that is in motion. I state the idea that I have with regard to space, & also time: the nature of these I explain much more fully in the Supplements. Further, I show that continuity itself is really a property of motions only, & that in all other things it is more or less a false assumption. Here I also consider some examples in which continuity at first sight appears to be violated, such as in some of the properties of light, & in certain other cases where things increase by addition of parts, and not by intussumption, as it is termed.

From Art. 153 onwards, I show how greatly these points of mine differ from object-souls. I consider how it comes about that continuous extension seems to be included in the very idea of a body; & in this connection, I investigate the origin of our ideas & I explain the prejudgments that arise therefrom. Finally, in Art. 165, I lightly sketch what might happen to enable points that are of no extent, & at a distance from one another, to coalesce into a coherent mass of any size, endowed with those properties that we experience in bodies. This, however, belongs to the third part; & there it will be much more fully developed. This finishes the first part.

PART II

In Art. 166 I state the theme of this second part; and in Art. 167 I declare what matters are to be considered more especially in connection with the curve of forces. Coming to the consideration of these matters, I first of all, as far as Art. 172, investigate the arcs of the curve, some of which are attractive, some repulsive and some asymptotic. Here a marvellous number of different cases present themselves, & to some of them there are noteworthy corollaries; such as that, since a curve of this kind is capable of possessing a considerable number of asymptotes, there can arise a series of perfectly similar cosmi, each of which will act upon all the others as a single inviolate elementary system. From Art. 172

to Art. 179, I consider the areas included by the arcs; these, corresponding to different segments of the axis, may be of any magnitude whatever, either great or small; moreover they measure the increment or decrement in the squares of the velocities. Then, on as far as Art. 189, I investigate the approach of the curve to the axis; both when the former is cut by the latter, in which case there are transitions from repulsion to attraction and from attraction to repulsion, which I call 'limits,' & use very largely in every part of my Theory; & also when the former is touched by the latter, & the curve once again recedes from the axis. I consider, too, as a case of approach, recession to infinity along an asymptotic arc; and I investigate what transitions, or limits, may arise from such a case, & whether such are admissible in Nature.

In Art. 189, I pass on from the consideration of the curve to combinations of points. First, as far as Art. 204, I deal with a system of two points. I work out those things that concern their mutual forces, and motions, whether they are left to themselves or projected in any manner whatever. Here also, having explained the connection between these motions & the distances of the limits, & different cases of oscillations, whether they are affected by external action of other points, or are not so disturbed, I make an anticipatory note of the great use to which this will be put in the third part, for the purpose of explaining various kinds of cohesion, fermentations, conflagrations, emissions of vapours, the properties of light, elasticity and flexibility.

There follows, from Art. 204 to Art. 239, the much more fruitful consideration of a system of three points. The forces connected with them can in general be easily determined for any given positions of the points; but, when any position & velocity are given, the motions have not yet been obtained by geometricians in such a form that the general calculation can be performed for every possible case. So I proceed to consider the forces, & the huge variation that different combinations of the points beget, although they are only three in number, as far as Art. 209. From that, on to Art. 214, I consider certain things that have to do with the forces that arise from the action, on each of the points, of the other two together, & how these urge the third point not only to approach, or recede from, themselves, but also in a direction at right angles; in this connection there comes forth an analogy with solidity, & a truly immense difference between the several cases when the distances are very small, & the greatest conformity possible at very great distances such as those at which gravity acts; & I point out what great use will be made of this also in explaining the constitution of Nature. Then up to Art. 221, I give ocular demonstrations of the huge differences that there are in the laws of forces with which two points act upon a third, whether it lies in the right line joining them, or in the right line that is the perpendicular which bisects the interval between them; this I do by constructing, from the primary curve, curves representing the composite forces. Then in the two articles that follow, I consider the case, a really important one, in which, by merely changing the position of the two points, the third point, at any and the same definite interval situated at the same distance from the middle point of the interval between the two points, will be either continually attracted, or continually repelled, or neither attracted nor repelled; & since a difference of this kind should hold to a much greater degree in masses, I point out, in Art. 222, the great use that will be made of this also in Physics.

At this point then, in Art. 223, I pass from the forces derived from two points to the consideration of a whole system of them; and, as far as Art. 228, I study three points situated in a right line, from the mutual forces of which there arise certain relations, which I return to later in much greater generality; in this connection also are outlined, for three points only, matters that have to do with rods, either rigid, flexible or elastic, and with the lever, as well as many other things; these, too, are treated much more generally later on, when I consider masses. Then right on to Art. 238, I consider three points that do not lie in a right line, whether they are in equilibrium, or moving in the perimeters of certain ellipses or other curves. Here we come across a marvellous analogy between certain limits and the limits which two points lying on the axis of the primary curve have with respect to each other; & here also a much greater variety of cases for masses is shown, & an example is given of the application to solidity, & liquefaction, on account of a quick internal motion being impressed on the points of the body. Moreover, in the two articles that then follow, I state some general propositions with regard to a system of four points, together with their application to solid rods, both rigid and flexible; I also give an illustration of various classes of particles by means of pyramids, each of which is formed of four points in the most simple case, & of four of such pyramids in the more complicated cases.

From Art. 240 as far as Art. 264, I pass on to masses & consider matters pertaining to the centre of gravity; & I prove that in general there is one, & only one, in any given mass. I show how it can in general be determined, & I set forth in clear terms the point that is lacking in the usual method, when it comes to a question of rigorous proof; this deficiency

I supply, & I bring forward a certain example of the same sort, that deals with the multiplication of numbers, & to the composition of forces by the parallelogram law; the latter I prove by another more general method, analogous to that which I use in the general investigation for the centre of gravity. Then by its help I prove very expeditiously & with extreme rigour that well-known theorem of Newton, in which he affirmed that the state of the centre of gravity is in no way altered by the internal mutual forces.

I gather several good results from this method of treatment. In Art. 264, the conservation of the same quantity of motion in the Universe in one plane; in Art. 265 the equality of action and reaction amongst masses; then the collision of solid bodies, and the communication of motions in direct impacts & the laws that govern them, & from that, in Art. 276, oblique impacts; in Art. 277 I reduce the theory of these from resolution of motions to compositions, & in the article that follows, Art. 278, I pass to impact on to a fixed plane; from Art. 279 to Art. 289 I show that there can be no real resolution of forces or of motions in Nature, but only a hypothetical one; & in this connection I consider & explain all sorts of cases, in which at first sight it would seem that there must be resolution.

From Art. 289 to Art. 297, I state the laws for the composition & resolution of forces; here also I give the explanation of that well-known fact, that force decreases in composition, increases in resolution, but always remains equal to the sum of the parts acting in the same direction as itself in the first, the rest being equal & opposite cancel one another; whilst in the second, all that is done is to suppose that two equal & opposite forces are added on, which supposition has no effect on the phenomena. Thus it comes about that nothing can be deduced from this in favour of the Theory of living forces, since everything can be explained without them; in the same connection, I explain also many of the phenomena, which are usually brought forward as evidence in favour of these 'living forces.'

In Art. 297, I seize the opportunity offered by the results just mentioned to attack certain matters that relate to the law of continuity, which in all cases of motion is strictly observed; & I show that, in the collision of solid bodies, & in reflected motion, the laws, as usually stated, are therefore only approximately followed. From this, as far as Art. 307, I make out the various relations between the angles of incidence & reflection, whether the forces, as the bodies approach one another, continually attract, or continually repel, or attract at one time & repel at another. I also consider what will happen if the roughness of the acting surface is very slight, & what if it is very great. I also state the first principles, derived from mechanics, that are required for the explanation & determination of the reflection & refraction of light; also the relation of the absolute to the relative force in the oblique descent of heavy bodies; & some theorems that are requisite for the more accurate theory of oscillations; these, though quite elementary, I explain with great care.

From Art. 307 onwards, I investigate the system of three bodies; in this connection, as far as Art. 313, I evolve several theorems dealing with the direction of the forces on each one of the three compounded from the combined actions of the other two; such as the theorem, that these directions are either all parallel to one another, or all pass through some one common point, when they are produced indefinitely on both sides. Then, as far as Art. 321, I make out several other theorems dealing with the ratios of these same resultant forces to one another; such as the following very simple & elegant theorem, that the accelerating forces of two of the masses will always be in a ratio compounded of three reciprocal ratios; namely, that of the distance of either one of them from the third mass, that of the sine of the angle which the direction of each force makes with the corresponding distance of this kind, & that of the mass itself on which the force is acting, to the corresponding distance, sine and mass for the other: also that the motive forces only have the first two ratios, that of the masses being omitted.

I then collect the results to be derived from these theorems, deriving from them, as far as Art. 328, theorems relating to the equilibrium of forces diverging in any manner, & the centre of equilibrium, & the pressure of the centre on a fulcrum. I extend the theorem relating to preponderance to the case also, in which the masses do not mutually act upon one another in a direct manner, but through others intermediate between them, which connect them together, & supply the place of rods joining them; and also to any number of masses, each of which I suppose to be connected with the centre of rotation & some other assumed mass, & from this I derive the principles of moments for all machines. Then I consider all the different kinds of levers; one of the theorems that I obtain is, that, if a lever is suspended from the centre of gravity, then there is equilibrium; but a force should be felt in this centre from the fulcrum or sustaining point, equal to the sum of the weights of the whole system; from which there follows most clearly the reason, which is everywhere assumed without proof, why the whole mass can be supposed to be collected at its centre of gravity, so long as the system is in a state of rest & all motions of its parts are prohibited by equilibrium.

From Art. 328 to Art. 347, I deduce from these same theorems, others that relate to the centre of oscillation of any number of masses, whether they are in the same right line, or anywhere in a plane perpendicular to the axis of rotation; this theory wants to be worked somewhat more carefully with a system of four bodies, to be gone into more fully, & to be extended so as to include the general case of a system of solid bodies; having stated this, I evolve from it the centre of percussion, & I show the analogy between it & the centre of oscillation.

I obtain all such results from theorems relating to three masses. After that, in Art. 347, I intimate the matters in which I agree with all others, & especially with the followers of Newton, concerning sums of forces, acting on a point, or an attracted or repelled mass, due to the separate points of another mass. Then, from Art. 348 to the end of this part, i.e., as far as Art. 359, I expound certain theorems that belong to the theory of fluids; & first of all, theorems with regard to pressure, in connection with which I mention that one which was proved by Newton, namely, that, if the compression of a fluid is proportional to the compressing force, then the repulsive forces between the points are in the reciprocal ratio of the distances, & conversely. Moreover, I show that, if the same force is insensible, then the matter can be represented by the logistic & other curves; also that in fluids subject to our terrestrial gravity pressures should be found proportional to the depths. After that, I touch upon those things that relate to the velocity of a fluid issuing from a vessel; & I show what is necessary in order that this should be equal to the velocity which would be acquired by falling through the depth itself, just as it is seen to happen in the case of an efflux of water. These things in some part being explained, & in some part merely indicated, I bring this second part to an end.

PART III

In Art. 358, I state the theme of this third part; in it I derive all the general & most of the special, properties of matter from my Theory. Then, as far as Art. 371, I deal somewhat more at length with the subject of impenetrability, which I remark is of a twofold kind in my masses of non-extended points; in this connection also, I deal with a certain apparent case of compenetrability, & the passage of light through the innermost parts of bodies without real compenetration; I also explain in a very summary manner several striking phenomena relating to the above. From here on to Art. 375, I deal with extension; this in my opinion is not continuous either in matter or in solid bodies, & yet it yields the same phenomena to the senses as does the usually accepted idea of it; here I also deal with geometry, which conserves all its power under my Theory. Then, as far as Art. 383, I discuss figurability, volume, mass & density, each in turn; in all of these subjects there are certain special points of my Theory that are not unworthy of investigation. Important theorems on mobility & continuity of motions are to be found from here on to Art. 388; then, as far as Art. 391, I deal with the equality of action & reaction, & my conclusions with regard to the subject corroborate in a wonderful way the hypothesis of those forces, upon which my Theory depends. Then follows divisibility, as far as Art. 398; this principle I admit only to the extent that any existing mass may be made up of a number of real points that are finite only, although in any given mass this finite number may be as great as you please. Hence for infinite divisibility, as commonly accepted, I substitute infinite multiplicity; which comes to exactly the same thing, as far as it is concerned with the explanation of the phenomena of Nature. Having considered these subjects I add, in Art. 398, that of the immutability of the primary elements of matter; according to my idea, these are quite simple in composition, of no extent, they are everywhere unchangeable, & hence are splendidly adapted for explaining a continually recurring set of phenomena.

From Art. 399 to Art. 406, I derive gravity from my Theory of forces, as if it were a particular branch on a common trunk; in this connection also I explain how it can happen that the fixed stars do not all coalesce into one mass, as would seem to be required under universal gravitation. Then, as far as Art. 419, I deal with cohesion, which is also as it were another branch; I show that this is not dependent upon quiescence, nor on motion that is the same for all parts, nor on the pressure of some fluid, nor on the idea that the attraction is greatest at actual contact, but on the limits between repulsion and attraction. I propose, & solve, a general problem relating to this, namely, why masses, once broken, do not again stick together, why the fibres are stretched or contracted before fracture takes place; & I intimate which of my ideas relative to cohesion are the same as those held by other philosophers.

In Art. 419, I pass on from cohesion to particles which are formed from a number of cohering points; & I consider these as far as Art. 426, & investigate the various distinctions between them. I show how it is possible for various shapes of all sorts to be assumed, which offer great resistance to rupture; & how in a given shape they may differ very greatly in the number & disposition of the points forming them. Also that from this fact there arise very different forces for the action of one particle upon another, & also for the action of different parts of this particle upon other different parts of it, or on the same part of another particle. For that depends solely on the number & distribution of the points, so that one given particle either attracts, or repels, or is perfectly inert with regard to another given particle, the distances between them and the positions of their surfaces being also given. Then I state in addition that the smaller the particles, the greater is the difficulty in dissociating them; moreover, that they ought to be quite uniform as regards gravitation, no matter what the disposition of the points may be; but in most other properties they should be quite different from one another (which we observe to be the case); & that this difference ought to be much greater in larger masses.

From Art. 426 to Art. 446, I consider solids & fluids, the difference between which is also a matter of different kinds of cohesion. I explain with great care the difference between solids & fluids; deriving the nature of the latter from the greater freedom of motion of the particles in the matter of rotation about one another, this being due to the forces being nearly equal; & that of the former from the inequality of the forces, and from certain lateral forces which help them to keep a definite position with regard to one another. I distinguish between various kinds of fluids also, & I cite the distinction between rigid, flexible, elastic & fragile rods, when I deal with viscosity & humidity; & also in dealing with organic bodies & those solids bounded by certain fixed figures, of which the formation presents no difficulty; in these one particle can only attract another particle in certain parts of the surface, & thus urge it to take up some definite position with regard to itself, & keep it there. I also show that the whole system of the Atomists, & also of the Corpuscularians, can be quite easily derived by my Theory, from the idea of particles of definite shape, offering a high resistance to deformation; so that it comes to nothing else than another single branch of this so to speak most fertile trunk, breaking forth from it on account of a different manner of cohesion. Lastly, I show the reason why it is that not every mass, in spite of its being constantly made up of homogeneous points, & even these in a high degree capable of rotary motion about one another, is a fluid. I also touch upon the resistance of fluids, & investigate the laws that govern it.

From Art. 446 to Art. 450, I deal with those things that relate to the different kinds of solidity, that is to say, with elastic bodies, & those that are soft. I attribute the nature of the former to the existence of a large interval between the consecutive limits, on account of which it comes about that points that are far removed from their natural positions still feel the effects of the same kind of forces, & therefore return to their natural positions; & that of the latter to the frequency & great closeness of the limits, on account of which it comes about that points that have been moved from one limit to another, remain there in relative rest as they were to start with. Then I deal with ductile and malleable solids, pointing out how they differ from fragile solids. Moreover I show that all these differences are in no way dependent on density; so that, for instance, a body that is much more dense than another body may have either a much greater or a much less solidity and cohesion than another; in fact, any of the properties set forth may just as well be combined with any density either greater or less.

In Art. 450 I consider what are commonly called the "four elements"; then from Art. 451 to Art. 467, I treat of chemical operations; I explain solution in Art. 452, precipitation in Art. 453, the mixture of several substances to form a single mass in Art. 454, 455, liquefaction by two methods in Art. 456, 457, volatilization & effervescence in Art. 458, emission of effluvia (which from a constant mass ought to be approximately constant) in Art. 461, ebullition & various kinds of evaporation in Art. 462, deflagration & generation of gas in Art. 463, crystallization with definite forms of crystals in Art. 464; & lastly, I show, in Art. 465, how it is possible for fermentation to cease, & in Art. 466, how it is that any one thing does not ferment when mixed with any other thing.

From fermentation I pass on, in Art. 467, to fire, which I look upon as a fermentation of some substance in light with some sulphureal substance; & from this I deduce several propositions, up to Art. 471. There I pass on from fire to light, the chief properties of which, from which all the phenomena of light arise, I set forth in Art. 472; & I deduce & fully explain each of them in turn as far as Art. 503. Thus, emission in Art. 473, velocity in Art. 474, rectilinear propagation in homogeneous media, & a compenetration that is merely apparent, from Art. 475 on to Art. 483, pellucidity & opacity in Art. 483, reflection at equal angles to Art. 484, & refraction to Art. 487, tenuity in Art. 487, heat & the great internal motions arising from the smooth passage of the extremely tenuous light in Art. 488, the greater action of oleose & sulphurous bodies on light in Art. 489. Then I show, in Art. 490, that it suffers no real resistance, & in Art. 491 I explain the origin of bodies emitting light, in Art. 492 the reason why light that falls with greater obliquity is reflected more strongly, in Art. 493, 494 the origin of different degrees of refrangibility, & in Art. 495, 496 I deduce that there are two different dispositions recurring at equal intervals; hence, in Art. 497, I bring out those alternations, discovered by Newton, of easier reflection & easier transmission, & in Art. 498 I deduce that some rays should be reflected & others transmitted in the passage to a fresh medium, & that the greater the obliquity of incidence, the greater the number of reflected rays. In Art. 499, 500 I state the origin of the difference between the lengths of the intervals of the alternations; upon this alone depends the whole of the Newtonian theory of natural colours. Finally, in Art. 501, I touch upon the wonderful property of Iceland spar & its cause, & in Art. 502 I explain diffraction, which is a kind of imperfect refraction or reflection.

After light derived from fire, which has to do with vision, I very briefly deal with taste & smell in Art. 503, & of sound in the three articles that follow next. Then, in the next four articles, I consider touch, & in connection with it, cold & heat also. After that, as far as Art. 514, I deal with electricity; here I explain the whole of the Franklin theory by means of my principles; I reduce this theory to two principles only, & these are derived from my general Theory of forces in almost the same manner as I have already derived precipitations & solutions. Finally, in Art. 514, 515, I investigate magnetism, explaining both magnetic direction & attraction.

These things being expounded, all of which relate to special properties, I once more consider, in the articles from 516 to the end, the general nature of bodies, what matter is, its form, what things ought to be considered as essential, & what as accidental, attributes; and also the nature of transformation and alteration are investigated, each in turn; & thus I bring to a close the third part of my Theory.

I will mention here but this one thing with regard to the appendix on Metaphysics; namely, that I there expound more especially how greatly different is the soul from matter, the connection between the soul & the body, & the manner of its action upon it. Then with regard to God, I prove that He must exist by many arguments that have a close connection with this Theory of mine; I especially mention, though but slightly, His Wisdom and Providence, from which there is but a step to be made towards revelation. But I think that I have, so to speak, given my preliminary foretaste quite sufficiently.

6

Reprinted from *Philos. Mag.* n.s., 4:161-169 (1828)

A BRIEF ACCOUNT OF MICROSCOPICAL OBSERVATIONS MADE IN THE MONTHS OF JUNE, JULY, AND AUGUST, 1827 ON THE PARTICLES CONTAINED IN THE POLLEN OF PLANTS: AND ON THE GENERAL EXISTENCE OF ACTIVE MOLECULES IN ORGANIC AND INORGANIC BODIES

Robert Brown

[We have been favoured by the Author with permission to insert the following paper, which has just been printed for private distribution.—Ed.]

THE observations, of which it is my object to give a summary in the following pages, have all been made with a simple microscope, and indeed with one and the same lens, the focal length of which is about $\frac{1}{32}$nd of an inch*.

The examination of the unimpregnated vegetable Ovulum, an account of which was published early in 1826†, led me to attend more minutely than I had before done to the structure of the Pollen, and to inquire into its mode of action on the Pistillum in Phænogamous plants.

In the Essay referred-to, it was shown that the apex of the

* This double convex Lens, which has been several years in my possession, I obtained from Mr. Bancks, optician, in the Strand. After I had made considerable progress in the inquiry, I explained the nature of my subject to Mr. Dollond, who obligingly made for me a simple pocket microscope, having very delicate adjustment, and furnished with excellent lenses, two of which are of much higher power than that above mentioned. To these I have often had recourse, and with great advantage, in investigating several minute points. But to give greater consistency to my statements, and to bring the subject as much as possible within the reach of general observation, I continued to employ throughout the whole of the inquiry the same lens with which it was commenced.

† In the Botanical Appendix to Captain King's Voyages to Australia, vol. ii. p. 534. *et seq.*

nucleus of the Ovulum, the point which is universally the seat of the future Embryo, was very generally brought into contact with the terminations of the probable channels of fecundation; these being either the surface of the placenta, the extremity of the descending processes of the style, or more rarely, a part of the surface of the umbilical cord. It also appeared, however, from some of the facts noticed in the same Essay, that there were cases in which the Particles contained in the grains of pollen could hardly be conveyed to that point of the ovulum through the vessels or cellular tissue of the ovarium; and the knowledge of these cases, as well as of the structure and economy of the antheræ in *Asclepiadeæ*, had led me to doubt the correctness of observations made by Stiles and Gleichen upwards of sixty years ago, as well as of some very recent statements, respecting the mode of action of the pollen in the process of impregnation.

It was not until late in the autumn of 1826 that I could attend to this subject; and the season was too far advanced to enable me to pursue the investigation. Finding, however, in one of the few plants then examined, the figure of the particles contained in the grains of pollen clearly discernible, and that figure not spherical but oblong, I expected, with some confidence, to meet with plants in other respects more favourable to the inquiry, in which these particles, from peculiarity of form, might be traced through their whole course: and thus, perhaps, the question determined whether they in any case reach the apex of the ovulum, or whether their direct action is limited to other parts of the female organ.

My inquiry on this point was commenced in June 1827, and the first plant examined proved in some respects remarkably well adapted to the object in view.

This plant was *Clarckia pulchella*, of which the grains of pollen, taken from antheræ full grown, but before bursting, were filled with particles or granules of unusually large size, varying from nearly $\frac{1}{4000}$th to about $\frac{1}{5000}$th of an inch in length, and of a figure between cylindrical and oblong, perhaps slightly flattened, and having rounded and equal extremities. While examining the form of these particles immersed in water, I observed many of them very evidently in motion; their motion consisting not only of a change of place in the fluid, manifested by alterations in their relative positions, but also not unfrequently of a change of form in the particle itself; a contraction or curvature taking place repeatedly about the middle of one side, accompanied by a corresponding swelling or convexity on the opposite side of the particle. In a few instances the particle was seen to turn on its longer axis. These motions were such as to satisfy me, after frequently repeated observation, that they arose neither from currents in the fluid, nor from its gradual evaporation, but belonged to the particle itself.

Grains of pollen of the same plant taken from antheræ immediately after bursting, contained similar subcylindrical particles, in reduced numbers however, and mixed with other particles, at least as numerous, of much smaller size, apparently spherical, and in rapid oscillatory motion.

These smaller particles, or Molecules as I shall term them, when first seen, I considered to be some of the cylindrical particles swimming vertically in the fluid. But frequent and careful examination lessened my confidence in this supposition; and on continuing to observe them until the water had entirely evaporated, both the cylindrical particles and spherical molecules were found on the stage of the microscope.

In extending my observations to many other plants of the same natural family, namely *Onagrariæ*, the same general form and similar motions of particles were ascertained to exist, especially in the various species of *Œnothera*, which I examined. I found also in their grains of pollen taken from the antheræ immediately after bursting, a manifest reduction in the proportion of the cylindrical or oblong particles, and a corresponding increase in that of the molecules, in a less remarkable degree, however, than in *Clarckia*.

This appearance, or rather the great increase in the number of the molecules, and the reduction in that of the cylindrical particles, before the grain of pollen could possibly have come in contact with the stigma,—were perplexing circumstances in this stage of the inquiry, and certainly not favourable to the supposition of the cylindrical particles acting directly on the ovulum; an opinion which I was inclined to adopt when I first saw them in motion. These circumstances, however, induced me to multiply my observations, and I accordingly examined numerous species of many of the more important and remarkable families of the two great primary divisions of Phænogamous plants.

In all these plants particles were found, which in the different families or genera varied in form from oblong to spherical, having manifest motions similar to those already described; except that the change of form in the oval and oblong particles was generally less obvious than in *Onagrariæ*, and in the spherical particle was in no degree observable*.

* In *Lolium perenne*, however, which I have more recently examined, though the particle was oval and of smaller size than in *Onagrariæ*, this change of form was at least as remarkable, consisting in an equal contraction in the middle of each side, so as to divide it into two nearly orbicular portions.

In a great proportion of these plants I also remarked the same reduction of the larger particles, and a corresponding increase of the molecules after the bursting of the antheræ: the molecule, of apparently uniform size and form, being then always present; and in some cases indeed, no other particles were observed, either in this or in any earlier stage of the secreting organ.

In many plants belonging to several different families, but especially to *Gramineæ*, the membrane of the grain of pollen is so transparent that the motion of the larger particles within the entire grain was distinctly visible; and it was manifest also at the more transparent angles, and in some cases even in the body of the grain in *Onagrariæ*.

In *Asclepiadeæ*, strictly so called, the mass of pollen filling each cell of the anthera is in no stage separable into distinct grains; but within, its tessellated or cellular membrane is filled with spherical particles, commonly of two sizes. Both these kinds of particles when immersed in water are generally seen in vivid motion; but the apparent motions of the larger particle might in these cases perhaps be caused by the rapid oscillation of the more numerous molecules. The mass of pollen in this tribe of plants never bursts, but merely connects itself by a determinate point, which is not unfrequently semitransparent, to a process of nearly similar consistence, derived from the gland of the corresponding angle of the stigma.

In *Periploceæ*, and in a few *Apocineæ*, the pollen, which in these plants is separable into compound grains filled with spherical moving particles, is applied to processes of the stigma, analogous to those of *Asclepiadeæ*. A similar economy exists in *Orchideæ*, in which the pollen-masses are always, at least in the early stage, granular; the grains, whether simple or compound, containing minute, nearly spherical particles, but the whole mass being, with very few exceptions, connected by a determinate point of its surface with the stigma, or a glandular process of that organ.

Having found motion in the particles of the pollen of all the living plants which I had examined, I was led next to inquire whether this property continued after the death of the plant, and for what length of time it was retained.

In plants, either dried or immersed in spirit for a few days only, the particles of pollen of both kinds were found in motion equally evident with that observed in the living plant; specimens of several plants, some of which had been dried and preserved in an herbarium for upwards of twenty years, and others not less than a century, still exhibited the molecules or smaller spherical particles in considerable numbers, and in evident motion, along with a few of the larger particles, whose motions were much less manifest, and in some cases not observable*.

In this stage of the investigation having found, as I believed, a peculiar character in the motions of the particles of pollen in water, it occurred to me to appeal to this peculiarity as a test in certain families of Cryptogamous plants, namely Mosses, and the genus *Equisetum*, in which the existence of sexual organs had not been universally admitted.

In the supposed stamina of both these families, namely, in the cylindrical antheræ or pollen of Mosses, and on the surface of the four spathulate bodies surrounding the naked ovulum, as it may be considered, of *Equisetum*, I found minute spherical particles, apparently of the same size with the molecule described in *Onagrariæ*, and having equally vivid motion on immersion in water; and this motion was still observable in specimens both of Mosses and of *Equiseta*, which had been dried upwards of one hundred years.

The very unexpected fact of seeming vitality retained by these minute particles so long after the death of the plant, would not perhaps have materially lessened my confidence in the supposed peculiarity. But I at the same time observed, that on bruising the ovula or seeds of *Equisetum*, which at first happened accidentally, I so greatly increased the number of moving particles, that the source of the added quantity could not be doubted. I found also that on bruising first the floral leaves of Mosses, and then all other parts of those plants, that I readily obtained similar particles, not in equal quantity indeed, but equally in motion. My supposed test of the male organ was therefore necessarily abandoned.

Reflecting on all the facts with which I had now become acquainted, I was disposed to believe that the minute spherical particles or Molecules of apparently uniform size, first seen in the advanced state of the pollen of *Onagrariæ*, and most other Phænogamous plants,—then in the antheræ of Mosses and on

* While this sheet was passing through the press I have examined the pollen of several flowers which have been immersed in weak spirit about eleven months, particularly of *Viola tricolor*, *Zizania aquatica*, and *Zea Mais*; and in all these plants the peculiar particles of the pollen, which are oval or short oblong, though somewhat reduced in number, retain their form perfectly, and exhibit evident motion, though I think not so vivid as in those belonging to the living plant. In *Viola tricolor*, in which, as well as in other species of the same natural section of the genus, the pollen has a very remarkable form, the grain on immersion in nitric acid still discharged its contents by its four angles, though with less force than in the recent plant.

the surface of the bodies regarded as the stamina of *Equisetum*,—and lastly in bruised portions of other parts of the same plants, were in reality the supposed constituent or elementary molecules of organic bodies, first so considered by Buffon and Needham, then by Wrisberg with greater precision, soon after and still more particularly by Müller, and, very recently by Dr. Milne Edwards, who has revived the doctrine and supported it with much interesting detail. I now therefore expected to find these molecules in all organic bodies: and accordingly on examining the various animal and vegetable tissues, whether living or dead, they were always found to exist; and merely by bruising these substances in water, I never failed to disengage the molecules in sufficient numbers to ascertain their apparent identity in size, form, and motion, with the smaller particles of the grains of pollen.

I examined also various products of organic bodies, particularly the gum-resins, and substances of vegetable origin, extending my inquiry even to pit-coal; and in all these bodies Molecules were found in abundance. I remark here also, partly as a caution to those who may hereafter engage in the same inquiry, that the dust or soot deposited on all bodies in such quantity, especially in London, is entirely composed of these molecules.

One of the substances examined, was a specimen of fossil wood, found in Wiltshire oolite, in a state to burn with flame; and as I found these molecules abundantly, and in motion in this specimen, I supposed that their existence, though in smaller quantity, might be ascertained in mineralized vegetable remains. With this view a minute portion of silicified wood, which exhibited the structure of *Coniferæ*, was bruised, and spherical particles, or molecules in all respects like those so frequently mentioned, were readily obtained from it; in such quantity, however, that the whole substance of the petrifaction seemed to be formed of them. But hence I inferred that these molecules were not limited to organic bodies, nor even to their products.

To establish the correctness of the inference, and to ascertain to what extent the molecules existed in mineral bodies, became the next object of inquiry. The first substance examined was a minute fragment of window-glass, from which, when merely bruised on the stage of the microscope, I readily and copiously obtained molecules agreeing in size, form, and motion with those which I had already seen.

I then proceeded to examine, and with similar results, such

minerals as I either had at hand or could readily obtain, including several of the simple earths and metals, with many of their combinations.

Rocks of all ages, including those in which organic remains have never been found, yielded the molecules in abundance. Their existence was ascertained in each of the constituent minerals of granite, a fragment of the Sphinx being one of the specimens examined.

To mention all the mineral substances in which I have found these molecules, would be tedious; and I shall confine myself in this summary to an enumeration of a few of the most remarkable. These were both of aqueous and igneous origin, as travertine, stalactites, lava, obsidian, pumice, volcanic ashes, and meteorites from various localities*. Of metals I may mention manganese, nickel, plumbago, bismuth, antimony, and arsenic. In a word, in every mineral which I could reduce to a powder, sufficiently fine to be temporarily suspended in water, I found these molecules more or less copiously; and in some cases, more particularly in siliceous crystals, the whole body submitted to examination appeared to be composed of them.

In many of the substances examined, especially those of a fibrous structure, as asbestus, actinolite, tremolite, zeolite, and even steatite, along with the spherical molecules, other corpuscules were found, like short fibres somewhat moniliform, whose transverse diameter appeared not to exceed that of the molecule, of which they seemed to be primary combinations. These fibrils, when of such length as to be probably composed of not more than four or five molecules, and still more evidently when formed of two or three only, were generally in motion, at least as vivid as that of the simple molecule itself; and which from the fibril often changing its position in the fluid, and from its occasional bending, might be said to be somewhat vermicular.

In other bodies which did not exhibit these fibrils, oval particles of a size about equal to two molecules, and which were also conjectured to be primary combinations of these, were not unfrequently met with, and in motion generally more vivid than that of the simple molecule; their motion consisting in turning usually on their longer axis, and then often appearing to be flattened. Such oval particles were found to be numerous and extremely active in white arsenic.

As mineral bodies which had been fused contained the

* I have since found the molecules in the sand-tubes, formed by lightning, from Drig in Cumberland.

moving molecules as abundantly as those of alluvial deposits, I was desirous of ascertaining whether the mobility of the particles existing in organic bodies was in any degree affected by the application of intense heat to the containing substance. With this view small portions of wood, both living and dead, linen, paper, cotton, wool, silk, hair, and muscular fibres, were exposed to the flame of a candle or burned in platina-forceps, heated by the blowpipe; and in all these bodies so heated, quenched in water, and immediately submitted to examination, the molecules were found, and in as evident motion as those obtained from the same substances before burning.

In some of the vegetable bodies burned in this manner, in addition to the simple molecules, primary combinations of these were observed, consisting of fibrils having transverse contractions, corresponding in number, as I conjectured, with that of the molecules composing them; and those fibrils, when not consisting of a greater number than four or five molecules, exhibited motion resembling in kind and vivacity that of the mineral fibrils already described, while longer fibrils of the same apparent diameter were at rest.

The substance found to yield these active fibrils in the largest proportion and in the most vivid motion, was the mucous coat interposed between the skin and muscles of the haddock, especially after coagulation by heat.

The fine powder produced on the under surface of the fronds of several Ferns, particularly of *Acrostichum calomelanos*, and the species nearly related to it, was found to be entirely composed of simple molecules and their primary fibre-like compounds, both of them being evidently in motion.

There are three points of great importance which I was anxious to ascertain respecting these molecules, namely, their form, whether they are of uniform size, and their absolute magnitude. I am not, however, entirely satisfied with what I have been able to determine on any of these points.

As to form, I have stated the molecule to be spherical, and this I have done with some confidence; the apparent exceptions which occurred admitting, as it seems to me, of being explained by supposing such particles to be compounds. This supposition in some of the cases is indeed hardly reconcileable with their apparent size, and requires for its support the further admission, that, in combination, the figure of the molecule may be altered. In the particles formerly considered as primary combinations of molecules, a certain change of form must also be allowed; and even the simple molecule itself has sometimes appeared to me when in motion to have been slightly modified in this respect.

My manner of estimating the absolute magnitude and uniformity in size of the molecules, found in the various bodies submitted to examination, was by placing them on a micrometer divided to five thousandths of an inch, the lines of which were very distinct; or more rarely on one divided to ten thousandths, with fainter lines, not readily visible without the application of plumbago, as employed by Dr. Wollaston, but which in my subject was inadmissible.

The results so obtained can only be regarded as approximations, on which perhaps, for an obvious reason, much reliance will not be placed. From the number and degree of accordance of my observations, however, I am upon the whole disposed to believe the simple molecule to be of uniform size, though as existing in various substances and examined in circumstances more or less favourable, it is necessary to state that its diameter appeared to vary from $\frac{1}{15,000}$th to $\frac{1}{20,000}$th of an inch*.

I shall not at present enter into additional details, nor shall I hazard any conjectures whatever respecting these molecules, which appear to be of such general existence in inorganic as well as in organic bodies; and it is only further necessary to mention the principal substances from which I have not been able to obtain them. These are oil, resin, wax, and sulphur; such of the metals as I could not reduce to that minute state of division necessary for their separation; and finally, bodies soluble in water.

[*Editor's Note:* Material has been omitted at this point.]

* While this sheet was passing through the press, Mr. Dollond, at my request, obligingly examined the supposed pollen of *Equisetum virgatum* with his compound achromatic microscope, having in its focus a glass divided into 10,000ths of an inch, upon which the object was placed; and although the greater number of particles or molecules seen were about $\frac{1}{20,000}$th, yet the smallest did not exceed $\frac{1}{30,000}$th of an inch.

7

Reprinted from pp. 161–165 of *A Source Book in Physics,* W. F. Magie, ed.
Cambridge, Mass.: Harvard University Press, 1935, 620pp.

DAVY

Sir Humphry Davy was born in Penzance on Decenber 17, 1778. He worked as assistant and student with a surgeon who was also a druggist. He became interested in chemistry and was early distinguished for his discoveries in chemistry and electricity. He was for many years professor of chemistry at the Royal Institution and for seven years was president of the Royal Society. He isolated sodium and potassium by means of the electric current and was the first to observe the electric arc. He died in Geneva on May 29, 1829.

The paper which follows, entitled "An Essay on Heat, Light, and the Combinations of Light," was published in a book entitled *Contributions to Physical and Medical Knowledge, Principally from the West of England,* Collected by Thomas Beddoes, M.D., 1799. It also may be found in Davy's Collected Works. In it Davy describes his experiments on the production of heat by friction.

Heat Produced by Friction

Matter is possessed of the power of attraction. By this power the particles of bodies tend to approximate, and to exist in a state of contiguity. The particles of all bodies with which we are acquainted, can be made to approach nearer to each other, by peculiar means, that is, the specific gravity of all bodies can be increased by diminishing their temperatures. Consequently (on the supposition of the impenetrability of matter) the particles of bodies are not in actual contact. There must then act on the corpuscles of bodies some other power, which prevents their actual contact; this may be called repulsion. The phaenomena of repulsion have been supposed, by the greater part of chemical philosophers, to depend on a peculiar elastic fluid; to which the names of latent heat, and caloric, have been given. The peculiar modes of existence of bodies, solidity, fluidity, and gazity, depend (according to the calorists) on the quantity of the fluid of heat entering into their composition; this substance insinuating itself between their corpuscles, separating them from each other, and preventing their actual contact, is, by them, supposed to be the cause of repulsion.

Other philosophers, dissatisfied with the evidences produced in favour of the existence of this fluid, and perceiving the generation of heat by friction and percussion, have supposed it to be motion.

Considering the discovery of the true cause of the repulsive power as highly important to philosophy, I have endeavoured to investigate this part of chemical science by experiments: from these experiments (of which I am now about to give a detail), I conclude that heat, or the power of repulsion, is not matter.

. . . .

The Phaenomena of Repulsion Are Not Dependent on a Peculiar Elastic Fluid for Their Existence, or Caloric Does Not Exist

Without considering the effects of the repulsive power on bodies, or endeavouring to prove from these effects that it is motion, I shall attempt to demonstrate by experiments that it is not matter; and in doing this, I shall use the method called by mathematicians, *reductio ad absurdum.*

Let heat be considered as matter, and let it be granted that the temperature of bodies cannot be increased, unless their capacities are diminished from some cause, or heat added to them from some bodies in contact.

Now the temperatures of bodies are uniformly raised by friction and percussion. And since an increase of temperature is consequent on friction and percussion, it must consequently be generated in one of these modes. First, either from a diminution of the capacities of the acting bodies from some change induced in them by friction, a change producing in them an increase of temperature.

Secondly, or from heat communicated, from the decomposition of the oxygen gas in contact by one or both of the bodies, and then friction must effect some change in them (similar to an increase of temperature), enabling them to decompose oxygen gas, and they must be found after friction, partially or wholly oxydated.

Thirdly, or from a communication of caloric from the bodies in contact, produced by a change induced by friction in the acting bodies, enabling them to attract caloric from the surrounding bodies.

Now first let the increase of temperature produced by friction and percussion be supposed to arise from a diminution of the capacities of the acting bodies. In this case it is evident some change must be induced in the bodies by the action, which lessens their capacities and increases their temperatures.

. . . .

Experiment II

I procured two parallelopipedons of ice, of the temperature of 29°, six inches long, two wide, and two-thirds of an inch thick: they were fastened by wires to two bars of iron. By a peculiar mechanism, their surfaces were placed in contact, and kept in a continued and violent friction for some minutes. They were almost entirely converted into water, which water was collected, and its temperature ascertained to be 35°, after remaining in an atmosphere of a lower temperature for some minutes. The fusion took place only at the plane of contact of the two pieces of ice, and no bodies were in friction but ice. From this experiment it is evident that ice by friction is converted into water, and according to the supposition its capacity is diminished; but it is a well-known fact, that the capacity of water for heat is much greater than that of ice; and ice must have an absolute quantity of heat added to it, before it can be converted into water. Friction consequently does not diminish the capacities of bodies for heat.

From this experiment it is likewise evident, that the increase of temperature consequent on friction cannot arise from the decomposition of the oxygen gas in contact, for ice has no attraction for oxygen. Since the increase of temperature consequent on friction cannot arise from the diminution of capacity, or oxydation of the acting bodies, the only remaining supposition is, that it arises from an absolute quantity of heat added to them, which heat must be attracted from the bodies in contact. Then friction must induce some change in bodies, enabling them to attract heat from the bodies in contact.

Experiment III

I procured a piece of clock-work so constructed as to be set to work in the exhausted receiver; one of the external wheels of this machine came in contact with a thin metallic plate. A considerable degree of sensible heat was produced by friction between the wheel and plate when the machine worked uninsulated from bodies capable of communicating heat. I next procured a small piece of ice; round the superior edge of this a small canal was made and filled with water. The machine was placed on the ice, but not in contact with the water. Thus disposed, the whole was placed under the receiver, (which had been previously filled with carbonic acid,) a quantity of potash (i.e. caustic vegetable alkali) being at the same time introduced.

The receiver was now exhausted. From the exhaustion, and from the attraction of the carbonic acid gas by the potash, a vacuum nearly perfect was, I believe, made.

The machine was now set to work. The wax rapidly melting, proved the increase of temperature.

Caloric then was collected by friction; which caloric, on the supposition, was communicated by the bodies in contact with the machine. In this experiment, ice was the only body in contact with the machine. Had this ice given out caloric, the water on the top of it must have been frozen. The water on the top of it was not frozen, consequently the ice did not give out caloric. The caloric could not come from the bodies in contact with the ice; for it must have passed through the ice to penetrate the machine, and an addition of caloric to the ice would have converted it into water.

Heat, when produced by friction, cannot be collected from the bodies in contact, and it was proved by the second experiment, that the increase of temperature consequent on friction cannot arise from diminution of capacity, or from oxydation. But if it be considered as matter, it must be produced in one of these modes. Since (as is demonstrated by these experiments) it is produced in neither of these modes, it cannot be considered as matter. It has then been experimentally demonstrated that caloric, or the matter of heat, does not exist.

Solids, by long and violent friction, become expanded, and if of a higher temperature than our bodies, affect the sensory organs with the peculiar sensation known by the common name of heat.

Since bodies become expanded by friction, it is evident that their corpuscles must move or separate from each other. Now a motion or vibration of the corpuscles of bodies must be necessarily generated by friction and percussion. Therefore we may reasonably conclude that this motion or vibration is heat, or the repulsive power.

Heat, then or that power which prevents the actual contact of the corpuscles of bodies, and which is the cause of our peculiar sensations of heat and cold, may be defined as a peculiar motion, probably a vibration, of the corpuscles of bodies, tending to separate them. It may with propriety be called the repulsive motion.

8

Reprinted from pages 340–346 and 401–403 of *Ann. Philos.* n.s., 1:340–351, 401–416 (1821)

MATHEMATICAL INQUIRY INTO THE CAUSES, LAWS, AND PRINCIPAL PHENOMENA OF HEAT, GASES, GRAVITATION, ETC.

John Herapath

[*Editor's Note:* In the first of the three papers under the same title and in the same journal and year (pages 273–293), Herapath advances his hypotheses concerning the atomic nature of matter. These includes the assumption that heat arises from the motion of atoms and the erroneous one that it is proportional to their individual momentum. This paper is not printed here.]

Of the Laws of Gaseous Bodies.

Definitions.

Def. 1.—The homogeneity of a gas is the perfect equality of its atoms, or particles, throughout, in quantity of matter. It is of no consequence whether the atoms are similar in figure or not; their figures, might be very different, provided, however, the quantity of solid matter in each is the same.

Def. 2.—Density is the quantity of matter in a given space, when the atoms, or particles, whatever be their relative magnitudes, are uniformly disposed in the medium.

Cor. 1.—The mean density of any body is, therefore, proportional to the whole quantity of matter, divided by the whole bulk, or magnitude of the body.

Cor. 2.—In a homogeneous body, the density is equal to the mass of an atom drawn into the number contained in a given space.

Def. 3.—Numeratom is a term I have employed to express the number of atoms, or particles, distributed throughout a given space, without respect to the density or homogeneity of the body.

Cor.—Hence the numeratom of a homogeneous body is proportional to the density directly, and the mass of a particle, or atom, inversely.

This conclusion might also have been drawn from Cor. 2 of the preceding definition.

Def. 4.—The elasticity of a gas is the force with which it endeavours to expand itself, or with which it resists compression; and is estimated by the amount of its action against similar and equal portions of the containing bodies.

General Considerations on Force and the Constitution of Gaseous Bodies.

Force is the cause of any change, or is that power, by which the change is produced. If, therefore, a body be perfectly free to yield to any change of state, and none take place, it follows, that the body is not acted on by any force; or if it be, it must be by an accumulation of opposing forces, which, in the aggregate, destroy one another. Mathematicians usually estimate force by the effect it produces, or would produce, in a given time, and consider the intensity of the force as proportional to the effect. Thus, if one of two forces produce twice the effect of the other, it is considered to be doubly as powerful; if three times the effect, trebly; and so on. But this is rather estimating the effects than the forces; since it is possible for the same primitive force, by successive actions, to produce very unequal effects in equal times.

Forces have likewise been distinguished into two kinds, pressive and impulsive, or those which act by pressure, and those which act by impulsion. The former generate changes in bodies by a continued unceasing action, from the beginning to the end of the stroke, or fit, of action; and, consequently, always consume time. The latter act either by an instantaneous impulse, or by a succession of those impulses; but each impulse, individually considered, occupies not the smallest portion of time. It is an action that in one moment might be said not to have come into existence; and in the next to have ended. This is the kind of force we have now to consider.

If we take single impulses, it is plain that the forces and the changes they produce are proportional, under the same circumstances of action. But if a succession of impulses takes place, the effects are no longer proportional to the forces, but to the collected actions of the forces in equal times. When, therefore, the forces are equal, the effects are proportional to the numbers of their respective repetitions; and when those forces are unequal, but uniformly intense in their actions, the effects are in the compound ratio of those numbers and the forces. It is hence manifestly possible for a weak force, by a greater number of successive actions, to produce effects equal to those of a stronger by a less number of actions; and if the impulses of the two forces be at regular intervals, and yet the intervals of the slower be not sensible to observation, the effects will always have the same ratio, whatever be the length of time in which they are compared.

These impulses may also be opposed to a pressive force, and effect precisely the same things. For let us suppose a hard body to be acted on from a state of quiescence by a continued force, such, for instance, as what we commonly conceive gravity to be, urging it in any direction; then, after it has been impelled forward for any length of time by this force, and has acquired a certain momentum, let it be met by another hard body, moving uniformly with an equal momentum in the opposite direction; and by our laws of collision, it will receive such a change in its motion by the contact as will give it an equal and opposite momentum in the path it has already described. It will, therefore, being acted on still by the soliciting force, retrace this path, until, having ascended to the point whence it set out, it has lost all its velocity, and begins to descend again as before. As soon as it has arrived again to the place where it before was met by the other body, let it receive a second equal shock, and again it will begin to ascend; and so on as long as we please. Thus it might be continually kept oscillating between these two points by impulses properly regulated. And if the distance between the points be diminished, the number of oscillations will be increased, and the intervals between the shocks diminished, and the body consequently approach nearer to a state of apparent rest.

This is how things will happen when the affected body comes in contact with only one other body, and receives the strokes in the direction of the track of its centre of gravity. But the same would evidently take place, if, instead of one body striking it in this identical manner, it were struck by several in different parts of its surface, either all at once, or in any order; the aggregate direction and intensity of whose strokes, however, should be similar to those of the single body under similar circumstances. Then might the body acted on by the force remain in a state of rest, if not absolute, at least so nearly so, as to differ from it insensibly. Now the tendency of this body to move in the direction in which it is solicited is precisely the same thing as mathematicians understand by the force, which they call pressure; and, therefore, the opposition of the other smaller bodies is also equivalent; and, as to effect, the same as this force of pressure. And because the value of this pressure might be increased or diminished; by increasing or diminishing the soliciting force, or the pressing body; so also might the value of the resisting force of the smaller bodies be increased or diminished, by increasing or diminishing their momentum, the number of their collisions, or the number of the bodies.

From these considerations it follows, that if a number of small bodies be inclosed in any hollow body, and be continually impinging on one another, and on the sides of the enclosing body; and if the motions of the bodies be conserved by an equivalent action in the sides of the containing body, then will these small bodies compose a medium, whose elastic force will be like that of our air and other gaseous bodies; for if the bodies be exceedingly small, the medium might, like any aeriform body, be compressed into a very small space; and yet, if it had no other tendency than what would arise from the internal collision of its atoms, it would, if left to itself, extend to the occupation of a space of almost indefinite greatness. And its temperature remaining the same, its elasticity would also be greater when occupying a less, and less when occupying a greater space; for in a condensed state the number of atoms striking against a given portion of the containing vessel must be augmented; and the space in which the atoms have to move being less, their returns, or periods, must be shorter; and the number of them, in a given time, consequently greater, on both of which accounts the elasticity is greater, the greater the condensation. Besides, when other things are the same, the elastic force augments with an augmentation of temperature, and diminishes with a diminution; for an increase of temperature, according to our theory, must necessarily be attended with an increase of velocity; and, therefore, with an increase in the number of collisions. But these things will be more accurately treated of presently.

Whether all pressive forces be not the same as this gaseous action, is a question we do not at present intend to consider. It might come under our cognizance hereafter, when we shall have had more opportunities of collating our principles with experiments. In the mean time it is sufficient to anticipate that our inquiries, whatever be the possibility of the existence of another kind of pressive force, lead us to conclude that nature operates by no other.

It might be asked by what means is it, if the parts of the gases are absolutely hard, that they are reflected back into the medium from the sides of the containing vessel? This question is easily answered, if we allow heat to consist in an intestine motion of the parts of the bodies; for then the parts of a solid, of equal temperature with a gas, must have the same quantity, though they have not the same freedom of motion as the parts of a gas have; and hence the parts of the gas will have the same reflection from the sides of the vessel that they have from one another.

Prop. VI.

The figure of the vessel, in which a gas is contained, has no influence on the elastic force of the gas, such a constitution of things being granted as we have supposed.

For since the particles of the gas receive their reflection from the superficial particles of the containing body; and since these particles have not only their intestine motion, but likewise their figure and arrangement, independent of the superficial figure of the containing body, it follows, that the direction of their action on the particles of the gas is also independent of the figure of the body in which they are enclosed. But other things being alike, this action is equal to the contrary action of the particles of the gas, which constitutes its elastic force; therefore, its elastic force is not affected by the figure of the body in which it is contained.

Cor.—Hence we gather that in any bodies of equal capacity and of the same temperature, the same, or an equal portion of the same gas, would be equally elastic; for the motion of the particles being independent of the figure of the containing body must depend entirely upon the temperature of that body; and this being the same, the motion is the same. But the motion, quantity of gas, and capacity, being the same, the elasticity must be the same.

Scholium.

This theorem and its cor. agree with the opinion of philosophers respecting the elasticity of gaseous bodies, though I am not aware that they have ever been made the subject of direct experiment. Indeed it seems to have been taken for granted that wherever the temperatures, spaces occupied, gases, and quantities of gas, were the same, the elasticities were the same; but it would be worth while to be more certain of it. An experiment to settle this point, taking every circumstance into consideration, would, perhaps, require as much care and skill as almost any of those that have been made on gases.

PROP. VII.

If a given portion of a fluid gas, composed of particles mutually impinging on one another and the sides of the containing body, in the manner that has been described, has its temperature the same; and if the particles be indefinitely small, its elastic force, under different compressions, is reciprocally proportional to the space it occupies.

Let us suppose that equal portions of the same gas be enclosed in two vessels of unequal capacity. Then, by the last Prop. it is immaterial whether these vessels be of the same or of different figures; the difference of figures would have no influence upon the ratio of the elasticities; but, for the sake of simplicity, we will suppose the two figures similar. Now because the only change that is supposed to take place is in the space which the gas occupies, the motions and collisions of a particle in the one will be similar to those of a corresponding particle in the other; and the temperature, that is, in this case, the velocity being the same in each, the numbers of revolutions that two corresponding particles in the two media make in a given time must be inversely proportional to the paths the particles describe; that is, these paths being alike and described with equal velocities in the inverse subtriplicate ratio of the spaces occupied by the equal portions of gases. But because the elasticity of a gas is proportional to the action of its particles against a given portion of the surface of the containing body, the ratio of the elastic forces, arising from the repeated actions of equal numbers of corresponding particles in the two media, will likewise, their velocities being the same in both media, be inversely as the subtriplicate of the spaces occupied. And if we conceive the two gases to be divided into strata, parallel to the sides of the bodies on which the elastic forces are measured, and of one, two, or any number of particles thick, it is manifest, since the motions of the particles are alike in each medium, that if the elasticity in one medium arises from the action of the particles of the first stratum alone, so it does also in the other medium; and if it arises from the action of the particles of the two first, three first, or *n* first strata in one medium, the same holds true with the elasticity in the other medium. But the number of particles of any one stratum that strike against a given portion of the containing surface of one medium, is to the number of particles of the corresponding stratum that strike against an equal and similar portion of the other medium, in the duplisubtriplicate ratio of the numeratoms directly; that is, in the duplisubtriplicate ratio of the spaces occupied by equal portions of the gases inversely. Therefore as the whole elastic forces of these corresponding strata are in a ratio compounded of the ratios of the numbers of particles that strike against equal portions of the sides of the containing bodies, and of the numbers of returns which they make to the sides in a given time, that ratio must be equal to one compounded of the inverse duplisubtriplicate and of the inverse subtriplicate ratios of the spaces occupied by the two gases; it must, therefore, be equal to the simple inverse ratio of the spaces occupied. And since the same number of strata affects the elasticity of the one gas as of the other; and since the inverse ratio of the spaces is the ratio of the elastic force of any two corresponding strata, it is consequently the ratio of the united elastic forces of all the strata that affect the elasticity; and is, therefore, the ratio of the elastic forces of the two gases.

Cor.—Because the numeratoms are reciprocally proportional to the spaces occupied, it follows that the elasticities are, under equal temperatures, directly as the numeratoms.

Scholium.

We have in the two preceding theorems and their corollaries supposed the atoms, or particles, to be perfectly hard; but the same consequences would follow if they were either perfectly or imperfectly elastic, and the containing vessel either elastic or hard. For the temperature being invariable, the intensity of the collisions, and consequently of the reflections, would remain the same in a rare as in a denser medium. The law, therefore, that the elasticities and compressions are proportional, under equal temperatures, is true not only in permanent airs or gases, but in all kinds of vapours, which is conformable to experience.

Prop. VIII.

The same things remaining, the elasticity of a gas under a variable temperature and compression, is proportional to its numeratom and the square of its temperature conjointly; or the elasticity varies as the square of the temperature directly, and the simple of the space inversely.

If we first suppose in two portions of the same gas the numeratoms to be equal, the elasticities of those portions will have the same ratio as the elasticities arising from the actions of corresponding particles in the two media; for the change of temperature does not alter the manner in which the corresponding particles act in the media, but only the intensity of action. This being the case, the elasticities due to the actions of corresponding particles are to one another as their momenta and the number of their revolutions or returns in a given time; that is, as their temperatures and velocities. But the masses of the corresponding particles being the same, the velocities are as the temperatures; therefore, the elasticities due to corresponding particles, and consequently the elasticities of the media, are as the squares of the temperatures. And by the cor. to the preceding prop. the temperatures being the same, the elasticities are as the numeratoms. Whence, if neither the temperatures, nor the numeratoms are the same, the elasticities are in a ratio compounded of the ratio of the numeratoms, and that of the squares of the temperatures, or, which is the same, in a ratio compounded of the inverse ratio of the volumes and the duplicate direct of the temperatures.

Cor. 1.—Hence the elasticities are also in a ratio which is equal to that compounded of the simple ratio of the numeratoms, and the duplicate of the velocities.

Cor. 2.—And hence also the elasticities are in a ratio compounded of these three simple ratios; namely, the ratio of the numeratoms, the ratio of the temperatures, and the ratio of the velocities.

[*Editor's Note:* Material has been omitted at this point.]

General Scholium.

Sudden condensation in all gaseous or aeriform bodies produces heat, and sudden rarefaction cold; but if the condensation or rarefaction be made slowly, no perceptible change in the temperature takes place. These are natural consequences of our theory of the constitution of gases. For if it be a condensation, by the motion inward, for instance, of one of the sides of the containing body, it is evident that the particles which strike against this side and are reflected back among the rest, will no longer be reflected by the same force only, with which they were previous to the commencement of the condensation, but by a force, which will augment their velocity individually by a quantity equal to the velocity of the moving side. And this excess of velocity being distributed to the rest of the particles, or communicated to them by the continuance of the stroke, will generate an excess of temperature throughout the medium proportional chiefly to the velocity of condensation. On the contrary, in the case of rarefaction, those particles, instead of returning among the rest with the celerity with which they did before the side began to move, will now return with a diminution of their velocity, equal to the velocity of the moving side. This diminution must, therefore, cause a decrease of temperature in the medium, in the same manner as the excess of velocity would cause an increase. In both these cases, the excess or defect of temperature will be proportional to the celerity of condensation or rarefaction, supposing the time occupied by the change of volume to be of sufficient length for the stroke to be propagated to all parts of the medium. Therefore, if the celerity of the stroke be great, a great change in the temperature will ensue; and if it be small, there will be little or none.

Our theory of gases likewise explains another phænomenon, which has usually been attributed to a different cause. It is well known that if two gases be placed in any position with respect to each other, and have ever so small a communication, they will intermix, and become equally diffused the one throughout the other. This has been accounted for on the principle that all gases have a slight affinity for one another, or in other words, that though the parts of all gaseous bodies individually considered mutually repel each other, yet the parts of different gases, however great or however small that difference may be, mutually attract each other. It is repugnant to my intention to attempt any refutation of this idea, or to make any observations on the confusion that would result from thus burthening nature with hypotheses; I shall, therefore, merely show that the phænomenon in question is an easy consequence of our theory. Suppose that portions of the surfaces of two media are exposed to each other, and that parallel to these portions the media are divided into strata. Then because a difference in the numeratoms, a difference in the motions of the particles of the two strata, or a cornered irregular figure in the particles, would render it impossible for the particles of one stratum to be so reflected by the particles of the other, that each stratum would retain its particles entire and unmixed with those of the other, the two strata would intermix; and would not arrive to a state of equilibrious action, until the particles were uniformly and equally disposed in each stratum. For the same reasons, these newly compounded strata would mix with their next; and thus it would go on stratum intermixing with stratum, until an equilibrious action throughout was restored, by the equal and uniform intermixture of the two gases.

By the same principles it is easily perceived, why gases so readily acquire the temperature of the surrounding bodies; why they transmit temperature so rapidly and so feebly in right lines; and why, according to the manner in which Mr. Leslie, Sir H. Davy, and MM. Dulong and Petit, have made their experiments, it is owing to the gravitation of the gases alone, that the lighter gases cool bodies immersed in them quicker than the heavier. When I first turned my attention to the cooling power of gases, I was surprised to find from my theory, that all gases, under the same circumstances, ought to cool equally fast. This inference for some time perplexed me very much. As soon, however, as I had read the ingenious treatise of Prof. Leslie on Heat and Moisture, and had considered anew the manner in which Sir H. Davy had conducted his experiments, which were about that time published, I saw that the theory and phænomena still agreed; and that the circumstances of my investigation precisely coincided with Mr. Leslie's cooling by gaseous pulsation; while the error that I had committed, with respect to Sir H. Davy's experiments, rested wholly on my not having taken into account the earth's attraction.

Their temperatures and elasticities being the same, the ratio of the numeratoms of two homogeneous gases appears by our theory to be equal to the subduplicate ratio of the weights or specific gravities of equal volumes.* Supposing, therefore, that oxygen and hydrogen are homogeneous (the truth of which in oxygen I much doubt), the numeratom of the former will be quadruple that of the latter. So that if two in volume of hydrogen unite with one in volume of oxygen to form water, the atoms of oxygen will be double in number those of hydrogen; and the numeratom of the compound gas before being condensed will be the geometrical mean between the numeratoms of the two simples. It has commonly been conceived that two atoms of hydrogen and one of oxygen form a particle of water; but whether this, or whether the result of our theory, or whether neither of them be true, it is out of our power to determine. Such kind of speculations transcends the corroboration of any experiment yet devised. I simply mention this theorem, which is one among several that I have investigated, relative to the mixture of gases, to give some small idea of the powers of our theory of the universe for unravelling the secret operations of nature.

[*Editor's Note:* Material has been omitted at this point.]

* This theorem was published in a more general form in the *Annals of Philosophy*, for July, 1816.

9

Reprinted from pages 5–26, 78, and 79 of *Roy. Soc. London Philos. Trans.* A, **183**:1–79 (1892)

ON THE PHYSICS OF MEDIA THAT ARE COMPOSED OF FREE AND PERFECTLY ELASTIC MOLECULES IN A STATE OF MOTION

J. J. Waterston

[*Editor's Note:* In the original, material precedes this excerpt.]

[*Author's Introduction.*]

Of the physical theories of heat that have claimed attention since the time of Bacon, that which ascribes its cause to the intense vibrations of the elementary parts of bodies has received a considerable accession of probability from the recent experiments of Forbes and Melloni. It is admitted that these have been the means of demonstrating that the mode of its radiation is identical with that of light in the quantities of refraction and polarization. The evidence that has been accumulated in favour of the undulatory theory of light has thus been made to support with a great portion of its weight a like theory of the phenomena of heat; and we are, perhaps, justified in expecting that the complete development of this theory will have a much more important influence on the progress of science, because of its more obvious connection and intimate blending with almost every appearance of Nature. Heat is not only the subject of direct sensation and the vivifier of organic life, but it is manifested as the accompaniment of mechanical force. It is related to it both as cause and effect, and submits itself readily to measurement by means of the mechanical changes that are among the most prominent indications of its change of intensity. The undulatory theory at once leads us to the conclusion that, inasmuch as the temperature of a body is a persistent quality due to the motion of its molecules, its internal constitution must admit of it retaining a vast amount of living force. Indeed, it seems to be almost impossible now to escape from the inference that heat is essentially molecular *vis viva*. In solids, the molecular oscillations may be viewed as being restrained by the intense forces of aggregation. In vapours and gases these seem to be overcome; vibrations can no longer be produced by the inherent *vis insita* of the molecules struggling with attractive and repellant forces; the struggle is over and the molecules are free; but they, nevertheless, continue to maintain a certain temperature; they are capable of heating and being heated; they are endowed with the quality heat, which, being of itself motion, compels us to infer that a molecule in motion without any force to restrain or qualify it, is in every respect to be considered as a free projectile. Allow such free projectiles to be endowed with perfect elasticity, and likewise extend the same property to the elementary parts of all bodies that they strike against, and we immediately introduce the principle of the conservation of *vis viva* to regulate the general effects of their fortuitous encounters. Whether gases do consist of such minute elastic projectiles or not, it seems worth while to enquire into the physical attributes of media so constituted, and to see what analogy they bear to the elegant and symmetrical laws of aeriform bodies.

Some years ago I made an attempt to do so, proceeding synthetically from this fundamental hypothesis, and have lately obtained demonstration of one or two points where the proof was then deficient. The results have appeared so encouraging, although derived from very humble applications of mathematics, that I have been led to hope a popular account of the train of reasoning may not prove unacceptable to the Royal Society.—Sept. 1, 1845.

SECTION I.—OF A HOMOGENEOUS MEDIUM AND THE LAWS OF ITS ELASTICITY.

§ 1. The term medium is, perhaps, not quite appropriate to what is here intended to be signified. We speak of a resisting medium. of the medium of light, and in each expression something is referred to as intervening between bodies, and it is the quality of interposition that entitles it to the name. Here, for want of better, it is employed to denote a certain hypothetical condition of matter which it is the object of this Paper to show has physical properties that resemble those that have been found to belong to aeriform bodies. Inasmuch, therefore, as the word may be applied to a simple unmixed gas so as to speak of it as an oxygen medium or a hydrogen medium, &c., so far we may be allowed the use of it in treating of a hypothetical medium, which we have carefully to refrain from assimilating to any known form of matter until, by synthetical reasoning, circumstantial evidence has been accumulated sufficient to prove or render probable its identity.

To have a proper conception of what the medium is that forms the subject of speculation, we must imagine a vast multitude of small particles of matter, perfectly alike in every respect, perfectly elastic as glass or ivory—but of size, form and texture that requires not to be specified further than that they are not liable to change by mutual action—to be enclosed by elastic walls or surfaces in a space so much greater than their aggregate bulk as to allow them freely to move amongst each other in every direction. As all consideration of attractive forces is left out at present, it is obvious that each particle must proceed on a straight line until it strikes against another, or against the sides of the enclosure; that it must then be reflected and driven into another line of motion, traversing backwards and forwards in every direction, so that the intestine condition of the multitude of these that form the medium may be likened to the familiar appearance of a swarm of gnats in a sunbeam.

The quality of perfect elasticity being common to all the particles, the original amount of *vis viva*, or living, acting force, of the whole multitude must for ever remain the same. If undisturbed by external action it cannot, of itself, diminish or increase, but must for ever remain as unchanged as the matter that is associated with it and that it endows with activity. Such is the case if we view the whole mass of moving particles as one object, but each individual of the multitude must at every encounter give or receive, according to the ever-changing angle and plane of impact, some portion of its force, so that, considered separately, they are for ever continually changing the velocity and direction of their individual motions; striking against and rebounding from each other, they run rapidly in their zig-zag conflict through every possible mode of concurrence, and *at each point of the medium we may thus conceive that particles are moving in every possible direction and encountering each other in every possible manner during so small an elapsed interval of time that it may be viewed as infinitesimal in respect to any sensible period.* The medium must in this way become endowed with a permanent state of elastic energy or disposition to expand, uniformly sustained in every part and communicating to it the physical character of an elastic fluid.

The simplicity of this hypothesis facilitates the application of mathematics in ascertaining the nature and properties of such media, and the study acquires much interest from the analogies that it unfolds. For if the reasoning is correct, the physical laws common to all gases and vapours—those laws, namely, that concern heat and pressure—do actually belong to such media, and may be synthetically deduced from the constitution which has now been assigned to them.

The characteristic which renders a medium susceptible of mathematical treatment is that of its being composed of particles perfectly alike in every respect, but it is chiefly their identity in weight or mass that is the important point of distinction. A particle thus conforms to the definition that the eminent physicist AMPÈRE has given to the term molecule, which we may therefore adopt as a more significant name for the element of a medium.

The first department of the subject must naturally be devoted to the consideration of the circumstances that determine the equilibrium of such a homogeneous medium considered by itself. Its density, by which is to be understood not its specific gravity but the number of molecules in a constant volume,* may be supposed to vary without disturbing its homogeneity. The mean square velocity of the molecules (which in any infinitesimal portion of the medium may be assumed as uniform) we also have to consider as a variable quantity, and the physical qualities of a medium being dependent on these two elements of its constitution, it is necessary to determine clearly their mathematical relations.

§ 2. It is evident from the definition of the hypothesis, that the medium must exert some expansive force on the surface that encloses it; but the nature of the force is not strictly continuous, it is composed of a multitude of successive strokes. Nevertheless, their succession is certainly continuous, and it is not difficult to conceive how they may be sufficient to counterbalance and support a superincumbent weight. To obtain an exact idea of this, let us suppose that a small elastic plane whose weight is n times that of a molecule, is supported by a regular succession of such molecules striking its centre of gravity with a velocity v. We seek to know the condition of their mutual action when an equilibrium is maintained.

The following are the well-known equations that express the law of elastic collision. They are necessarily the foundation of all reasoning on the effects of the mutual action of elastic bodies by impact.

1. *The Meeting Impact.*

Two molecules, B and D, meet directly in an intermediate point and strike each other with the respective velocities β and δ. The velocities after impact are respectively :—

$$\beta_0 = -\beta + \frac{2(\delta+\beta)\mathrm{D}}{(\mathrm{B}+\mathrm{D})}; \quad \text{and} \quad \delta_0 = \delta - \frac{2(\delta+\beta)\mathrm{B}}{(\mathrm{B}+\mathrm{D})};$$

the direction of D's motion being reckoned positive.

2. *The Overtaking Impact.*

The two molecules, B and D, with the same velocities, β and δ, move in the same direction and D overtakes B; the velocities after impact are respectively :—

$$\beta_1 = \beta + \frac{2(\delta-\beta)\mathrm{D}}{(\mathrm{B}+\mathrm{D})}; \quad \text{and} \quad \delta_1 = \delta - \frac{2(\delta-\beta)\mathrm{B}}{(\mathrm{B}+\mathrm{D})};$$

the direction of D's motion being reckoned positive.

* [Attention should be directed to this use of the word "density."—R.]

In the first of these let $\beta_0 = \beta$, $\delta = v$, $B = nD$; then shall

$$\beta = -\beta + \frac{2(v+\beta)}{n+1} ; \quad \text{or} \quad \beta = \frac{v}{n},*$$

which evidently expresses the upward velocity given to the plane by the impulse of one molecule when the velocity of incidence and reflexion is the same. The plane ascends and descends the height due to this velocity, and then encounters the next in the succession of molecular impacts without any transference of force taking place between them; and n being taken an indefinitely great number, β is infinitesimal in respect to v, and the height through which the plane traverses is also infinitesimal, so that it is supported as if by a continuous force of upward pressure. The time between each impact is, according to the law of falling bodies, equal to the time taken by the force of gravity to destroy and reproduce the infinitesimal velocity v/n. This is $2v/gn$: the velocity which a free body gains or loses in a unit of time by the force of gravity being represented by g. The number of impacts in a unit of time is therefore $gn/2v = A$. This, then, is the relation between the weight of the plane, in terms of that of the molecule unity, and the rapidity of the succession of impacts necessary to support it in a condition of statical equilibrium. Now, if the plane forms part of the surface that encloses the medium and that counterbalances by its weight the effect of the impacts of the confined molecules, such effect must correspond with the succession represented by A; and we deduce that *the elastic force of a medium, as represented by the weight or pressure required to confine it, is directly proportional to he number of molecular impacts that take place against a unit surface in a unit time with a constant velocity* (or $e \doteqdot A$, if v is constant) . . . . I.

§ 3. Such being the nature of the elastic force, it will not be difficult to prove that it increases exactly as the density of the medium. The proposition stands thus: if the number of molecules in a volume of the medium be doubled, the number of impacts that take place on a constant surface in a constant time will also be doubled, the velocity being unchanged.

Suppose the number octupled, the mean distance is reduced to one-half. If they were equidistant and moving in one direction with the constant velocity, it is evident that eight times the previous number would pass the same imaginary plane in the same time, and if the plane were solid that eight times the previous number would impinge against it. Now, although all do not move in one direction, yet in both cases the same proportion of the whole must in each case do so. Whatever may be the density no preference can be assigned to one direction more than to another in the molecular movements; they must in every case be equally distributed in every

* [The case is that where the particle (mass 1) and the plane (mass n) both reverse their velocities at impact. The conservation of *vis viva* is thereby secured, and the condition of momentum gives at once $n\beta = v$. —R.]

direction, and if the number is increased eight times in any one direction it must be so in every other.

This may be viewed in another light. Suppose in both cases, the density being 1 and 8 respectively, that the molecules are arrested in their motion. It is evident that opposite a unit of surface in density 8 there will, in the first row, be four times as many molecules as in density 1, and that the average distance between the rows is only one-half. Suppose the molecules to resume their motion, and compare density 8 with density 1, it is obvious that in half the time four times the number will impinge on the unit of surface, and in the same time eight times the number. Now it has been shown (§ 2) that the elastic force is proportional to the number of molecular impacts made with a constant velocity against a unit of surface in a unit of time, hence we deduce that *the elastic force* (e) *of a medium with a constant mean molecular velocity* (v) *is proportional to its density* (Δ^3) (or $e \doteqdot \Delta^3$, if v or v^2 is constant). . II.

§ 4. Hitherto the molecular velocity has been supposed constant. We have now to enquire how the elasticity of the medium is affected by a change in the velocity from v to mv. The intestine action of the medium may be viewed as the traversing of a certain mean distance, L, by the molecules in a given time, t; and in this time a certain mean number, A, of impacts take place against a unit of surface. If the velocity is increased m times, the distance L is traversed in $1/m$th the time t, or t/m, and in this reduced time the same number of impacts must take place as before took place in the time t; for there is nothing in the change of velocity simply that can alter the ratio that subsists between the mean distance traversed and the mean number of impacts, unless that ratio were subject to change without any change whatever in the medium, which is absurd; hence, in the original time, t, there is m times the original number of impacts, A.

It was shown in § 2 that if the weight of each of the molecules were represented by 1, their mean velocity by v, and weight of plane supported by their impinging action n, the number of impacts in a second or unit of time required to support the plane is $\frac{gn}{2v} = \mathrm{A}$, or $n = \frac{2}{g}\mathrm{A}v$, and this equation must evidently be maintained in altering the value of the terms. Now, it has been shown that in changing v to mv in a medium that does not alter its density we cause A to become $m\mathrm{A}$, and $\frac{2}{g}\mathrm{A}v$ becomes $\frac{2}{g}\mathrm{A}mvm = \frac{2}{g}\mathrm{A}vm^2 = nm^2$. Hence n, the weight of the plane, or measure of tension, must be increased m^2 times so that it may continue to equilibrate the impinging action. Thus, we deduce that while the molecular velocity increases from v to mv, the elasticity increases from n to m^2n or *the elasticity of a medium having a constant density is proportional to the mean square molecular velocity or* vis viva *of the medium* (or $e \doteqdot v^2$, when Δ^3 is constant)* . . . . III.

* [II. and III. were given by D. BERNOULLI. See 'POGG. Ann.,' vol. 107, p. 490, 1859.—R.]

§ 5. Combining II. with III. it appears that when both the density and the *vis viva* are subject to change that the elasticity is equal to their product, or $e \doteqdot \Delta^3 v^2$, and this is the law that includes all the conditions of equilibrium of an enclosed homogeneous medium. One other condition only remains to be specified. *Under a constant pressure the density is inversely as the* vis viva *or mean square molecular velocity* ($\Delta^3 \doteqdot \frac{1}{v^2}$, if e is constant) . . . . IV.

§ 6. In concluding this part of the subject, we cannot fail of being sensible of the analogies that subsist between these synthetical deductions and the chief properties that distinguish aeriform fluids.

The first point that was inductively established is MARIOTTE'S law, viz.: *at the same temperature the density of air is as its compression.* This is analogous to the second deduction:—The square of the velocity being constant, the elastic force of a medium is proportional to its density. The accordance appears as complete as could be desired, and there is a residual evidence in favour of v^2 being identical with temperature, or being a quality that varies simultaneously with it.

The second point is DALTON and GAY-LUSSAC'S law of expansion. By experimenting upon the same weight of air at different temperatures under a constant pressure, these philosophers found that an increment of one degree caused always the same augmentation of bulk, and that this amounted to $\frac{1}{480}$th part of the space that it occupied at 32°. Thus, if the same law hold good at all temperatures, 480 cubic inches of air at this temperature should diminish one inch in bulk for every degree it was lowered in temperature, and would become zero in bulk at 480° below the freezing point of water, or — 448° on Fahrenheit scale.

Now in IV. we had $\Delta^3 \doteqdot \frac{1}{v^2}$, or $v^2 \doteqdot \frac{1}{\Delta^3}$, when e is constant; but $\frac{1}{\Delta^3}$ is the volume occupied by a constant number of molecules; hence with the same constant number of molecules the volume is as the mean square molecular velocity, and a constant increment of *vis viva* is followed by the same increment of volume under a constant pressure, and as the constant increment of volume (1 cubic inch) is to the constant increment of *vis viva* (1°) so is the volume (480 cubic inches) corresponding to a certain *vis viva* (32° Fahr.) to that *vis viva* (480°).

The analogy therefore still holds good, and the evidence continues in favour of the absolute temperature being represented by v^2.

When air is not allowed to expand and heat is applied, the elastic force increases with the temperature, and a rise of 1° causes an absolute increase in the elasticity, which is the same at all temperatures, and corresponds with the increase of bulk it would assume if allowed freely to expand. This is analogous to III., where, the density being constant or bulk unchanged, the elasticity is shown to be proportional to the mean square molecular velocity.

Thus, the laws of MARIOTTE and of DALTON and GAY-LUSSAC are represented by the formula $(448 + t)\Delta^3 = e$; in which t = temperature, Fahrenheit scale; Δ^3 = density, and e = elasticity.

The law of elasticity in the hypothetical medium is represented by the formula $v^2\Delta^3 = e$; in which v^2 is the mean square molecular velocity; Δ^3 = density, and e = elasticity.

The *first* expresses physical laws that have been *found* to belong to a certain *existent* form of matter.

The *second* expresses physical laws that have been *proved* to belong to a certain *possible* form of matter.

The cause of the effect represented by $(448 + t)$ in the first is unknown, but has, at various times, by eminent authorities, been referred to *molecular motion*.

The corresponding term, v^2, of the second represents *molecular motion*.

SECTION II.—ON THE PHYSICAL RELATIONS OF MEDIA THAT DIFFER FROM EACH OTHER IN THE SPECIFIC WEIGHT OF THEIR MOLECULES.

§ 7. The synthetical deductions of last section apply to a homogeneous medium without respect to the absolute weight of its molecules, if the weight of each molecule is the same. This weight, common to all, may be viewed as the specific molecular weight of the medium, and distinguishes it from any other medium with a different specific molecular weight. We have now to enquire into the relations that subsist between the density and molecular velocity of two such media that have the same elasticity, or that are in equilibrium of pressure and also of *vis viva*.

We deduced from the law of impinging elastic bodies that if v represents the mean molecular velocity in feet per second, A the number of molecular impacts in a second upon a small elastic plane which is equal in weight to n molecules, then $n = \frac{2}{g}\text{A}v$. Let ω represent the specific weight of the molecules, we have $\omega n = \frac{2\omega}{g}\text{A}v = e$ = the elastic force exerted by the medium on a unit of surface;* and as this must in the present enquiry be assumed constant, we may easily remark how a change in ω affects v and A.

It is evident that since $\frac{2\omega}{g}\text{A}v$ is a constant quantity and ω, A, and v variable, we have $\text{A}v \doteqdot \frac{1}{\omega}$; but $e = \omega\Delta^3v^2$ (§ 5) $= \frac{2\omega}{g}\text{A}v$, and, therefore, $\Delta^3 v = \frac{2}{g}\text{A}$, or $\text{A} = \frac{g}{2}\Delta^3 v$, and $\text{A}v \doteqdot \frac{1}{\omega} \doteqdot \Delta^3 v^2$. Hence it is obvious that if Δ^3, the density or number of molecules in a constant volume, as well as e, the tension, are constant, while the molecular

* e is the absolute weight of the small elastic plane that is supported by the succession of A number of molecular impacts per second, the weight of each of which is ω, and their common impinging velocity v feet per second.

velocity and specific weight are variable, these variables are bound by the relation expressed by $v^2 \doteqdot 1/\omega$, which signifies that if two media are in equilibrium of pressure and have in the same volume of each the same number of molecules, the squares of their molecular velocities must be inversely as their specific molecular weights. Hence, we deduce that *if any number of separate media have equal density and tension, the molecular velocity of each must be proportional to the inverse square root of their specific molecular weight, or to the inverse square root of the specific gravity of the media respectively* * . . . . V.

§ 8. But media may be in equilibrium of pressure without being of equal density, for a deficiency of density may be compensated by an excess in the molecular velocity. It is plain that if e and v^2 are constant in any two media they may still be in equilibrium of pressure if ω is proportional to $1/\Delta^3$, or if the molecular weight of each medium is inversely as its density. If the specific molecular weights are in a constant ratio to each other and the tension and velocity also constant, the media must be kept in equilibrium of pressure if the density of each is reciprocally proportional to the specific molecular weight of the other.

We have supposed hitherto that the media are separate while their respective elasticities are compared. Let us now enquire into the effects of allowing them to have access to each other. The united media immediately obtain a heterogeneous character, for it requires no demonstration to convince us that the molecules of each will permeate through the volume occupied by the other, the vacuities in the space occupied by each presenting no more obstacle to the motion of one set of molecules than it does to the other; and as collision must take place amongst them in every possible manner and direction, the common space of the united media are free alike to each individual molecule of both to range through in its zig-zag course. Consequently, *media in contact with each other become gradually equally diffused through their common volume* . . . . VI.

The internal condition of the mixture must after a time become settled so that in any infinitesimal portion the same mean velocity will be found proper to the molecules of each medium respectively.

But as each of the two sets of molecules, although completely mixed together, preserve their specific weights, so must they have corresponding specific velocities that remain intact, notwithstanding that they as often impinge on molecules of the other set as on the molecules of their own kind. It is of consequence to settle what the ratio of these specific velocities is, for upon this point depends the nature of the *vis viva* equilibrium of different media, and we have to determine the relative condition of two media when they are in equilibrium both of pressure and of *vis viva.*

* [The deduction of V. appears to be correct, though much embarrassed by the irrelevant g. In his first memoir on the Theory of Gases ('Pogg. Ann.,' vol. 100, 1857), Clausius arrives at the same conclusion. His assumption that the density (in Waterston's sense) of various gases is the same, appears to have been made upon chemical grounds.—R.]

§ 9. We must now refer back to the equations of impact (§ 2). It is apparent that the sum of the impinging *vis viva* of both molecules does not alter in either the meeting or overtaking impact; what is gained by one is lost by the other, or $\beta^2 B + \delta^2 D = \beta_0{}^2 B + \delta_0{}^2 D = \beta_1{}^2 B + \delta_1{}^2 D$. But in every case except one a transference of *vis viva* must take place from the one to the other.

The exception is found in the meeting impact when $\beta \doteqdot 1/B$ and $\delta \doteqdot 1/D$; then shall $\beta_0 = \beta$ and $\delta_0 = \delta$, but at the same time β_1 is not equal to β, or δ_1 to δ; in every other case β_0 is not equal to β nor to β_1, nor is δ_0 equal to δ or to δ_1.

It can seldom happen that the molecules strike each other directly. In taking account of the collective result of their fortuitous concourse we must view the position of the plane of concurrence and the respective inclinations of the line of motion of each molecule to it as three independent variables. The incident velocity of each is the absolute velocity resolved perpendicular to the plane, and the equations apply to this portion only of the *vis viva* of the molecules.

Although the variety in the mode of impact is infinite, it is certain that one direction of motion is as likely as any other, and hence, that the opposite of any direction is equally probable to the direction itself.

Let us confine our attention to any single case of impact and suppose that the directions of the motions of the two impinging molecules lie on one side of the plane of concurrence, then it appears that the nature of the impact must be *overtaking*. Again, let us suppose that they lie similarly disposed on the other side of the plane; the nature of the impact is again *overtaking*. Now, instead of having the opposite of both the original lines of motion, suppose the opposite of one only is taken; it is clear that the nature of the impact is in this case of the *meeting* kind; and the opposite of the other line of motion being taken while the first is in its original position, the impact is again of the meeting kind.

Each of these four cases are equally probable, and the resolved velocities, or the values of β and δ, are the same in all, but two are meeting impacts and two are overtaking, each couple having perfectly distinct numerical equations to define the relation between the incident and reflected *vis viva*.

We are thus obliged to infer that the intestine action of the medium must be viewed in this manner as divided into two kinds of impacts specifically distinct in the numerical relation that subsists between the velocity before and after concurrence, and when employing the equations for summing up the results of the whole indefinitely great multitude that take place in mixed media, the effect of any one *meeting* impact must be considered along with its counterpart overtaking impact with the same velocities.

§ 10. We have remarked that it is only the resolved portion of the whole *vis viva* of a molecule that is dealt with by the equations—that forms the force of impact—and it may be questioned whether the mean of these forces in each kind of molecules bears the same proportion to each other as the whole *vis viva* of each. That the ratio is the same is best seen by the *reductio ad absurdum* method of reasoning. If the ratio is different, the motions of the heavier molecules must be resolved in a different way from those of the lighter, or the plane of concurrence must incline to one set of molecules in a different manner to that of the other set. Now any such effect is quite inconsistent with the fundamental hypothesis, and would require us to admit the influence of a modifying power whose nature and mode of bringing about the effect in question is unknown.

The ratio ρ of the resolved to the absolute *vis viva* is actually one-third, and will become obvious in the next section, but it seems needless to require the demonstration in this place as all that we have to be assured of is the constancy of the ratio, whatever its actual value may be.

In seeking to demonstrate the nature of the *vis viva* equilibrium, the solitary condition that we have to reason from is that the mean value of $(\beta_0^2 + \beta_1^2)$ is equal to $2\beta^2$ (§ 2) and the mean value of $(\delta_0^2 + \delta_1^2) = 2\delta^2$. That this is a necessary condition is obvious, because if either were less there would be a continual transfer from the molecules B to the molecules D, or from D to B and *vice versâ*.

By squaring the equations in § 2 and adding, we have the following:—

$$\beta_0^2 + \beta_1^2 = 2\{\beta^2 - \frac{4D}{B+D}\cdot\beta^2 + \frac{4D^2}{(B+D)^2}\cdot(\delta^2 + \beta^2)\}$$

$$\delta_0^2 + \delta_1^3 = 2\{\delta^2 - \frac{4D}{B+D}\cdot\delta^2 + \frac{4B^2}{(B+D)^2}\cdot(\delta^2 + \beta^2)\}.$$

If in any case it happens that $\beta_0^2 + \beta_1^2 = \beta^2$, we shall have

$$\beta^2\left(\frac{4D}{B+D}\right) = \frac{4D^2}{(B+D)^2}\cdot(\delta^2 + \beta^2),$$

or $\beta^2 B = \delta^2 D$, and $\delta_0^2 + \delta_1^2 = 2\delta^2$. Hence, if the squares of the impinging velocities happen to be in the inverse ratio of the molecular weights, then in either molecule the sum of the *vis viva* of the twofold encounter (one *meeting*, the other *overtaking* with the same impinging velocities) before impact, or $2\beta^2$, is equal to the sum after impact, or to $\beta_0^2 + \beta_1^2$.

But this is only one case out of an infinite number where the ratio is different. Generally, we may express the equation thus:

$$\beta_0^2 + \beta_1^2 = 2\beta^2 + p, \text{ and } \delta_0^2 + \delta_1^2 = 2\delta^2 + q.$$

Now, suppose that in an indefinitely great multitude of impacts the sum of all the individual values of $\beta_0^2 + \beta_1^2$ and $\delta_0^2 + \delta_1^2$ are taken, we shall have the mean of the values of the first equal to $\beta_m^2\rho$, in accordance with the necessary condition of permanence noticed above (by β_m^2 we mean to denote the mean molecular *vis viva* or mean square velocity of the B molecules, &c.). The mean value of $\beta^2 + p$ is, therefore, $\beta_m^2\rho$, but the mean value of β^2 is also evidently equal to $\beta_m^2\rho$, as above; hence, the mean value of p is 0, or the positive values of p balance the negative values. In the same way, it may be shown that the mean of the values of q is also 0. Hence, we deduce that $\beta_m^2 B = \delta_m^2 D$, or that *in mixed media the mean square molecular velocity is inversely proportional to the specific weight of the molecules* * . . . VII.

This is the law of the equilibrium of *vis viva.*

§ 11. Thus, it appears that the inverse ratio of the specific molecular weight is that which is naturally assumed by the mean square molecular velocity of media in contact, and according to the foregoing reasoning (§ 10), this is also the ratio that ensures an equilibrium of pressure between media of the same density, or which have the same number of molecules contained in the same volume. Thus, by combining V. with VII. we deduce that *media in equilibrium of pressure and* vis viva *are of equal density, or have specific gravities respectively proportional to their specific molecular weights* . . . . VIII.

§ 12. We may likewise remark that as the mean value of the product $\beta^2 B$ is equal to the mean $\delta^2 D$, or $B_m^2 = \delta_m^2 D$, *there is the same amount of* vis viva *or mechanical force contained in equal volumes of all media that are in equilibrium of pressure and* vis viva . . . . IX.

§ 13. If different media are placed in contact they must diffuse themselves through their common volume with velocities proportional to their mean molecular velocity; but this velocity being in each inversely as the square root of its specific molecular weight, which is equal to the square root of its specific gravity, we may deduce, by combining VI. with VII., that *media in equilibrium of pressure and* vis viva *diffuse themselves through their common volume with velocities inversely proportional to the square root of their specific gravity* . . . . X.

§ 14. Such are the principal points by which different media are related to each other. Their analogies to the properties of gases may be stated as follows:

(1.) The specific gravities of gases of the same *temperature* and pressure are respectively proportional to their atomic weight. [The combining equivalents or proportions may be viewed as simple multiples or divisors of the atomic weight or specific gravity.]

This is analogous to the VIII. deduction. Media in equilibrium of *vis viva* and pressure have specific gravities proportional to their molecular weight. It will be remarked that here again we have temperature represented by *vis viva.*

(2.) It is considered as almost proved that gases in equilibrium of pressure and temperature have, in equal volumes, the same absolute quantity of heat.

* [This is the first statement of a very important theorem. (See also 'Brit. Assoc. Rep.,' 1851). The demonstration, however, of § 10 can hardly be defended. It bears some resemblance to an argument indicated and exposed by Professor TAIT ('Edinburgh Trans.,' vol. 33, p. 79, 1886). There is reason to think that this law is intimately connected with the Maxwellian distribution of velocities, of which WATERSTON had no knowledge.—R.]

We have deduced, in reference to media, that when they are in equilibrium of pressure and *vis viva* they have, in equal volumes, the same amount of *vis viva*.

(3.) It is considered probable that the quantity of heat contained in equal volumes of all gases in equilibrium of pressure and temperature is proportional to their absolute temperature; and, in general, that equal increments of heat cause equal increments of temperature or expansion.

In all media in equilibrium of pressure and *vis viva* the same increment of *vis viva* causes the same expansive effect. (See also § 6.)

(4.) Gases, however different in specific gravity, when placed in connection diffuse themselves equally through each other; and according to the elegant induction of Dr. GRAHAM ('Edin. Trans.,' 1831) the velocity of diffusion is inversely as the square root of their specific gravity.

This is very exactly responded to by the hypothesis. DALTON'S law of diffusion belongs to media as a necessary and the most obvious consequence of the constitution assigned to them (VI.) DR. GRAHAM'S law of diffusive velocity or volume applies also to media, because if placed in similar circumstances such an effect must depend on the molecular velocity, which, singularly enough, bears exactly the like ratio to their specific gravity (X.).

§ 15. In the first point of analogy it was stated the atomic weight of a gas corresponded with its specific gravity, but with the reservation that the combining proportions are simple multiples or quotients of the same. This appears to me to be the fair statement of the remarkable connection that is always found between the combining volume and combining weight. It seems impossible that the fact of a volume of every gas containing the same number of molecules can ever be inductively established, but all analogy leads us to this conclusion. One volume of oxygen combines with two volumes of hydrogen to form two volumes of aqueous vapour. If we inferred from this that one molecule of oxygen combined with two molecules of hydrogen to form one of steam, we must admit that the molecule of steam occupies double the volume of a molecule of hydrogen or oxygen. If it is admitted, on the other hand, that the oxygen molecule is capable of disintegration, and that half a molecule combines with one of hydrogen to form one of steam, the bulk of the three molecules are equal. This last is the view that is responded to by the hypothetical media. The objection to it is plausible from the natural repugnance to the idea of dividing what has been considered as an ultimate element into parts, and of supposing it possible that an element should have a strong affinity to itself—for this point is also involved. Half molecules of oxygen must have a powerful attraction to each other as they never appear separate. It is the same with all the other simple gases and vapours that combine in half or other fractional volumes. We have an analogous example among the hydrocarburets. Methyl is half a molecule of olefiant gas, and enters into distinct combinations similar in every respect to those of olefiant gas, but it never appears in a separate form. Olefiant gas enters into combination with itself in various proportions, forming etherine, ceten, elaen, &c., all of which are isomeric but have different specific gravities as vapours. These the chemist is unable to form directly; they are organic products and show a great molecular capability that is undoubtedly common to all bodies.

There has certainly as yet been no recognised example of a simple element appearing in the gaseous form with two specific gravities, but it would be rash to affirm that such was either impossible or absurd. Indeed, chlorine, bromine, iodine, and fluorine are so similar in their chemical qualities that it is not very unlikely they may be examples of isomeric combinations of one base in which condensation of atomic volume takes place simultaneously with increment of atomic mass. In the hydrocarburets, on the other hand, there is little, if any, condensation of atomic volume, their specific gravity in the liquid form being all nearly the same.

A strong argument in favour of the physical origin of the law of volumes is to be obtained from the combination of acids with ether. Some of these rise up into vapour in the same state of chemical union as when liquid, while others quite similar are decomposed in the act of rising and occupy double the volume of the others. Even those that are fully combined when they first rise, by an increase of temperature not very considerable, are decomposed and immediately assume a double volume. Oxalic and nitrous ethers are examples of the last kind. Sulphate and nitrate of pyroxilic ether are examples of disunion in the act of rising.

If the hypothetical law of volumes is true in all cases, we should have to distinguish between atomic weight and combining proportion. Thus, if the atomic weight of oxygen is 1 (or specific gravity of gas), its combining proportions are $\frac{1}{2}$, 1, $1\frac{1}{2}$, 2, $2\frac{1}{2}$, 3, $3\frac{1}{2}$ (?). If the atomic weight of hydrogen is 1, its ascertained combining proportions are $\frac{1}{2}$, 1, $1\frac{1}{2}$, 2, 3, 4, 5, 6, 8, 9, 16. If the atomic weight of nitrogen is 1, its combining proportions are $\frac{1}{2}$ and 1. The same for chlorine is $\frac{1}{2}$, 1, $1\frac{1}{2}$, 2, $2\frac{1}{2}$; for bromine they are $\frac{1}{2}$, 1, $1\frac{1}{2}$; for iodine, $\frac{1}{2}$, 1, $1\frac{1}{2}$; for arsenic, $\frac{1}{4}$, 1; for sulphur, $\frac{1}{6}$, $\frac{1}{3}$; for phosphorus, $\frac{1}{4}$, &c. These are necessarily derived from the specific gravity of the simple gases and of their compounds.

The labours of DUMAS, MITSCHERLICH, REGNAULT, and BINEAU, have extended the list of gases and vapours, whose specific gravities have been accurately measured, to nearly 150. Such determinations throw a light upon the atomic constitution of compounds which it is impossible to obtain from their mere chemical analysis, and form an important guide to theoretical discussion, when the arithmetic of volumes is properly applied.

SECTION III.—ON THE PHENOMENA THAT ATTEND THE CONDENSING AND DILATING OF MEDIA, AND THE MECHANICAL VALUE OF THEIR MOLECULAR VIS VIVA.

§ 16. In the first section a distinct idea of the elastic force of the medium was obtained by viewing it as a rapid succession of impacts on the lower surface of a gravitating elastic plane ; and the equilibrium as being maintained when the upward velocity given to the plane by the shock of one molecule was equal to the downward velocity given to it by gravity acting through half the infinitesimal portion of time that elapses between two successive impacts. During the first half of this time gravity acts in destroying the upward velocity : during the second half it acts and generates the same velocity downwards, and by applying the equation for the meeting impact (§ 2) we found the relation between the pressure and the number of impacts in a given time. This relation is expressed by $gn/2w = A$, in which n is the number of molecules whose aggregate weight is the weight of the plane supported by A number of molecular impacts in a second of time : the impinging velocity being w, and g the accelerating force of gravity. It was also shown that the upward velocity given to the plane by one impact is w/n, and this is likewise the descending velocity with which it encounters the molecular shock.

We have now to examine the case where the encounter takes place with the plane at rest. Applying the equation for the meeting impact as in § 2, and putting $\epsilon = \beta = 0$; $n = B$; $D = 1$; $w = \delta$; we have

$$\epsilon_0 = \beta_0 = \frac{2w}{n+1} = \text{velocity upwards of the plane after the shock ;}$$

$$w_0 = \delta_0 = w - \frac{2wn}{n+1} = -w + \frac{2wn}{n+1} = w - \frac{w}{n+1}$$

$$= \text{velocity downwards of the molecules after the shock.*}$$

Thus, n being an indefinitely great number, we have the ascending velocity of the plane $\epsilon_0 = 2w/n$, being double what it was in the former case when the result of the impacts was statical equilibrium: and the decrement of molecular velocity $= w - w_0 = w/n$, which is a new and important feature. In the former case there was no decrement of molecular velocity : the molecule and plane continually meeting and retreating with velocity of impact and reflexion the same, and inversely proportional to their respective weight.

With its velocity $\frac{2w}{n}$ the gravitating plane ascends to the height $\left(\frac{2w}{gn}\right)^2 \frac{1}{2}g = \frac{2w^2}{gn^2}$, so that the weight of n molecules is raised through this height by the decrement w/n of the impinging velocity w of one molecule. Employing the differential notation,

* [It is easy to see that in the case supposed $n\epsilon_0 = 2w$, when n is great, so that the velocity of the plane is $2w/n$; but in the next step there is an unfortunate error which runs through many of the subsequent deductions.

$$w_0 = \delta_0 = w - \frac{2wn}{n+1} = -\left(-w + \frac{2wn}{n+1}\right) = -\left(w - \frac{2w}{n+1}\right),$$

not $-\left(w - \frac{w}{n+1}\right)$. The *vis viva* expended in raising n to the height $2w^2/gn^2$ is thus $4w^2/n$, not $2w^2/n$.—R.]

$dw = w/n$; and $2dw\,w = 2w^2/n = dw^2$, the *vis viva* expended in raising n to the height $2w^2/gn^2$.

§ 17. If we recur to § 4 we may remark the necessity of considering molecular velocity in two points of view when applying the arguments of §§ 2 and 16 to an enclosed volume of a medium. The first point is that upon the molecular velocity of impact depends the intensity of the shock on the plane, the ascending velocity given to it, and therefore also the time between the impacts, if the weight of the plane is considered constant. Thus as any one velocity is to the time of ascent and descent of the plane caused by that velocity, so is the mean of the impinging velocities to the mean of the time intervals, or inversely as the number of impacts in a unit of time; and the equilibrium does not require that the succession of impacts should be regular; the rapidity of the succession may fluctuate, but the average time and velocity must be constant. The second point is that any augmentation of velocity causes an increase in the frequency of the encounters (§ 4). In the equation $ng/2w = A$, if A were not a function of w it would remain unchanged, if n and w increased or diminished in the same proportion; but it was shown in § 4 that it was proportional to w, when Δ^3, the density, is constant, and to Δ^3 or n or c when w is constant; hence $A = wc\,\Delta^3$, in which c is a constant factor that has to be determined. We have also to determine the ratio between w^2, the mean square impinging velocity, and v^2, the mean square absolute molecular velocity, in the equation $ng/2w = A = wc\,\Delta^3$, or $n = \frac{2}{g}\,wc\,\Delta^3$.

Suppose the unit of volume in which the medium is confined to be a cube, the upper side of which is the plane n, and let v^2 be the mean square velocity of the molecules, so that if the squares of the respective velocities of all the molecules be added together, the sum will at all times be equal to $\Delta^3\,v^2$. Resolve the motion of each molecule at any instant into the six rectangular directions parallel to the side of the cube and add up the squares of the resolved velocities that are perpendicular to one side; it is evident that the sum must be $\frac{1}{6}\,\Delta^3\,v^2$, as the force is equally distributed in every direction, and in the stratum of the medium next the plane n one-sixth of the force of the molecules that happen to be in the stratum at any given instant is directed perpendicularly upon the plane. Suppose the breadth of the stratum is $1/\Delta$, the number of molecules that at all times are to be found moving in it is Δ^2, and half of these are diminishing their distance from the plane, and half increasing their distance with the mean square velocity $\frac{1}{3}\,v^2$.

The molecules moving equally in every direction must necessarily impinge equally in every possible direction on the plane, so that if their lines of motion were brought from every point of the surface of the plane where they impinge and made to issue from one central point, they would radiate equally to every part of the hemisphere; and as soon as those belonging to any one direction have impinged and thus withdrawn from forming part of the constant aggregate force $\frac{1}{6}v^2\,\Delta^2$, their place is immediately supplied by others of the same entering the stratum. The time taken by the set whose velocity is u and inclination to the plane θ to traverse the breadth of the stratum is evidently as $\frac{1}{u \sin \theta}$, and in a unit of time the number of the impacts in the succession of these belonging to the set is proportional to $u \sin \theta$. But this is the value of the resolved velocity of the set. Referring back to the reasoning in §§ 2, 16, the supporting effect of each impact on the heavy plane n was shown to be proportional also to the velocity of impact or molecular velocity resolved perpendicular to the plane. The supporting effect of each set in a unit of time is therefore as the square of the resolved or impinging velocity of the set. But the mean of all the square impinging velocities is $\frac{1}{3}v^2$, and half the molecules in the stratum are continually approaching the plane; the supporting effect of their continuous action is therefore the same as would be derived from the medium reduced to half density advancing against the plane with the uniform velocity $\sqrt{\frac{1}{3}v^2}$. Now as Δ^3 represents the number of molecules in the cubical unit of volume, the side of the cube being the unit of length, and $\sqrt{\frac{1}{3}v^2}$ the number of such units traversed in a unit of time, the supporting effect of the medium on the heavy plane n in the unit of time, is the same as that derived from $\frac{1}{2}\Delta^3\sqrt{\frac{1}{3}v^2}$ molecules impinging with the velocity $\sqrt{\frac{1}{3}v^2}$. Hence it is obvious that $A = \frac{1}{2}\Delta^3\sqrt{\frac{1}{3}v^2}$, and $w = \sqrt{\frac{1}{3}v^2}$, or $c = \frac{1}{2}$, and $3w^2 = v^2$. Also $n = \frac{2}{g}\, w^2 c\, \Delta^3 = \frac{v^2\, \Delta^3}{3g}$; or $v^2\, \Delta^3 = 3gn$; or $v = \sqrt{3g\frac{n}{\Delta^3}}$.

Thus we obtain an expression for the square root of the mean square molecula velocity in terms of the height of a uniform atmosphere $\frac{n}{\Delta^3}$, or what is the same, in terms of the ratio of the number of molecules in a column of the medium of the height of a uniform atmosphere to the number in the column of a unit height: and since $\sqrt{2gh}$ expresses the velocity acquired by a body in falling through the height h, we arrive at the following deduction. *The mean square molecular velocity of a medium is equal to the square of the velocity that a body acquires in falling through one-and-a-half times the height of a uniform atmosphere;* if the pressure of the medium is estimated from the effects of the molecular impacts on a perfectly rigid and elastic surface. If it were estimated by the effect on molecular elastic surfaces, there is reason to believe that the mean square velocity is double the amount specified (see Sec. 4) . . . . XI.*

§ 18. Suppose that the cubical volume of the medium receives such an increment of *vis viva* that under the constant pressure n the volume from 1 becomes $1 + \frac{2w^2}{gn^2}$; it has been shown in § 5 that the molecular *vis viva* must also from 1 have become

* [The author here arrives at the correct conclusion,

$$v^2 = 2g \cdot \tfrac{3}{2}h. \text{—R.]}$$

$1 + \frac{2w^2}{gn^2}$ that it may sustain the same pressure with the reduced density. But the molecular *vis viva* in the cubical volume is $\Delta^3 v^2$: hence we have the proportion

$$1 : 1 + \frac{2w^2}{gn^2} :: \Delta^3 v^2 : \Delta^3 v^2 \left(1 + \frac{2w^2}{gn^2}\right);$$

the increment of *vis viva* in the contained medium is thus $\Delta^3 v^2 \frac{2w^2}{gn^2}$. But it has been shown in § 17 that $w^2 = \frac{1}{3}v^2$, and $gn = \frac{1}{3}v^2 \Delta^3$, hence $\Delta^3 v^2 \frac{2w^2}{gn^2} = \frac{2v^2}{n}$.

Now in § 16 it was shown that to raise the weight of n molecules to the height $2w^2/gn^2$ the *vis viva* expended was $2w^2/n = \frac{2}{3} v^2/n$; comparing this, the *vis viva* expended in the act of increasing the volume, with $2v^2/n$, the increment of *vis viva* required to support the increased volume, it is evident that the ratio is exactly one-third. Thus we deduce that *to effect one increment of expansion in the volume of a medium subsisting under a constant pressure, four-thirds of an increment of vis viva are required: one of which thirds is expended in the act and does not appear in the medium: the remaining three-thirds, or one increment, appears in the medium and assists in sustaining its augmented volume** . . . . XII.

§ 19. This result is a necessary consequence of the perfect conservation of *vis viva* in the impinging action of perfectly elastic bodies. To enlarge a volume that is pressed upon is to raise a weight; is to expend mechanical force; is to expend molecular *vis viva*, and the last train of reasoning has led us to the relation between the molecular force expended and the work performed by it.

The mechanical value of the whole of the *vis viva* of the medium may be ascertained by the following proportion

$$\frac{2v^2}{3n} : \frac{2v^2}{3n^2} :: \Delta^3 v^2 : \frac{\Delta^3 v^2}{gn}. \quad \text{But } gn = \tfrac{1}{3}\Delta^3 v^2, \text{ therefore } \frac{\Delta^3 v^2}{gn} = 3.$$

Thus the *vis viva* expended in raising n, or the constant pressure, through the increment of the unit of volume, is to that increment as the whole *vis viva* of the medium is to three units of volume. Hence we deduce that if a medium is supposed to expand to four times its original volume with its original tension undiminished, it will in doing so expend as much as the whole of its original molecular force. This may be more concisely expressed by: *The molecular vis viva of a medium is equal to its tension acting through three times its volume*† . . . . XIII.

* [The corrected version of XII. will be "To effect one increment of expansion in the volume of a medium subsisting under a constant pressure, *five*-thirds of an increment of *vis viva* are required; *two* of which thirds is expended in the act, and does not appear in the medium; the remaining three-thirds, or one increment, appears in the medium, and assists in sustaining its augmented volume."—R.]

† [This is the virial equation applicable under the supposed conditions.—R.]

§ 20. If a medium is not allowed to increase in volume while its *vis viva* is increasing, no force will of course be expended, and each increment of *vis viva* engenders a like increment of tension. Thus *if we compare the amount of vis viva required to produce an increment of molecular vis viva in the medium, in the two cases of constant pressure and constant volume; it is manifest that the ratio is* 4 *to* 3, *or* 4/3* . . . . XIV.

§ 21. If we suppose the heavy plane n instead of being raised by the medium to descend upon it through the same differential height $2w^2/gn^2$, it is obvious that the same differential *vis viva* that was formerly abstracted is now communicated to the molecules of the unit volume. Force is exerted by the descending weight upon the medium and is transferred to its molecules.† Thus it is evident that the conversion of mechanical force into molecular *vis viva* is subject to the same law as the conversion of molecular *vis viva* into mechanical force. This law is expressed in XII. and XIV. The following is another form of annunciation which refers to an experimental method of ascertaining it if such media were actual existents. *The ratio of the increment of vis viva evolved by a small condensation of a medium to the diminution of molecular vis viva required to maintain the same condensation under a constant pressure is* $\frac{1}{3}$‡ . . . . XV.

§ 22. If a medium is compressed or dilated and the molecular *vis viva* evolved in it or given out from it by the act of condensation and dilatation be retained, let us enquire into the ratio of the density to the pressure. The preceding reasoning has shown that the increment or decrement of *vis viva* is equal to one-third of the increment or decrement of density, or $\frac{dv^2}{v^2} = \frac{d\Delta^3}{3\Delta^3}$; hence $\frac{2dv}{v} = \frac{d\Delta}{\Delta}$, which being integrated gives $v^2 \doteqdot \Delta$. But $v^2 \Delta^3 = e$, therefore $\Delta^4 \doteqdot e$, and $v^2 \doteqdot \sqrt[4]{e}$. Thus we deduce that *if a medium is compressed or dilated from a given condition of density and vis viva, the mean square molecular velocity varies as the fourth root of the tension or as the cube root of the density*§ . . . . XVI.‖

§ 23. The tendency of media to have their *vis viva* augmented when being forced into smaller volume is very similar to the rise of temperature that appears in air when being condensed. Thus tinder may be inflamed by the sudden compression of a small quantity of air, and on charging an air gun the condenser and force pump become so hot as to be painful to touch. Again, mercury may be frozen if exposed to a jet of air escaping from a state of high compression and expanding against the

* [The ratio of specific heats, commonly called γ, should be 5 : 3, not 4 : 3.—R.]

† Note A (motion indestructible as matter).

‡ [This result also requires correction.—R.]

§ Note B (vapours).

‖ [The corrected argument is:—Since $dv^2/v^2 = \frac{2}{3}\, d\Delta^3/\Delta^3$, we get $dv/v = d\Delta/\Delta$, or $v \doteqdot \Delta$. Accordingly $v^2 \doteqdot \Delta^2 \doteqdot (\Delta^3)^{\frac{2}{3}}$. But $v^2\Delta^3 = e$; therefore, $\Delta^5 \doteqdot e$, and $v^2 \doteqdot e^{\frac{2}{5}}$. Also $e \doteqdot (\Delta^3)^{\frac{5}{3}}$, or $p \doteqdot \rho^{\gamma}$, where $\gamma = \frac{5}{3}$.—R.]

atmospheric pressure. Media also lose their *vis viva* if allowed to expand against pressure.

Thus by XVI., if at 60° Fahr. a vessel containing air of double density is allowed to empty itself into the atmosphere, the decrement of temperature in the air that remains ought to be − 48° by the XVI. deduction. If on the contrary its density is suddenly increased from 1 to 2, its temperature ought to rise to 196°.*

Thus the analogy between media and gases and between *vis viva* and temperature is still maintained, and the phenomena of latent heat in gases appear also in media as the transference of force during a change of volume: *out* of the medium when it expands, and exerts, or gives out, mechanical force; *into* the medium when it is compressed and acted upon, or receives mechanical force.

The phenomena of latent heat thus appear to be the conversion of mechanical force into molecular *vis viva;* the visible into the invisible, as in condensation: and molecular *vis viva* into mechanical force; the invisible into the visible, as in expansion.

It is a necessary consequence of the conservation of *vis viva* or indestructibity of force among perfectly elastic bodies.

§ 24. Several experiments have been made on the *ratio* of the increment of temperature evolved by a small condensation of a volume of air to the diminution of temperature required to produce the same condensation under a constant pressure, and found it to be $\frac{3}{8}$. Mr. IVORY ('Phil. Mag.,' 1827) has proved that this ratio is constant under every change of temperature or density so long as DALTON and GAY-LUSSAC's law is maintained, or the air thermometer is an exact measure of heat. MM. GAY-LUSSAC and WELTER have also proved this experimentally for a considerable range. Mr. IVORY has also expressed his opinion that the nascent value of this ratio will be found to be $\frac{1}{3}$, and that the cause is probably connected with the proportion that subsists between the linear and the solid increments of expansion.

The value of this ratio in all media, whatever may be their condition of density or *vis viva*, is $\frac{1}{3}$ (§ 21); and the synthetical demonstration rests on the same fundamental principle that determines the proportion of linear to solid increments of expansion.

This ratio $\frac{1}{3}$ applies only to infinitesimal changes of volume, and it slowly increases with the amount of condensation. When the medium is compressed from 1 to 1·20 the ratio becomes $\frac{3}{8}$.†

In the experiments of MM. GAY-LUSSAC and WELTER referred to in the 'Mécanique Céleste,' the condensation did not exceed $\frac{1}{40}$th part of an atmosphere. The discrepancy may be exactly ascertained by performing the same experiments as it were on the medium by computation. The absolute temperature is denoted by v^2

* Note C (temperature of condensed air).

† Note D (to find the compression that corresponds with a given ratio between latent and sensible heat.)

and the absolute zero corresponds to $-461°$ of Fahr. scale (taking RUDBERG's constant of expansion, see § 6). By employing the equation $e = v^2 \Delta^3$ and $\left(\frac{e_0}{e_1}\right)^{\frac{1}{3}} = \frac{v_0^2}{v_1^2}$; (XVI), it is easy, by substituting the barometric height for e, to compute the result of any given experiment as if it had been performed on the hypothetical medium. This I have done in the case of MM. CLÉMENT and DÉSORMES' experiments. Referring to the account given of them in the 'Mécanique Céleste,' the minus interior pressure of the medium would have been 3·42 millims.; the mean result is given as 3·61, the difference being only about $\frac{1}{130}$th part of an inch of mercury. If the experiment gave the nascent ratio, the minus pressure at the end would have been one-fourth of the minus pressure at the beginning, or 3·45 millims. In MM. GAY-LUSSAC and WELTER's experiments, the difference of pressure at the end was 4·44 millims.; the nascent ratio in a medium would in a like experiment be 4·09 millims., the difference being about $\frac{1}{70}$th of an inch of mercury.

The evidence afforded by these experiments may be summed up as follows :—

The initial ratio of the increment of *vis viva* under a constant volume to the increment required to effect the same change of *vis viva* under a constant pressure in the medium is . . . . $\frac{1}{1\cdot33}$.

In air the same ratio of the increments of heat under the same circumstances by MM. CLÉMENT and DÉSORMES' experiments, is . . . . $\frac{1}{1\cdot35}$.

In air the same ratio, by MM. GAY-LUSSAC and WELTER's experiments, is . $\frac{1}{1\cdot37}$.

A difference in the reading of the height of the mercury in the manometer of $\frac{1}{130}$th and $\frac{1}{70}$th part of an inch would bring the respective experiments to coincide with the theory.*

§ 25. In the more recent of Mr. JOULE's physical researches that gentleman has applied mechanical force to the compressing of air surrounded with water, to collect the heat evolved, and has found that about 800 lbs. descending through the height of one foot increases the temperature of a pound of water one degree. The same result nearly was obtained by forcing water through narrow tubes. Mechanical force was expended, and the same proportionate amount of heat was produced in the water. It is remarkable that the same mechanical value was found for the heat generated by the magneto-electric machine. Such accordance in the results, as Mr. JOULE remarks, seems strongly to favour the *vis viva* or vibratory theory of heat.

If air is similar in its constitution to a medium we may employ the deductions of this section to determine the mechanical value of any quantity of heat applied to it,

* [The fair agreement of the erroneously deduced value of γ, viz., $\frac{4}{3}$, with observation, was doubtless the reason of the author not discovering his mistake of calculation. We know that upon his principles the calculated value should be $\frac{5}{3}$, which accords much less well with the results observed for ordinary gases than does $\frac{4}{3}$. It should be borne in mind that the observed value, $\gamma = 1\cdot405$, has not, even at the present time, been reconciled with theory, although reasons may be given for a departure from $\gamma = \frac{5}{3}$. —R.]

and assuming the specific heat of air to be 0·238 that of water,* we may ascertain the mechanical value of 1° applied to 1 lb. of water, which is equal to $\frac{1}{0\cdot238}$ degree applied to 1 lb. of air. Since 820 cubic feet of air at 60° and 30 in. tension weigh as much as one cubic foot of water, we have $\frac{820}{0\cdot238} = 3444$ cubic feet of air which, heated one degree without being allowed to change its volume, requires as much heat as one cubic foot of water to raise it one degree. The absolute temperature at 60° is 520°, and one degree added augments by $\frac{1}{521}$ part the absolute heat or molecular *vis viva* of the air. But the whole *vis viva* in 3444 cubic feet of air at the temperature 60° and pressure 30 in., is equal to the whole pressure of the atmosphere on a square foot, acting through three times 3444 feet in height, or 10,332 feet (XIII.). The pressure of a column of 30 inches of mercury on a base of 1 square foot or 144 square inches is equal to 14·722 lbs. × 144 = 2120 lbs. This weight raised through 10,332 ft., corresponds to 21,904,000 lbs. raised one foot high, and $\frac{1}{521}$ part of this, or 42,043 lbs., raised one foot high represents the absolute mechanical effect of 1° of heat applied to one cubic foot of water. Dividing this by 62½, the number of lbs. in a cubic foot of water, we get 673 lbs. raised one foot high equal to the mechanical effect corresponding to 1° of heat applied to 1 lb. of water. This compared with Mr. Joule's result is not unsatisfactory considering the difficulties that attend the experiments that afford the data.†‡

[*Editor's Note:* Material has been omitted at this point.]

* Note E (specific heat of air).

† Note F (M. Clapeyron's view of the motive power of heat examined).

‡ [This is an independent calculation of the mechanical equivalent of heat, quite distinct from that of Mayer.—R.]

APPENDIX I.

EXTRACT from the Proceedings of the Royal Society (vol. 5, p. 604—March 5, 1846).

On the Physics of Media that are composed of free and perfectly elastic Molecules in a state of Motion. By J. J. WATERSON (*sic*), *Esq. Communicated by Captain* BEAUFORT, *R.N., F.R.S.*

THIS memoir contains the enunciation of a new theory of heat, capable of explaining the phenomena of its radiation and polarization, and the elasticity of various bodies, founded on the hypothesis of a medium consisting of a vast multitude of particles of matter endowed with perfect elasticity, and enclosed in elastic walls, but moving in all directions within that space, with perfect freedom, and in every possible direction. In the course of these motions, the particles must be supposed to encounter one another in every possible manner, during an interval of time so small as to allow of their being considered infinitesimal in respect to any sensible period; still, however, preserving the molecular *vis viva* constant and undiminished.

The author then enters into extensive analytical investigations; first, of the conditions that determine the equilibrium of such a homogeneous medium as is implied by the hypothesis, and of the laws of its elasticity; secondly, of the physical relations of media that differ from each other in the specific weight of their molecules; thirdly, of the phenomena that attend the condensing and dilating of media, and of the mechanical value of their molecular *vis viva;* fourthly, of the resistance of media to a moving surface; fifthly, of the vertical equilibrium of a medium surrounding a planet and constituting its atmosphere; and lastly, of the velocity with which impulses are transmitted through a medium so constituted.

In an Appendix, the author enters into a full explanation of a table of gases and vapours, drawn up with reference to the subjects discussed in his paper.

Appendix II.

Extract from the Report of the 21st Meeting of the British Association, Ipswich, 1851. (Transactions of the Sections, p. 6.)

On a General Theory of Gases. By J. J. Waterston, *Bombay.*

The author deduces the properties of gases, with respect to heat and elasticity, from a peculiar form of the theory which regards heat as consisting in small but rapid motions of the particles of matter. He conceives that the atoms of a gas, being perfectly elastic, are in continual motion in all directions, being restrained within a limited space by their collisions with each other, and with the particles of surrounding bodies. The *vis viva* of those motions in a given portion of gas constitutes the quantity of heat contained in it.

He shows that the result of this state of motion must be to give the gas an elasticity proportional to the mean square of the velocity of the molecular motions, and to the total mass of the atoms contained in unity of bulk; that is to say, to the density of the medium. This elasticity, in a given gas, is the measure of temperature. Equilibrium of pressure and heat between two gases takes place when the number of atoms in unity of volume is equal, and the *vis viva* of each atom equal. Temperature, therefore, in all gases, is proportional to the mass of one atom multiplied by the mean square of the velocity of the molecular motions, being measured from an *absolute zero* 491° below the zero of Fahrenheit's thermometer.

If a gas be compressed, the mechanical power expended in the compression is transferred to the molecules of the gas, increasing their *vis viva;* and conversely, when the gas expands, the mechanical power given out during the expansion is obtained at the expense of the *vis viva* of the atoms. This principle explains the variations of temperature produced by the expansion and condensation of gases—the laws of their specific heat under different circumstances, and of the velocity of sound in them. The fall of temperature found on ascending in the atmosphere, if not disturbed by radiation and other causes, would correspond with the *vis viva* necessary to raise the atoms through the given height.

The author shows that the velocity with which gases diffuse themselves is proportional to that possessed by their atoms according to his hypothesis.

Part III

REFINEMENTS OF THE ENERGETIC DESCRIPTION OF ATOMIC MOTIONS

Editor's Comments on Papers 10 Through 14

10 JOULE
Some Remarks on Heat, and the Constitution of Elastic Fluids

11 KRÖNIG
Basic Features of a Theory of Gases

12 CLAUSIUS
On the Nature of the Motion Which We Call Heat

13 CLAUSIUS
On the Mean Length of the Paths Described by the Separate Molecules of Gaseous Bodies on the Occurrence of Molecular Motion; Together with Some Other Remarks Upon the Mechanical Theory of Heat

14A MAXWELL
Illustrations of the Dynamical Theory of Gases. Part I. On the Motions and Collisions of Perfectly Elastic Spheres

14B MAXWELL
Excerpts from *Illustrations of the Dynamical Theory of Gases. Part II. On the Process of Diffusion of Two or More Kinds of Moving Particles Among One Another*

We have seen that even though significant steps had, by 1850, been taken toward the development of an analytical kinetic theory of matter, the reception of these approaches by the influential segments of the scientific community was something less than cordial. Part of this was due to the continued strong bias in favor of the caloric theory of heat, reinforced by the lack of sufficiently convincing experiments showing how heat could be associated with molecular motion. But as Boorse and Motz have pointed out (see "The World of the Atom," ed. by Henry A. Boorse and Lloyd Motz, Basic Books, Vol. I, p. 256, 1966), a stamp of respectability on kinetic theory was badly needed. The person who finally helped break the stubborn resistance of the adherents to the caloric theory and confer that much needed stamp through careful experimentation and keen intellectual effort was the distinguished English scientist James Prescott Joule (1818–1889). Joule was born near Manchester, England, to a family of brewers. He did not attend a university but acquired an adequate mathematical and scientific training from tutors including John Dalton, who at that time was at the close of his own illustrious career. Joule became an imaginative and ingenious experimenter, and during the 1840s he conducted a series of quantitative studies on the heat evolved in electrical and mechanical processes. By 1845 he had determined experimentally the mechanical equivalent of heat, with a value close to that accepted today, namely 4.18605 joules of mechanical energy per calorie of heat. (Some of these experiments are described in Benchmark volume *Energy: Historical Development of the Concept*, ed. R. Bruce Lindsay, Papers 30A and 30B.) The reports on these experiments did not contain any reference to a theory of heat. However, in a lecture delivered at Saint Ann's Church (Manchester) Reading Room in May, 1847 (see Joule in Lindsay 1975) Joule presented his views on the nature of heat and laid down in qualitative fashion the foundations for a kinetic theory of matter. The following quotation from this lecture is relevant to our theme: "We may conceive, then, that the communication of heat to a body consists, in fact, in the communications of impetus, or living force, to its particles. It will perhaps appear to some of you something strange that a body apparently quiescent should in reality be the seat of motions of great rapidity; but you will observe that the bodies themselves, considered as a whole, are not supposed to be in motion. The constituent particles, or atoms, being far too small to be seen even by the help of the most powerful microscopes, it is no wonder that we cannot observe their motion. There is therefore reason to suppose that the particles of all bodies, their constituent atoms, are in a state of motion almost too rapid for us to conceive, for the phenomena cannot be otherwise explained. . . ."

Joule felt certain that the original Bernoulli-Herapath model of a gas, attributing the pressure of a gas to elastic collisions of the constituent molecules with the confining surface, was qualitatively correct (see Paper 8). Joule proceeded to calculate the average speed of the molecules in hydrogen gas by using the known weight of a volume of gas and assuming one third of the particles travel back and forth in each of three perpendicular directions. An implicit step in the analysis was the identification of the pressure with the *square* of the molecular velocity, which led to the proportionality of absolute temperature to *vis viva.* Joule's results were first published in 1851 in the *Memoirs of the Manchester Literary and Philosophical Society* and later printed in the *Philosophical Magazine*. The latter article is reproduced here as Paper 10. It proved to be "seminal" in the fullest scientific sense of the word. Joule's signature provided the necessary endorsement of scientific legitimacy on the new kinetic theory and opened the field to vigorous analytical exploitation by others.

Among those whose work was to make a quick impact as a consequence of Joule's interest in the subject were the Germans August Karl Krönig (1822–1879) and Rudolf Clausius (1822–1888). Krönig was born in Westphalia and earned his doctorate at the University of Berlin in 1845. He was a charter member of the Berlin Physical Society, and for many years served as professor at the Berlin Königliche Realschule, although he was forced to retire prematurely in 1861 because of ill health. Krönig's only major contribution, aside from several elementary texts, was his 1856 paper "Grundzüge einer Theorie der Gase," which is reproduced here in translation as Paper 11. Krönig was apparently not aware of Joule's work. He used a somewhat more formal mathematical approach than Joule, assumed that the molecules move in equal numbers with equal velocities along three perpendicular directions, and explicitly derived the expression for the product of pressure and volume of gas in terms of the square of the molecular velocity in its modern form. He showed formally that the general gas law for an ideal gas is obtained if the absolute temperature is taken as proportional to the molecular *vis viva.* There have been suggestions advanced that Krönig may have actually read an abstract of a paper by Waterston written in 1851 summarizing his proposed kinetic theory, and that perhaps Krönig was at least subconsciously influenced by Waterston's principal themes (Daub 1971). In that event the influence of Waterston becomes all the more pervasive in this history of the development of kinetic theory. Krönig's paper evidently served to stimulate his eminent fellow countryman, Clausius, to examine the problems of gas dynamics in a more general way.

Rudolf Clausius, one of the outstanding scientific figures of the nineteenth century, was born in Köslin, Germany (now Koszalin, Poland). After studying mathematics and physics at Berlin, he received his doctorate from the University at Halle in 1847. Following professorial appointments at Zurich and Würzburg, he went to Bonn in 1869 and remained there the rest of his life. After the appearance of Joule's and Krönig's works on kinetic theory, Clausius realized its important implications for the development of thermodynamics, a subject with which he was much concerned during his entire career. His key introductory paper in kinetics is reproduced here in translation as Paper 12. In this treatment he went beyond Krönig and Joule by showing that the ratio of the experimental specific heat at constant pressure to that at constant volume can not be accounted for by translational motion alone. The *vis viva* must be present in other forms. He inferred that collisions between molecules will transfer energy into these other forms and bring about a distribution of molecular speeds. Previously no one had considered gas models with individual molecules moving with different speeds at the same temperature. Clausius was also able to advance arguments in support of Avogadro's principle that equal volumes of all gases at the same temperature and pressure contain equal numbers of molecules. In a second paper (Paper 13 in this volume). Clausius invoked for the first time the idea of the mean free path of a molecule in a gas. He showed that contrary to earlier conjectures, the magnitude of the average distance between collisions for gases under standard conditions is quite small in comparison to the dimensions of familiar containers. A satisfactory explanation for such well-known visible effects as the slowness of dissipation of a smoke plume could thereby be given, with the relative shortness of the mean free path being the significant factor rather than the high average molecular speed. Together with Helmholtz, Kelvin, Carnot, Joule, and Tait, Clausius codified and interpreted the first and second laws of thermodynamics. In this connection, he introduced and defined the all-important functions of thermodynamic state—internal energy and entropy—the former in 1850 and the latter in 1865. Clausius was a scientist of strong convictions, and at various times he indulged in controversies with Tait, Maxwell, and Kelvin.

The revival of interest in kinetic theory as a consequence of the work of Krönig and Clausius soon attracted the attention of the Scot, James Clerk Maxwell (1831–1879). Maxwell was quick to see the import of Clausius' inferences concerning a distribution of molecular speeds and realized that a correct *a priori* calculation of the mean free path would require a general treatment of the motion of systems containing a large number of elastic particles moving with a wide range of speeds. His subsequent analysis was to pave the way for the introduction of statistical techniques into the solution of physical problems. Maxwell's creative efforts in the development of electromagnetic theory alone were probably sufficient to rank him as the greatest theoretical physicist of the classical nineteenth-century era. He was born in Edinburgh to a family of comfortable means. He obtained his formal education at the University of Edinburgh and in 1854 graduated from Cambridge University as the Second Wrangler in the mathematical Tripos. In 1856 he went to Marischal College in Aberdeen as the professor of natural philosophy. While there, he worked on his major memoir *Illustrations of the Dynamical Theory of Gases,* from which Paper 14 in this volume was taken. From 1860 to 1865 Maxwell was professor of natural philosophy at Kings College, London. He left London in 1865 and spent the next five years in semiretirement at his family's country home in Glenair, Scotland, where he completed his new formulation of electromagnetic theory. In 1871 the Cavendish Laboratory was established at Cambridge University with the aid of a generous gift from the Seventh Duke of Devonshire, and Maxwell was called to its Chair of Experimental Physics. Unfortunately, during the following years his health failed, and he died in 1879 at the premature age of forty-eight.

Maxwell's contributions to gas dynamics represented the coming into play of the full power and efficacy of the mathematical physicist's approach. His analytical skills were such that much more realistic features could be introduced into the model of a gas. From Maxwell's distribution of velocities, the mean free path could be calculated and an experimental confirmation obtained with data on thermal conductivity, gaseous diffusion, and gaseous viscosity. By his use of statistical methods, Maxwell removed one of the major obstacles facing those who were trying to apply mechanical theories to atoms, namely, the one of coping with systems containing enormously large numbers of particles in life-sized volumes. The application of statistics was to be exploited further in the following years by the Austrian Ludwig Boltzmann and the American J. Willard Gibbs. The interval from 1850 onward represented the coming of age of the modern methodology of science with powerful theoretical minds working in tandem with increasingly more precise experimentation.

Though we do not include in this book any paper of Ludwig Boltzmann (1844–1906), he played such an important part in the development of the role of energy in atomic physics that some reference to him is in order here. He was born in Vienna, educated at the University of Vienna, and received his doctorate from that institution in 1866. He held professorships in mathematics and physics at Vienna, Graz, Munich, and Leipzig during his professional career and had returned to Vienna when suicide brought a tragic end to his distinguished career. Boltzmann's important contribution to gas dynamics in 1868 was to account for the effect of an external force on a system of particles and to derive the distribution function giving the fraction of particles possessing a given total energy. He obtained the function in the form

$$f(v) = (\text{const})e^{-\frac{[\frac{1}{2}mv^2 + V(x)]}{h}}$$

In modern terminology the quantity $h = kT$, where k is the so-called Boltzmann constant, and T is the absolute temperature. The term $V(x)$ represents the potential energy. This equation was to become fundamental in the calculation of equilibrium values of thermodynamic quantities describing such diverse phenomena of atomic origin as dielectric polarization, paramagnetism, and specific heats. Boltzmann also linked together statistical physics and thermodynamics with the *H*-theorem in which he was able to show how a system with an originally nonequilibrium distribution evolves toward an equilibrium distribution as time passes. He identified the hitherto purely experimentally defined entropy-state function of Clausius as that statistical quantity characteristic of the system that increases as the system approaches equilibrium.

Throughout his scientific career Boltzmann felt a keen philosophical commitment to the atomic point of view and spent considerable time developing arguments to refute the criticisms of those who objected to atomism and preferred phenomenological theories of matter. Boltzmann's contemporary in Vienna, Ernst Mach (1838–1916), and the German physical chemist, Wilhelm Ostwald (1853–1932), were forceful advocates of a particular kind of phenomenological theory called "energetics." This school believed that through the use of general energy principles and analogies among descriptive mathematical equations, all natural phenomena could be explained without having to invoke atomic models as the starting point. Boltzmann's successful arguments against the energetics school had much to do with the ultimate triumph of the atomic point of view in modern science (Boltzmann 1897).

REFERENCES

Boltzmann, L., 1897. "Über die Unentbehrlichkeit der Atomistik in der Naturwissenschaft," *Ann. Phys.*, N.F., **60**, 237.

Daub, E. E., 1971. "Waterston's Influence on Krönig's Kinetic Theory of Gases," *Isis,* **62**, 512.

Joule, J. P., 1884*a*. "On the Calorific Effects of Magneto-Electricity, and on the Mechanical Value of Heat," *The Scientific Papers on Prescott Joule,* Vol. 1. London: Taylor & Francis, pp. 123–159. *In* R. Bruce Lindsay, 1975. *Energy: Historical Development of the Concept.* Stroudsburg, Pennsylvania: Dowden, Hutchinson & Ross, pp. 308–344.

Joule, J. P., 1884*b*. "On the Existence of an Equivalent Relation between Heat and the Ordinary Forms of Mechanical Power," *The Scientific Papers on Prescott Joule,* Vol. 1. London: Taylor & Francis, pp. 202–205. *In* R. Bruce Lindsay, 1975. *Energy: Historical Development of the Concept.* Stroudsburg, Pennsylvania: Dowden, Hutchinson & Ross, pp. 345–348.

Joule, J. P., 1884*c*. "On Matter, Living Force, and Heat," *The Scientific Papers on Prescott Joule,* Vol. 1. London: Taylor & Francis, pp. 265–276. *In* R. Bruce Lindsay, 1975. *Energy: Historical Development of the Concept.* Stroudsburg, Pennsylvania: Dowden Hutchinson & Ross, pp. 349–360.

10

Reprinted from *Philos. Mag.* ser. 4, 14:211-216 (1857)

SOME REMARKS ON HEAT, AND THE CONSTITUTION OF ELASTIC FLUIDS

J. P. Joule*

To the Editors of the Philosophical Magazine and Journal.

Oakfield, Moss Side, Manchester, August 22, 1857.

GENTLEMEN,

SHOULD you deem it to be of sufficient interest, I shall feel much obliged by your republishing the enclosed paper, referred to by Professor Clausius in his paper "On the Nature of the Motion which we call Heat," inserted in your last Number. I may observe in reference to the note at the foot of p. 109, that the Memoirs of the Literary and Philosophical Society of Manchester are at present regularly forwarded to the principal scientific societies of Europe and America.

I have the honour to remain,

Yours very respectfully,

JAMES P. JOULE.

In a paper "On the Heat evolved during the Electrolysis of Water," published in the seventh volume of the Memoirs of this Society, I stated that the magneto-electrical machine enabled us to convert mechanical power into heat; and that I had little doubt that, by interposing an electro-magnetic engine in the circuit of a voltaic battery, a diminution of the quantity of heat evolved, per equivalent of chemical reaction, would be observed, and that this diminution would be proportional to the mechanical power obtained.

The results of experiments in proof of the above proposition were communicated to the British Association for the Advancement of Science, in 1843†. They showed that whenever a current of electricity was generated by a magneto-electrical machine, the quantity of heat evolved by that current had a constant relation to the power required to turn the machine; and, on the other hand, that whenever an engine was worked by a voltaic battery, the power developed was at the expense of the calorific power of the battery for a given consumption of zinc, the mechanical effect produced having a fixed relation to the heat lost in the voltaic circuit.

The obvious conclusion from these experiments was, that heat and mechanical power were convertible into one another; and it became therefore evident that heat is either the *vis viva* of pon-

* Read at a Meeting of the Manchester Literary and Philosophical Society, October 3, 1848, and published in the Society's Memoirs, November 1851.

† Phil. Mag. vol. xxiii. pp. 263, 347, 435.

derable particles, or a state of attraction or repulsion capable of generating *vis viva*.

It now became important to ascertain the mechanical equivalent of heat with as much accuracy as its importance to physical science demanded. For this purpose the magnetic apparatus was not very well adapted, and therefore I sought in the heat generated by the friction of fluids for the means of obtaining exact results. I found, first, that the expenditure of a certain amount of mechanical power in the agitation of a given fluid uniformly produced a certain fixed quantity of heat; and, secondly, that the quantity of heat evolved in the friction of fluids was entirely uninfluenced by the nature of the liquid employed; for water, oil and mercury, fluids as diverse from one another as could have been well selected, gave sensibly the same result, viz. that the quantity of heat capable of raising the temperature of a pound of water 1°, is equal to the mechanical power developed by a weight of 770 lbs. in falling through one perpendicular foot*.

Believing that the discovery of the equivalent of heat furnished the means of solving several interesting phænomena, I commenced, in the spring of 1844, some experiments on the changes of temperature occasioned by the rarefaction and compression of atmospheric air†. It had long been known that air, when forcibly compressed, evolves heat; and that, on the contrary, when air is dilated, heat is absorbed. In order to account for these facts, it was assumed that a given weight of air has a smaller capacity for heat when compressed into a small compass than when occupying a larger space. A few experiments served to show the incorrectness of this hypothesis: thus, I found that by forcing 2956 cubic inches of air, at the ordinary atmospheric pressure, into the space of 136½ cubic inches, 13°·63 of heat per pound of water were produced; whereas by the reverse process, of allowing the compressed air to expand from a stopcock into the atmosphere, only 4°·09 were absorbed instead of 13°·63, which is the quantity of heat which ought to have been absorbed, according to the generally received hypothesis. I found, also, that when strongly compressed air was allowed to escape into a vacuum, no cooling effect took place on the whole, a fact likewise at variance with the received hypothesis. On the contrary, the theory I ventured to advocate‡ was in perfect agreement with the phænomena; for the heat evolved by compressing the

* The equivalent I have since arrived at is 772 foot-pounds. See Phil. Trans. 1850, part 1.—May 1851, J. P. J.

† Phil. Mag. vol. xxvi.

‡ I subsequently found that M. Mayer had previously advocated a similar hypothesis, without, however, attempting an experimental demonstration of its accuracy.—*Annalen* of Wöhler and Liebig for 1842.—May 1851, J. P. J.

air was found to be the equivalent of the mechanical power employed, and, *vice versâ*, the heat absorbed in rarefaction was found to be the equivalent of the mechanical power developed, estimated by the weight of the column of atmospheric air displaced. In the case of compressed air expanding into a vacuum, since no mechanical power was produced, no absorption of heat was expected or found. M. Seguin has confirmed the above results in the case of steam.

The above principles lead, indeed, to a more intimate acquaintance with the true theory of the steam-engine; for they have enabled us to estimate the calorific effect of the friction of the steam in passing through the various valves and pipes, as well as that of the piston in rubbing against the sides of the cylinder; and they have also informed us that the steam, while expanding in the cylinder, loses heat in quantity exactly proportional to the mechanical force developed*.

The experiments on the changes of temperature produced by the rarefaction and condensation of air give likewise an insight into the constitution of elastic fluids, for they show that the heat of elastic fluids is the mechanical force possessed by them; and since it is known that the temperature of a gas determines its elastic force, it follows that the elastic force, or pressure, must be the effect of the motion of the constituent particles in any gas. This motion may exist in several ways, and still account for the phænomena presented by elastic fluids. Davy, to whom belongs the signal merit of having made the first experiment absolutely demonstrative of the immateriality of heat, enunciated the beautiful hypothesis of a rotatory motion. He says, "It seems possible to account for all the phænomena of heat, if it be supposed that in solids the particles are in a constant state of vibratory motion, the particles of the hottest bodies moving with the greatest velocity and through the greatest space: that in fluids and elastic fluids, besides the vibratory motion, which must be considered greatest in the last, the particles have a motion round their own axes with different velocities, the particles of elastic fluids moving with the greatest quickness; and that in ætherial substances the particles move round their own axes, and separate from each other, penetrating in right lines through space. Temperature may be conceived to depend upon the velocity of the vibrations, increase of capacity on the motion being performed

* A complete theory of the motive power of heat has been recently communicated by Professor Thomson to the Royal Society of Edinburgh. In this paper the very important law is established, that the fraction of heat converted into power in any perfect engine, is equal to the range of temperature divided by the highest temperature above absolute zero.—May 1851, J. P. J.

in greater space; and the diminution of temperature during the conversion of solids into fluids or gases may be explained on the idea of the loss of vibratory motion, in consequence of the revolution of particles round their axes at the moment when the body becomes fluid or aëriform, or from the loss of rapidity of vibration in consequence of the motion of the particles through greater space*." I have myself endeavoured to prove that a rotary motion, such as that described by Sir H. Davy, will account for the law of Boyle and Mariotte, and other phænomena presented by elastic fluids†; nevertheless, since the hypothesis of Herapath, in which it is assumed that the particles of a gas are constantly flying about in every direction with great velocity, the pressure of the gas being owing to the impact of the particles against any surface presented to them, is somewhat simpler, I shall employ it in the following remarks on the constitution of elastic fluids; premising, however, that the hypothesis of a rotatory motion accords equally well with the phænomena.

Let us suppose an envelope of the size and shape of a cubic foot to be filled with hydrogen gas, which, at 60° temperature and 30 inches barometrical pressure, will weigh 36·927 grs. Further, let us suppose the above quantity to be divided into three equal and indefinitely small elastic particles, each weighing 12·309 grs.; and further, that each of these particles vibrates between opposite sides of the cube, and maintains a uniform velocity except at the instant of impact; it is required to find the velocity at which each particle must move so as to produce the atmospherical pressure of 14,831,712 grs. on each of the square sides of the cube. In the first place, it is known that if a body moving with the velocity of $32\frac{1}{6}$ feet per second be opposed, during one second, by a pressure equal to its weight, its motion will be stopped, and that if the pressure be continued one second longer, the particle will acquire the velocity of $32\frac{1}{6}$ feet per second in the contrary direction. At this velocity there will be $32\frac{1}{6}$ collisions of a particle of 12·309 grs. against each side of the cubical vessel in every two seconds of time; and the pressure occasioned thereby will be $12{\cdot}309 \times 32\frac{1}{6} = 395{\cdot}938$ grs. Therefore, since it is manifest that the pressure will be proportional to the square of the velocity of the particles, we shall have for the velocity of the particles requisite to produce the pressure of 14,831,712 grs. on each side of the cubical vessel,

$$v=\sqrt{\left(\frac{14,831,712}{395{\cdot}938}\right)}\,32\tfrac{1}{6}=6225 \text{ feet per second.}$$

* Elements of Chemical Philosophy, p. 95.

† Mr. Rankine has given a complete mathematical investigation of the action of vortices, in his paper on the Mechanical Action of Gases and Vapours.—Trans. Roy. Soc. Edinb. vol. xx. part 1.—May 1851, J. P. J.

The above velocity will be found equal to produce the atmospheric pressure, whether the particles strike each other before they arrive at the sides of the cubical vessel, whether they strike the sides obliquely, and thirdly, into whatever number of particles the 36·927 grs. of hydrogen are divided.

If only one-half the weight of hydrogen, or 18·4635 grs., be enclosed in the cubical vessel, and the velocity of the particles be as before, 6225 feet per second, the pressure will manifestly be only one-half of what it was previously, which shows that the law of Boyle and Mariotte flows naturally from the hypothesis.

The velocity above named is that of hydrogen at the temperature of 60°; but we know that the pressure of an elastic fluid at 60° is to that at 32° as 519 is to 491. Therefore the velocity of the particles at 60° will be to that at 32° as $\sqrt{519} : \sqrt{491}$, which shows that the velocity at the freezing temperature of water is 6055 feet per second.

In the above calculations it is supposed that the particles of hydrogen have no sensible magnitude, otherwise the velocity corresponding to the same pressure would be lessened.

Since the pressure of a gas increases with its temperature in arithmetical progression, and since the pressure is proportional to the square of the velocity of the particles, in other words, to their *vis viva*, it follows that the absolute temperature, pressure, and *vis viva* are proportional to one another, and that the zero of temperature is 491° below the freezing-point of water. Further, the absolute heat of the gas, or, in other words, its capacity, will be represented by the whole amount of *vis viva* at a given temperature. The specific heat may therefore be determined in the following simple manner:—

The velocity of the particles of hydrogen, at the temperature of 60°, has been stated to be 6225 feet per second, a velocity equivalent to a fall from the perpendicular height of 602,342 feet. The velocity at 61° will be $6225\sqrt{\frac{520}{519}} = 6230{\cdot}93$ feet per second, which is equivalent to a fall of 603,502 feet. The difference between the above falls is 1160 feet, which is therefore the space through which 1 lb. of pressure must operate upon each pound of hydrogen, in order to elevate its temperature one degree. But our mechanical equivalent of heat shows that 770 feet is the altitude representing the force required to raise the temperature of water one degree; consequently the specific heat of hydrogen will be $\frac{1160}{770} = 1{\cdot}506$, calling that of water unity.

The specific heats of the gases will be easily deduced from that of hydrogen; for the whole *vis viva* and capacity of equal bulks of the various gases will be equal to one another; and the velocity of the particles will be inversely as the square root of the specific gravity. Hence the specific heat will be inversely proportional to the specific gravity, a law which has been arrived at experimentally by De la Rive and Marcet.

In the following Table I have placed the specific heats of various gases determined in the above manner, in juxtaposition with the experimental results of Delaroche and Berard reduced to constant volume:—

	Experimental specific heat.	Theoretical specific heat.
Hydrogen . . .	2·352	1·506
Oxygen	0·168	0·094
Nitrogen	0·195	0·107
Carbonic oxide . .	0·158	0·068

The experimental results of Delaroche and Berard are invariably higher than those demanded by the hypothesis. But it must be observed, that the experiments of Delaroche and Berard, though considered the best that have hitherto been made, differ considerably from those of other philosophers. I believe, however, that the investigation undertaken by M. V. Regnault, for the French Government, will embrace the important subject of the capacity of bodies for heat, and that we may shortly expect a new series of determinations of the specific heat of gases, characterized by all the accuracy for which that distinguished philosopher is so justly famous. Till then, perhaps, it will be better to delay any further modifications of the dynamical theory, by which its deductions may be made to correspond more closely with the results of experiment*.

* If we assume that the particles of a gas are resisted uniformly until their motion is stopped, and that then their motion is renewed in the opposite direction, by the continued operation of the same cause, as in the projection upwards and subsequent fall of a heavy body; the maximum velocity of the particles will be to the uniform velocity required by the theory assumed in the text, as the square root of two is to one, and the comparison of the theoretical with the experimental specific heat will be as follows:—

	Experimental specific heat.	Theoretical specific heat.
Hydrogen	2·352	3·012
Oxygen	0·168	0·188
Nitrogen	0·195	0·214
Carbonic oxide . . .	0·158	0·136

I have just learned that the experiments of Regnault on the specific heat of elastic fluids are on the eve of publication, and doubt not that their accuracy will enable us to arrive at a decisive conclusion as to the correctness of the above hypothesis.—June 1851, J. P. J.

11

BASIC FEATURES OF A THEORY OF GASES

A. Krönig

This article was translated expressly for this Benchmark volume by R. Bruce Lindsay, Brown University, from "Grundzüge einer Theorie der Gase," Ann. Phys. (Leipzig) **99**:315–322 (1857).

The mechanical theory of heat maintains that the heat of a body consists of nothing but the motion of its smallest parts. There is lacking, however, a clear view of how this motion is really constituted. For gaseous bodies, which in relation to the theory of heat have received most consideration to date, I shall introduce in what follows a hypothesis that I think corresponds well with the facts. From this I believe we shall be able to deduce the explanation (not so far given) of the experimental laws of gases, i.e., those of Mariotte, Gay Lussac, etc. The hypothesis is as follows: Gases consist of atoms, which behave as solid, completely elastic spheres, moving with definite velocites in otherwise empty space. Solid and liquid bodies, as soon as they settle down to equilibrium or a steady state, are likewise absolutely elastic with respect to the collisions of gas atoms.

Let us now think of a box with absolutely elastic walls and imagine that it contains a number of absolutely elastic spheres, which when at rest take up a very small part of the space inside the box. The box is now shaken vigorously so that the spheres are put in motion. If the box is now allowed to return to rest, the spheres continue in the motion that has been communicated to them, although the direction and speed of motion of each individual sphere will change with each collision with another sphere or the walls. The gas atoms of a gas occupying a vessel will behave just like these spheres.

Thus a gas atom does not oscillate about a position of equilibrium but moves in a straight line with constant velocity until it hits another atom or strikes a solid or liquid wall. It is assumed that there is no mutual repulsion between gas atoms that are not in contact.

The smoothest wall is to be considered rough in comparison with the gas atoms, and the path of every atom must therefore be so irregular that it defies precise description. However, the laws of the theory of probability allow one to replace this complete irregularity with complete regularity.

Let there be n gas atoms in a vessel in the form of a rectangular paralle piped with plane walls and dimensions x, y, and z. Let the vessel be divided into $n/6$ equal cubes. At any given moment, let there be 6 atoms in each cube, moving in the directions $+x$, $-x$, $+y$, $-y$, $+z$, $-z$ with the common velocity c. Let us assume that the atoms have no mutual interaction whatever but rather move until they strike a wall (atoms whose midpoints move in the same straight line when they collide merely exchange velocities). We now calculate the pressure exerted by the gas on one of the yz walls.

This pressure is produced by the collisions of the atoms with the wall. If only one atom hits the wall, the pressure will be mca, where a is the number of collisions per unit time. An atom moving perpendicular to the yz wall, that is, parallel to the x

direction, hits the wall each time after traveling the distance $2x$ so that $a = c/2x$. To find the total pressure on the wall, we must multiply mca by the number of atoms moving in the x direction. Since out of every 6 atoms 2 move parallel to x, this number is $n/3$. Consequently,

$$P = mc \cdot \frac{c}{2x} \cdot \frac{n}{3}$$

If we call p the pressure on unit area of the wall, we have

$$p = mc \cdot \frac{c}{2x} \cdot \frac{n}{3} \cdot \frac{1}{yz}$$

[*Editor's Note:* In modern terminology P is the total force on the wall, while p, the force per unit area, is called the pressure.]

If we write $xyz = v$, the volume of the box, we have, leaving out the constant factor 1/6 [*Editor's Note:* Why does he do this? It seems unnecessary!], we have

$$p = \frac{nmc^2}{v}$$

This preceding expression shows that the pressure of the gas on each of the walls is equally great and that it is inversely proportional to the volume, as the law of Mariotte requires.

We must now add to the assumption previously made the additional one that the product mc^2, or the *vis viva* of an atom, is none other than the temperature of the gas reckoned from absolute zero.

According to the Gay-Lussac law, the gas exerts no pressure at –273°C. At a higher temperature t, however, it exerts a pressure proportional to t. Thus

$$p = nt/v$$

from the previous formula, if we put $mc^2 = t$.

[*Editor's Note:* There is a problem here. This formula seems to make temperature of the same dimension as energy. Modern molecular theory makes the kinetic energy proportional to the temperature and introduces Boltzmann's constant k as part of the proportionality factor.]

From $p_1 = p_2, t_1 = t_2$ and $v_1, = v_2$, we get n_1 and n_2, that is, different gases at the same temperature and pressure contain equal numbers of atoms per unit volume.

[*Editor's Note:* This is Avogadro's law.]

From this it also follows that the mass m of an atom is proportional to the specific gravity of the gas. Thus from $p = mnt/mv$, we get $p = P't/sv$, where $P' = nm$, or proportional to the mass of the gas, and s is the specific gravity.

Up to now I have considered the gas as contained in a closed vessel and have taken no account of the action of gravity on the atoms. I shall now determine the pressure that a single gas atom exerts on the earth. One might plausibly suppose that this pressure will increase as the temperature of the atom increases.

Let the atom leave the earth at time $t = 0$ with the velocity having vertical component c. After time c/g, the atom will reach the highest point in its path and have velocity zero there (g = acceleration of gravity). After time $2c/g$, it will reach the earth again with velocity c. In unit time it will hit the earth $g/2c$ times with the velocity c. Hence its pressure against the earth or its weight will be equal to $mc \cdot g/2c = mg/2$ and therefore independent of its velocity and hence also of its temperature.

When an atom oscillates between two horizontal walls whose vertical separation is z, we can find the difference between the pressure it exerts above and below. Let the atom leave the lower wall at time $t = 0$ with vertical component velocity equal to c. If the upper wall were not there, the atom would after time c/g reach the height $c^2/2g$, with velocity zero. If now $z > c^2/2g$, we obviously have the case just discussed. The atom then exerts on the lower wall the pressure $mg/2$ and on the upper wall the pressure zero. The difference between the two pressures is $mg/2$. However, for $z < c^2/2g$, the atom reaches the upper wall after time $(c - \sqrt{c^2 - 2gz})/g$ with a velocity having vertical component $\sqrt{c^2 - 2gz}$. If we substitute into the previous formula for the pressure mca the corresponding values, we get for the pressure of the atom on the lower wall

$$\frac{mcg}{2\,(c - \sqrt{c^2 - 2gz})}$$

The pressure on the upper wall is

$$\frac{mg\sqrt{c^2 - 2gz}}{2\,(c - \sqrt{c^2 - 2gz})}$$

The difference between the two pressures is again $mg/2$. Since this result is not affected by a possible horizontal component of the motion, a closed vessel containing n atoms each of mass m suffers an increase in weight of $nmg/2$.

We can readily deduce from this that Archimedes' law holds for gases and that a gas of lower specific gravity will rise above one with higher specific gravity.

To appreciate a surprising aspect of the assumption of the steady motion of gas atoms, we need to be reminded only of what a short time, for example, it takes for a very small amount of hydrogen sulphide to spread itself throughout a large room. From the assumption that the temperature is measured by mc^2 comes the result that a gas must diffuse the more quickly the smaller its specific gravity. If we allow m_w and c_w to correspond to hydrogen and m_s and c_s to correspond to oxygen, we then have $m_s = 16m_w$. From $m_s c_s^2 = m_w c_w^2$, it follows that $c_w = 4c_s$. Thus if hydrogen and oxygen are at the same temperature, it follows that the velocity of the hydrogen atoms is four times that of the oxygen atoms.

If the temperature of a gas is given by the *vis viva* of an atom of the gas, the quantity of heat in the gas Q is equal to the sum of the *vis viva* values of all its atoms. Accordingly $Q = nmc^2$, or $Q = nt$. If we represent the quantity of a gas by the number n of its atoms, it follows that equal amounts of any gas at a given temperature contain equal amounts of heat. In other words, for the same pressure and temperature, equal volumes of any gas contain equal quantities of heat.

The differential equation corresponding to the equation $Q = nt$ represents the fact that all gases measured by volume have the same constant specific heat.

It is well known that several of the laws just derived do not agree precisely with experimental results, when the experiments are performed with highest precision. One ought not to expect an unlimited validity of these laws. Suppose we have two plane parallel walls with separation equal to x, and suppose that two atoms move back and forth between these walls along the same straight line that is perpendicular to the walls. Assume that at time $t = 0$ an atom starts out from each wall with velocity c. In the foregoing, it has been assumed that each atom moves without influencing the other and that each atom moves back and forth along x. Then the number of collisions of both atoms against a wall per unit time is $2c/2x$. But in reality the two atoms cannot pass right through each other. They collide in the middle of their paths and bounce off each other with the same velocity in the opposite direction. Hence it is not true that both atoms strike both walls. Each atom then strikes only the one wall. Depending on the different time needed for the action of the elasticity, there are then three possible cases between which it is not possible to decide *a priori*. Either every atom uses exactly the time $(2 \cdot 1/2 \cdot x)/c$ to pass from a given wall to collision with another atom and then back to the wall, or it needs a smaller or a longer time. In the first case the formula previously derived for the pressure is correct. In the second case the pressure would be greater than nmc^2/v, and in the third case, smaller. Similar considerations hold in relation to several of the other laws that have been derived.

The temperature relations connected with volume changes in gases also follow easily from the assumptions that have been made.

Let us imagine two vessels of the same volume that have a wall in common. The first is assumed to be filled with n gas atoms each of mass m and velocity c, whereas the second is assumed to be empty (vacuum). Let a piece w be removed from the common wall. Those atoms that would have collided against the piece w must now go through the corresponding hole and be reflected from the rough wall of the second vessel in various directions. It will after a time develop that in unit time as many atoms pass through the hole in one direction as pass out in the other. Hence we shall ultimately have the same pressure in the two vessels. But during this process c has remained unchanged, and the same is true of mc^2. The result is that when a gas expands without having to overcome any resistance, it suffers no change in temperature.

If on the other hand, the collisions of the gas atoms take place against a movable piston, the atoms must lose as much of their energy (*vis viva*), i.e., of their heat, as they have contributed to the piston.

If, finally, a gas is compressed by a piston, the atoms must bounce off the moving piston with greater speed than they had on incidence. Hence the gas must become heated.

12

Reprinted from *Philos. Mag.* ser. 4, 14:108–127 (1857)

ON THE NATURE OF THE MOTION WHICH WE CALL HEAT

R. Clausius*

1. BEFORE writing my first memoir on heat, which was published in 1850, and in which heat is assumed to be a motion, I had already formed for myself a distinct conception of the nature of this motion, and had even employed the same in several investigations and calculations. In my former memoirs I intentionally avoided mentioning this conception, because I wished to separate the conclusions which are deducible from certain general principles from those which presuppose a particular kind of motion, and because I hoped to be able at some future time to devote a separate memoir to my notion of this motion and to the special conclusions which flow therefrom. The execution of this project, however, has been retarded longer than I at first expected, inasmuch as the difficulties of the subject, as well as other occupations, have hitherto prevented me from giving to its development that degree of completeness which I deemed necessary for publication.

A memoir has lately been published by Krönig, under the title *Grundzüge einer Theorie der Gase*†, in which I have recognized some of my own views. Seeing that Krönig has arrived at these views just as independently as I have, and has published them before me, all claim to priority on my part is of course out of the question; nevertheless, the subject having once been mooted in this memoir, I feel myself induced to publish those parts of my own views which I have not yet found in it. For the present, I shall confine myself to a brief indication of a

* From Poggendorff's *Annalen*, vol. c.

† This was first printed separately by A. W. Hayn in Berlin, and afterwards appeared in Poggendorff's *Annalen*, vol. xcix. p. 315.

few principal points, and reserve a more complete analysis for another time*.

2. Krönig assumes that the molecules of gas do not oscillate about definite positions of equilibrium, but that they move with constant velocity in right lines until they strike against other molecules, or against some surface which is to them impermeable. I share this view completely, and I also believe that the expansive force of the gas arises from this motion. On the other hand, I am of opinion that this is not the only motion present.

In the first place, the hypothesis of a rotatory as well as a progressive motion of the molecules at once suggests itself; for at every impact of two bodies, unless the same happens to be central and rectilineal, a rotatory as well as a translatory motion ensues.

I am also of opinion that vibrations take place within the several masses in a state of progressive motion. Such vibrations are conceivable in several ways. Even if we limit ourselves to the consideration of the atomic masses solely, and regard these as absolutely rigid, it is still possible that a molecule, which consists of several atoms, may not also constitute an absolutely rigid mass, but that within it the several atoms are to a certain extent moveable, and thus capable of oscillating with respect to each other.

I may also remark, that by thus ascribing a movement to the atomic masses themselves, we do not exclude the hypothesis that each atomic mass may be provided with a quantity of finer matter, which, without separating from the atom, may still be moveable in its vicinity.

By means of a mathematical investigation given at the end of the present memoir, it may be proved that the *vis viva* of the translatory motion alone is too small to represent the whole heat present in the gas; so that without entering into the probability of the same, we are thus compelled to assume one or more motions of another kind. According to this calculation, the excess of the whole *vis viva* over that of the translatory motion alone is particularly important in gases of a complicated chemical constitution, in which each molecule consists of a great number of atoms.

* I must not omit to mention here, that some time ago Mr. William Siemens of London, when on a visit in Berlin, informed me that Joule had also expressed similar ideas in the Memoirs of the Literary and Philosophical Society of Manchester. My views being consequently no longer completely new, this was an additional reason why I should hasten their publication less than I otherwise probably should have done. Hitherto I have not been able to procure the memoir of Joule in question, and therefore I am ignorant how far he has pursued the subject, and whether his views coincide with mine in all points. It is to be regretted that Joule did not publish his memoir in a more widely circulated periodical.

3. In one and the same gas the translatory motion of the whole molecules will always have a constant relation to the several motions which, in addition to the above, the constituents of the molecules likewise possess. For brevity I will call the latter the *motions of the constituents.*

Conceive a number of molecules whose constituents are in active motion, but which have no translatory motion. It is evident the latter will commence as soon as two molecules in contact strike against each other in consequence of the motion of their constituents. The translatory motion thus originated will of course occasion a corresponding loss of *vis viva* in the motion of the constituents. On the other hand, if the constituents of a number of molecules in a state of translatory motion were motionless, they could not long remain so, in consequence of the collisions between the molecules themselves, and between them and fixed sides or walls. It is only when all possible motions have reached a certain relation towards one another, which relation will depend upon the constitution of the molecules, that they will cease mutually to increase or diminish each other.

When two molecules whose constituents are in motion come into collision they will not rebound, like two elastic balls, according to the ordinary laws of elasticity; for their velocities and directions after collision will depend, not only upon the motion which the whole molecules had before impact, but also upon the motion of those constituents which are nearest each other at the moment of collision. After the equalization of the several motions, however, when the translatory motion is, on the whole, neither increased nor diminished by the motions of the constituents, we may, in our investigation of the total action of a great number of molecules, neglect the irregularities occurring at the several collisions, and assume that, in reference to the translatory motion, the molecules follow the common laws of elasticity.

4. The explanation of the expansive force of gases and its dependence upon volume and temperature, as given by Krönig, suffers no essential modification through the introduction of other motions. The pressure of the gas against a fixed surface is caused by the molecules in great number continually striking against and rebounding from the same. The force which must thence arise is, in the first place, by equal velocity of motion inversely proportional to the volume of the given quantity of gas; and secondly, by equal volume proportional to the *vis viva* of the translatory motion: the other motions do not here immediately come into consideration.

On the other hand, from Gay-Lussac's law we know that, under constant volume, the pressure of a perfect gas increases in the same ratio as the temperature calculated from −273° C., which we call the absolute temperature. Hence, according to the above, it follows that the absolute temperature is proportional to the *vis viva* of the translatory motion of the molecules. But as, according to a former remark, the several motions in one and the same gas bear a constant relation to each other, it is evident that the *vis viva* of the translatory motion forms an aliquot part of the total *vis viva*, so that the absolute temperature is also proportional to the whole *vis viva* in the gas.

These considerations, together with others connected therewith to be given hereafter, induced me, in my memoir "On the Moving Force of Heat," to express the opinion that the specific heat of gases was constant; which opinion was in opposition to the experiments then known*. The quantity of heat which must be imparted to a gas, under constant volume, in order to raise its temperature is to be considered as the increase of the *vis viva* in the gas, inasmuch as in this case no work is done whereby heat could be consumed. The specific heat *under constant volume*, therefore, is in a perfect gas the magnitude which Rankine calls the *true* specific heat. Now the assertion that the true specific heat of a gas is constant, is simply equivalent to the assertion that *the total* vis viva *in the gas has a constant ratio to the* vis viva *of the translatory motion which serves us as a measure of the temperature.* With respect to the specific heat under constant pressure, I have proved in the memoir before cited, and by means of a hypothesis proceeding from the same considerations, that it differs only by a constant magnitude from the true specific heat.

5. The foregoing is true for permanent gases only, and even for these only approximatively. In general, the small deviations which present themselves can be easily accounted for.

In order that Mariotte's and Gay-Lussac's laws, as well as others in connexion with the same, may be strictly fulfilled, the gas must satisfy the following conditions with respect to its molecular condition:—

(1) The space actually filled by the molecules of the gas must be infinitesimal in comparison to the whole space occupied by the gas itself.

(2) The duration of an impact, that is to say, the time required to produce the actually occurring change in the motion of a molecule when it strikes another molecule or a fixed surface, must be infinitesimal in comparison to the interval of time between two successive collisions.

(3) The influence of the molecular forces must be infinitesimal. Two conditions are herein involved. In the first place, it is requisite that the force with which all the molecules at their

* Poggendorff's *Annalen*, vol. lxxix. p. 393. Phil. Mag. vol. ii. pp. 1, 102.

mean distances attract each other, vanish when compared with the expansive force due to the motion. But the molecules are not always at their mean distances asunder; on the contrary, during their motion a molecule is often brought into close proximity to another, or to a fixed surface consisting of active molecules, and in such moments the molecular forces will of course commence their activity. The second condition requires, therefore, that those parts of the path described by a molecule under the influence of the molecular forces, when the latter are capable of altering appreciably the direction or velocity of the molecule's motion, should vanish when compared with those parts of its path with respect to which the influence of these forces may be regarded as zero.

If these conditions are not fulfilled, deviations in several ways from the simple laws of gases necessarily arise; and these deviations become more important the less the molecular condition of the gas fulfils the conditions in question.

On becoming acquainted with the celebrated investigations of Regnault on the deviations of gases from Mariotte's and Gay-Lussac's laws, I attempted, by means of the principles above intimated, to deduce some conclusions with respect to the molecular condition of several gases from the nature of the deviations which Regnault detected in the same. A description of this method, however, would be too prolix; and even the results, in consequence of the many difficulties encountered in actual calculation, are too uncertain to merit being here adduced.

Whenever, therefore, in the sequel a gas is spoken of, we shall, as before, conceive it to be one which *perfectly* fulfils the above conditions, and which Regnault calls an *ideal* gas, inasmuch as all known gases present but an approximation to this condition.

6. After these considerations on the *gaseous* condition, the question at once arises in what manner the *solid* and *liquid* conditions differ from the gaseous. Although a definition of these states of aggregation, in order to be satisfactory in all its details, would require a more complete knowledge than we at present possess of the condition of the individual molecules, yet it appears to me that several fundamental distinctions may be advanced with tolerable probability.

A motion of the molecules takes place in all three states of aggregation.

In the *solid* state, the motion is such that the molecules move about certain positions of equilibrium without ever forsaking the same, unless acted upon by foreign forces. In solid bodies, therefore, the motion may be characterized as a vibrating one, which may, however, be of a very complicated kind. In the first place, the constituents of a molecule may vibrate among themselves; and secondly, the molecule may vibrate as a whole: again, the latter vibrations may consist in oscillations to and fro of the centre of gravity, as well as in rotatory oscillations around this centre of gravity. In cases where external forces act on the body, as in concussions, the molecules may also be permanently displaced.

In the *liquid* state the molecules have no longer any definite position of equilibrium. They can turn completely around their centres of gravity; and the latter, too, may be moved completely out of its place. The separating action of the motion is not, however, sufficiently strong, in comparison to the mutual attraction between the molecules, to be able to separate the latter entirely. Although a molecule no longer adheres to definite neighbouring molecules, still it does not spontaneously forsake the latter, but only under the united actions of forces proceeding from other molecules, with respect to which it then occupies the same position as it formerly did with respect to its neighbouring molecules. In liquids, therefore, an oscillatory, a rotatory, and a translatory motion of the molecules takes place, but in such a manner that these molecules are not thereby separated from each other, but, even in the absence of external forces, remain within a certain volume.

Lastly, in the *gaseous* state the motion of the molecules entirely transports them beyond the spheres of their mutual attraction, causing them to recede in right lines according to the ordinary laws of motion. If two such molecules come into collision during their motion, they will in general fly asunder again with the same vehemence with which they moved towards each other; and this will the more readily occur, since a molecule will be attracted with much less force by another single molecule than by all the molecules which in the liquid or solid state surround it.

7. The phænomenon of *evaporation* appearing peculiarly interesting to me, I have attempted to account for the same in the following manner.

It has been stated above, that in liquids a molecule, during its motion, either remains within the sphere of attraction of its neighbouring molecules, or only leaves the same in order to take up a corresponding position with respect to other neighbouring molecules. This applies only to the mean value of the motions, however; and as the latter are quite irregular, we must assume that the velocities of the several molecules deviate within wide limits on both sides of this mean value.

Taking next the surface of a liquid into consideration, I assume that, amongst the varied motions to and fro, it happens that under the influence of a favourable cooperation of the translatory, oscillatory, and rotatory motions, a molecule separates itself with such violence from its neighbouring molecules that it has already receded from the sphere of their action before losing all its velocity under the influence of their attracting forces, and thus that it continues its flight into the space above the liquid.

Conceive this space to be enclosed, and at the commencement empty; it will gradually become more and more filled with these expelled molecules, which will now deport themselves in the space exactly as a gas, and consequently in their motion strike against the enclosing surfaces. The liquid itself, however, will form one of these surfaces; and when a molecule strikes against the same, it will not in general be driven back, but rather retained, and, as it were, absorbed in consequence of the renewed attraction of the other molecules into whose vicinity it has been driven. A state of equilibrium will ensue when the number of molecules in the superincumbent space is such, that on the average as many molecules strike against, and are retained by the surface of the liquid in a given time, as there are molecules expelled from it in the same time. The resulting state of equilibrium, therefore, is not a state of rest or a cessation of evaporation, but a state in which evaporation and condensation continually take place and compensate each other in consequence of their equal intensity.

The density of the vapour necessary for this compensation, depends upon the number of molecules expelled from the surface of the liquid in the unit of time; and this number is again evidently dependent upon the activity of the motion within the liquid, that is to say, upon its temperature. I have not yet succeeded in deducing from these considerations the law according to which the pressure of vapour must increase with the temperature.

The preceding remarks on the deportment of the surface of the liquid towards the superincumbent vapour, apply in a similar manner to the other surfaces which enclose the space filled with vapour. The vapour is in the first place condensed on these surfaces, and the liquid thus produced then suffers evaporation, so that here also a state must be attained in which condensation and evaporation become equal. The requisite quantity of condensed vapour on these surfaces depends upon the density of the vapour in the enclosed space, upon the temperature of the vapour and of the enclosing surfaces, and upon the force with which the molecules of vapour are attracted towards these surfaces. In this respect a maximum will occur when the enclosing surfaces are completely moistened with the condensed liquid; and as soon as this takes place, these surfaces deport themselves exactly like a single surface of the same liquid.

8. The reason why the presence of another gas above the liquid cannot impede the evaporation of the same may now be immediately explained.

The pressure of the gas on the liquid arises solely from the fact, that here and there single molecules of gas strike against the surface of the liquid. In other respects, however, inasmuch as the molecules of gas themselves actually fill but a very small part of the superincumbent space, the latter must be considered as empty, and as offering a free passage to the molecules of the liquid. In general these molecules will only come into collision with those of the gas at comparatively great distances from the surface, and the former will then deport themselves towards the latter as would the molecules of any other admixed gas. We must conclude, therefore, that the liquid also expels its molecules into the space filled with gas; and that in this case also the quantity of vapour thus mixed with the gas continues to increase until, on the whole, as many molecules of vapour strike against and are absorbed by the surface of the liquid as the latter itself expels; and the number of molecules of vapour to the unit of volume requisite hereto, is the same whether the space does or does not contain additional molecules of gas.

The pressure of the gas, however, exercises a different influence on the interior of the liquid. Here also, or at places where the mass of liquid is bounded by a side of the vessel, it may happen that the molecules separate from each other with such force that for a moment the continuity of the mass is broken. The small vacuum thus produced, however, is surrounded on all sides by masses which do not admit of the passage of the moved molecules; and hence this vacuum will only then become magnified into a bubble of vapour, and be able to continue as such, when the number of molecules expelled from its enclosing liquid walls is sufficient to produce an internal vapour-pressure capable of holding in equilibrium the pressure which acts externally and tends to compress the bubble again. Hence the expansive force of the enclosed vapour must be greater, the greater the pressure to which the liquid is exposed, and thus is explained the relation which exists between the pressure and the temperature of the boiling-point.

The relations will be more complicated when the gas above the liquid is itself condensable, and forms a liquid which mixes with the given one, for then of course the tendency of the two kinds of matter to mix enters as a new force. I shall not here enter into these phænomena.

As in liquids, so also in solids the possibility of an evaporation may be comprehended; nevertheless it does not follow from this that, on the contrary, an evaporation *must* take place on the surface of all bodies. It is, in fact, readily conceivable that the mutual cohesion of the molecules of a body may be so great, that, so long as the temperature does not exceed a certain limit, even the most favourable combination of the several molecular motions is not able to overcome this cohesion.

9. The explanation of the *generation* and *consumption* of heat accompanying changes of the state of aggregation and of the volume of a body, follows immediately, according to the above principles, from the consideration of the positive and negative work done by the active forces.

In the first place, let us consider the *interior* work.

When the molecules of a body change their relative positions, the change may be either in the same direction as that according to which the forces resident in the molecules tend to move the latter, or in a direction opposite to this. In the first case, a certain velocity is imparted by the forces to the molecules during their passage from one position to the other, and the *vis viva* of this velocity is immediately transformed into heat; in the second case, and disregarding for the present the action of extraordinary foreign forces, it is in virtue of heat that the molecules move in directions opposed to the interior forces, and the retardation which these molecules thereby suffer through the action of the opposing forces appears as a diminution of the motion of heat.

In the passage from the solid to the liquid state the molecules do not, indeed, recede beyond the spheres of their mutual action; but, according to the above hypothesis, they pass from a definite and, with respect to the molecular forces, suitable position, to other irregular positions, in doing which the forces which tend to retain the molecules in the former position have to be overcome.

In evaporation, the complete separation which takes place between the several molecules and the remaining mass evidently again necessitates the overcoming of opposing forces.

With respect to gaseous bodies, it is evident from what has been above remarked concerning the requisites of a perfect gaseous state, that as soon as a gas has once attained this state, molecular attractions have no longer to be overcome during its further expansion, so that during the changes of volume of an ideal gas no interior work has to be performed.

10. Besides the interior, we have in the next place to consider the *exterior* work, and the corresponding change in the quantity of heat.

In the first place, with respect to the changes in the heat of a permanent gas subjected to pressure or expanding by overcoming a counter-pressure, Krönig has already remarked at the conclusion of his memoir, that when one of the sides against which the molecules of gas strike is itself in motion, these molecules will not rebound with velocities equal to those they possessed when moving towards the side. If the side approaches the molecules, the velocity of the molecules on leaving the same is in general greater than before, if it recedes less. By means of special mathematical considerations, it may without difficulty be proved that the increase or decrease of the *vis viva* thereby produced must exactly correspond to the external work done by the expansive force of the gas; nevertheless it is not even necessary to give a special proof of this theorem, since the same is an immediate consequence of the general law of equivalence between *vis viva* and work.

If the side moves so slowly that the pressure of the gas against the moving side is just as great as against a stationary one, then, in determining the work, the velocity of the side no longer enters into consideration, but merely the whole path described by the same. On the contrary, if the velocity of the side is so great that in the vicinity of the same a sensible compression or rarefaction of the gas thereby ensues, then the pressure actually exercised by the gas during the motion must always be brought into calculation.

When an overflow takes place between two vessels filled with gases of different densities, or between a full and an empty vessel, on the whole no work will be performed, and therefore no change in the total quantity of heat can occur. It is not here asserted that no change in the quantity of heat takes place in either of the two vessels considered separately, for a mass of gas whose molecules move principally in a definite direction deports itself towards adjoining gaseous masses in the same manner as a moved side; and when the moved gaseous mass strikes against stationary walls, just as much motion of heat makes its appearance as *vis viva* is lost by the common translatory motion of the whole mass.

Just as in the changes of volume of gaseous bodies, so also in other cases the external work must be taken into consideration; as, for instance, the work which during the evaporation is employed to overcome external resistance, and thus to make room for the vapour as it is generated. In solid and liquid bodies which only suffer small changes of volume, the external work is also for the most part only small; nevertheless here also cases occur in which its influence becomes considerable.

11. Lastly, I must mention a phænomenon the explanation of which appears to me to be of great importance, viz. *when two gases combine with each other, or when a gas combines with another body, and the combination is also gaseous, the volume of the compound gas bears a simple ratio to the volumes of the single constituents, at least when the latter are gaseous.*

Krönig has already proved that the pressure exerted by a gas on the unit of its enclosing surface must be proportional to the number of molecules contained in the unit of volume, and to the *vis viva* of the several molecules arising from their translatory motion, the only one which Krönig considers.

If we apply this to simple gases, and assume that, when pressure and temperature are the same, equal volumes contain the same number of atoms—a hypothesis which for other reasons is very probable,—it follows that, in reference to their translatory motion, the atoms of different gases must have the same *vis viva*.

We will next examine in what manner this theorem remains true when applied to the molecules of compound gases.

12. In the first place, let us compare compound gases amongst themselves, *e. g.* two gases to form which the constituents have combined in ratios of volume respectively equal to 1 : 1 and 1:2. Nitric acid and nitrous acid may serve as examples.

With respect to these two gases, we know that quantities containing the same amount of oxygen occupy the same volume. Hence here, too, equal volumes contain the same number of molecules, although in the one gas each molecule consists of two, and in the other of three atoms; and we must further conclude, that even these differently constituted molecules have the same *vis viva* with respect to their translatory motion.

In most other compound gases we are led to the same conclusion; and in cases which do not submit themselves to this rule, it does not appear to me impossible that the discrepancy may be accounted for in one or both of two ways: either the gas was not sufficiently removed from its points of condensation when its volume was determined, or the chemical formula hitherto employed does not properly represent the manner in which the atoms are combined to form molecules.

On comparing compound and simple gases, however, an unmistakeable deviation from the foregoing rule shows itself, inasmuch as the space corresponding to an atom of the simple gas does not correspond to a molecule of the compound one. When two simple gases combine in equal volumes, it is well known that no change of volume takes place, whilst according to the above rule the volume ought to be diminished in the ratio of 2 : 1. Again, when a volume of one gas combines with two or three volumes of another, the combination is found to occupy two volumes, whereas according to rule it ought only to occupy one volume, and so on.

13. On seeking to explain these curious anomalies, and especially to find a common law governing the relations of volume in gases, I was led to adopt the following view as being most plausible. I beg to offer the same to the scientific public as a hypothesis which is at least worthy of further examination.

I assume that the force which determines chemical combination, and which probably consists in a kind of polarity of the atoms, is already active in simple substances, and that *in these likewise two or more atoms are combined to form one molecule.*

For instance, let equal volumes of oxygen and nitrogen be given. A mixture of these gases contains a certain number of molecules, which consist either of two atoms of oxygen or of two atoms of nitrogen. Conceive the mixture to pass into a chemical compound, and the latter then contains just as many molecules, which are merely constituted in a different manner, inasmuch as each consists of an atom of oxygen and an atom of nitrogen. Hence there is no reason why a change of volume should take place. If, on the other hand, one volume of oxygen and two of nitrogen are given, then in the mixture each molecule consists of two, and in the compound of three atoms. The chemical combination, therefore, has caused the number of molecules to diminish in the ratio of 3 : 2, and consequently the volume ought to diminish in the same ratio.

It is well known that some simple substances do not, in the gaseous form, occupy the volume which their atomic weights and the volumes of their combinations would lead us to anticipate, but another, and in most cases a smaller volume, which bears to the former a simple ratio. A special investigation of these substances would here be out of place, more especially as two of them, sulphur and phosphorus, deport themselves in other respects in so remarkable a manner, in consequence of the variety of conditions they are capable of assuming, that we may reasonably expect further discoveries from chemistry with respect to these bodies; and then, perhaps, besides other irregularities, those of the volumes of their vapours will be explained. Nevertheless I may here recall one circumstance which in some cases may possibly facilitate this explanation. I refer to the fact, that the above hypothesis, according to which the molecules of simple substances each consist of *two* atoms, may not be the only possible one.

On comparing with each other all cases of simple and compound gases, we must not expect to find immediately a perfect agreement throughout. I am of opinion, however, that, under the present uncertainty with respect to the inner constitution of several bodies, and particularly of those which possess a complicated chemical composition, too great weight ought not to be laid upon individual anomalies; and I deem it probable, that, by means of the above hypothesis respecting the molecules of simple substances, all relations of volume in gases may be referred back to the theorem, *that the several molecules of all gases possess equal* vis viva *in reference to their translatory motion.*

---

14. Proceeding to treat the subject mathematically, we will

first deduce the expression which shows in what manner the pressure of the gas on the sides of the vessel depends upon the motion of its molecules.

As the shape of the vessel is indifferent, we will select that which is most convenient for our purpose. We will assume the vessel to be very flat, and that two of its sides consist of parallel planes so close to one another that their distance asunder is infinitesimal when compared with the other dimensions of the vessel. Hence we need not consider the cases where the molecules strike against one of the narrow strips of sides, and we may assume that each moves in a right line until it either strikes against another molecule or against one of the large parallel sides. In fact, to take the small sides into consideration would change nothing in the final result, and would only make the development more prolix.

Let us consider one only of the two large sides; during the unit of time it is struck a certain number of times by molecules moving in all possible directions compatible with an approach towards the surface. We must first determine the number of such shocks, and how many correspond on the average to each direction.

15. Hereafter we shall always assume the gas to be an *ideal* one; in other words, we shall disregard the irregularities proceeding from an imperfect gaseous state, so that in determining the pressure we may, with Krönig, introduce certain simplifications in place of considering the motion exactly as it takes place.

The whole number of shocks received by the side remains unchanged when we assume that the molecules do not disturb each other in their motion, but that each pursues its rectilineal path until it arrives at the side.

Further, although it is not actually necessary that a molecule should obey the ordinary laws of elasticity with respect to elastic spheres and a perfectly plane side, in other words, that when striking the side, the angle and velocity of incidence should equal those of reflexion, yet, according to the laws of probability, we may assume that there are as many molecules whose angles of reflexion fall within a certain interval, *e.g.* between 60° and 61°, as there are molecules whose angles of incidence have the same limits, and that, on the whole, the velocities of the molecules are not changed by the side. No difference will be produced in the final result, therefore, if we assume that for each molecule the angle and velocity of reflexion are equal to those of incidence. According to this, each molecule would move to and fro between the large parallel sides, in the same directions as those chosen by a ray of light between two plane mirrors, until at length it would come in contact with one of the small sides; from this it would be reflected, and then commence a similar series of journeys to and fro, and so forth.

Lastly, there is no doubt that actually the greatest possible variety exists amongst the velocities of the several molecules. In our considerations, however, we may ascribe a certain mean velocity to all molecules. It will be evident from the following formulæ, that, in order to maintain an equal pressure, this mean velocity must be so chosen that with it the total *vis viva* of all the molecules may be the same as that corresponding to their actual velocities.

16. According to these assumptions, it is evident, that, during the unit of time, each molecule will strike the side under consideration just as often as during that time it can, by following its peculiar direction, travel from the side in question to the other and back again. Let h be the distance between the large parallel sides, and ϑ the acute angle between the normal and the direction of motion; then $\frac{h}{\cos\vartheta}$ is the length of the path from one side to the other, and

$$\frac{u.\cos\vartheta}{2h} \quad . \quad . \quad . \quad . \quad . \quad . \quad . \quad . \quad . \quad (1)$$

the number of impulses given to the side, u being the velocity of the molecule.

With respect to the directions of the several molecules, we must assume that on the average each direction is equally represented. From this it follows, that the number of molecules moving in directions which form with the normal angles included between ϑ and $\vartheta + d\vartheta$, has to the whole number of molecules the same ratio that the surface of the spherical zone, whose limiting circles correspond to the angles ϑ and $d\vartheta$, has to the surface of the hemisphere, that is, the ratio

$$2\pi \sin\vartheta d\vartheta : 2\pi.$$

Hence if n represents the whole number of molecules, the number which corresponds to the angular interval between ϑ and $\vartheta + d\vartheta$ will be

$$n \sin\vartheta d\vartheta,$$

and the number of shocks imparted by them will be

$$\frac{nu}{2h}\cos\vartheta \sin\vartheta d\vartheta. \quad . \quad . \quad . \quad . \quad . \quad . \quad (2)$$

17. In order to determine the intensity of a shock, the whole velocity must be resolved into two components, one parallel and the other perpendicular to the side. Of these components, the first will not be affected by the shock, and will not enter into

consideration in determining its intensity; the second, however, whose magnitude is represented by $u \cos \vartheta$, will be changed by the shock into an equal velocity in the opposite direction. The action of the side upon the molecule, therefore, consists in depriving it in one direction of the velocity $u \cos \vartheta$, calculated according to the normal, and of imparting to it an equal velocity in an opposite direction; in other words, of imparting to it a velocity of $2u \,.\, \cos \vartheta$ in the latter direction. Hence the quantity of motion imparted to the molecule will be

$$2mu \,.\, \cos \vartheta, \quad . \quad . \quad . \quad . \quad . \quad . \quad . \quad . \quad (3)$$

where m is the mass of the molecule.

Applying this to all molecules which correspond to the interval between ϑ and $\vartheta + d\vartheta$, we obtain during the unit of time,

$$\frac{nu}{2n} \cos \vartheta \sin \vartheta d\vartheta$$

times the same action, hence the quantity of motion imparted to these molecules during the unit of time is

$$\frac{nmu^2}{h} \cos^2 \vartheta \,.\, \sin \vartheta \,.\, d\vartheta. \quad . \quad . \quad . \quad . \quad . \quad . \quad (4)$$

Integrating this expression between the limits $\vartheta = 0$ and $\vartheta = \frac{\pi}{2}$, we find the motion imparted by the side to all the molecules which strike against it during the unit of time to be

$$\frac{nmu^2}{3h}. \quad . \quad . \quad . \quad . \quad . \quad . \quad . \quad . \quad . \quad (5)$$

Let us now conceive the side to be capable of moving freely; then in order that it may not recede before the shocks of the molecules, it must be acted upon on the other side by a counter force, which latter may in fact be regarded as continuous, in consequence of the great number of shocks and the feebleness of each. The intensity of this force must be such as to enable it, during the unit of time, to generate the quantity of motion represented by the above expression. Since all forces, however, are measured by the quantity of motion they can produce in the unit of time, the above expression at once represents this force as well as the pressure exerted by the gas, the latter being equilibrated by the former.

If α be the superficial area of the side, and p the pressure on the unit of surface, then

$$p = \frac{mnu^2}{3\alpha h}.$$

The product αh here involved gives the volume of the vessel or gas; hence representing the same by v, we have

$$p = \frac{mnu^2}{3v} \quad . \quad . \quad . \quad . \quad . \quad . \quad . \quad . \quad (6)$$

The same formula would have been obtained if, with Krönig, we had, for the sake of simplification, assumed that one third of the whole molecules move perpendicularly to the side under consideration, and the two remaining thirds in two other directions parallel to the side. Nevertheless I preferred deducing the formula for the pressure without using this simplifying hypothesis.

If we write the last equation in the form

$$\frac{3}{2}pv = \frac{nmu^2}{2}, \quad . \quad . \quad . \quad . \quad . \quad . \quad . \quad . \quad (6a)$$

the right-hand side then denotes *the* vis viva *of the translatory motion of the molecules**. But, according to Mariotte's and Gay-Lussac's laws,

$$pv = \text{T} \cdot \text{const.},$$

where T is the absolute temperature; hence

$$\frac{nmu^2}{2} = \text{T} \cdot \text{const.};$$

and, as before stated, the *vis viva* of the translatory motion is proportional to the absolute temperature.

18. We may now make an interesting application of the above equations by determining the velocity u with which the several molecules of gas move.

The product nm represents the mass of the whole given quantity of gas, whose weight we will call q. Then g being the force of gravity,

$$nm = \frac{q}{g};$$

and from equation (6) we deduce

$$u^2 = \frac{3gpv}{q} \quad . \quad . \quad . \quad . \quad . \quad . \quad . \quad . \quad (7)$$

Adopting the metre as unit of length, and the kilogramme as unit of weight, let us suppose a kilogramme of gas under the pressure of 1 atmosphere—10333 kilogrammes on the square

* In accordance with a practice lately become general, and with what I have myself done in former memoirs, I call the *semi*-product of the mass into the square of the velocity the *vis viva*, because it is only with this definition of the notion that we can, without the addition of a coefficient, equate the expressions representing a quantity of work and the increase or decrease of *vis viva* which corresponds to the same.

metre—to be given. Then

$$g=9^{m}\cdot 80896,$$
$$p=10333,$$
$$q=1.$$

To determine v, we know that, according to Regnault, a kilogramme of atmospheric air under the pressure of 1 atmosphere, and at the temperature of melting ice, occupies 0·7733 cubic metre. Hence ρ being the specific gravity of the gas under consideration, its volume at the temperature of melting ice will be

$$\frac{0\cdot 7733}{\rho};$$

and at the absolute temperature T, assuming 273° to be the absolute temperature of melting ice, it will be

$$\frac{0\cdot 7733}{\rho}\cdot\frac{T}{273}.$$

Hence we deduce

$$u^2=3\,.\,9\cdot 80896\,.\,10333\,.\,0\cdot 7733\,.\,\frac{T}{273\,.\,\rho}$$
$$=235130\frac{T}{273\,.\,\rho},$$

and consequently

$$u=485^{m}\sqrt{\frac{T}{273\,.\,\rho}}. \quad \dots\dots\dots \quad (8)$$

As particular cases, we obtain the following numbers corresponding to the temperature of melting ice:—

for oxygen . . . . 461^{m},
for nitrogen . . . . 492^{m},
for hydrogen . . 1844^{m}.

These numbers are the mean velocities which, for the totality of molecules, give the same *vis viva* as would their actual velocities. At the same time, however, it is possible that the actual velocities of the several molecules differ materially from their mean value.

19. By means of the equations above established, we will lastly examine *what relation exists between the* vis viva *of the translatory motion and the whole* vis viva, *or heat, in the gas.*

In doing so we will conceive the quantity of heat to be measured, not by the ordinary unit of heat, but by the mechanical unit of *vis viva*, or what is equivalent, by the unit of work. To this end we have only to divide the quantity of heat measured in the ordinary manner by the thermal equivalent of the unit of work, which as before I will denote by A. Let H be the quantity of heat thus determined.

Further, let c be the specific heat of the gas under constant volume, in other words, the true specific heat; then the increase of the quantity of heat in the quantity q of gas corresponding to an elevation of temperature dT is

$$dH = \frac{qc}{A} dT.$$

Integrating this equation, we have

$$H = \frac{qc}{A} T. \quad . \quad . \quad . \quad . \quad . \quad . \quad . \quad (9)$$

No constant need be added, since, as before remarked, the heat in the gas is proportional to the *vis viva* of the translatory motion, and hence also to the absolute temperature.

The expression on the right of this equation may be replaced by another which is very convenient for our present investigation.

The quantity of heat which must be imparted to the quantity of gas q in order to elevate its temperature by dT and its volume by dv is expressed thus,

$$\frac{qc}{A} dT + pdv,$$

wherein the first term represents the increase of the heat contained in the gas, and the second the quantity of heat consumed by work. If we assume the gas to be heated under constant pressure, the relation between dT and dv is thereby defined. For we have generally

$$pv = T \,.\, \text{const.};$$

and differentiating, under the supposition that p is constant, we obtain

$$pdv = dT \,.\, \text{const.};$$

whence the undetermined constant may be eliminated by means of the foregoing equation, and we have

$$dv = \frac{v}{T} dT.$$

Let us substitute this value of dv in the above equation, and at the same time note that, c' being the specific heat under constant pressure, the whole quantity of heat imparted to the gas in the case under consideration may be represented by $\frac{qc'}{A} dT$. In this manner we arrive at the equation

$$\frac{qc'}{A} dT = \frac{qc}{A} dT + \frac{pv}{T} dT,$$

whence we conclude that

$$\frac{q(c'-c)}{A} \cdot T = pv. \quad . \quad . \quad . \quad . \quad . \quad . \quad (10)$$

By means of this equation (9) becomes

$$H = \frac{c}{c'-c} \cdot pv. \quad . \quad . \quad . \quad . \quad . \quad . \quad . \quad (11)$$

20. Let us now return to the equation (6*a*) before established, and for brevity let us denote the *vis viva* of the translatory motion by K, then

$$K = \frac{3}{2} pv.$$

By combining this with the foregoing equation we obtain

$$\frac{K}{H} = \frac{3}{2}\left(\frac{c'}{c} - 1\right). \quad . \quad . \quad . \quad . \quad . \quad (12)$$

The ratio of the *vis viva* of the translatory motion to the whole *vis viva* is thus reduced to the ratio between the two specific heats.

In order to compare with each other the values of the ratio $\frac{K}{H}$ corresponding to different gases, it will be found convenient to introduce in the above formula, in place of the specific heats calculated with reference to the unit of weight, those calculated according to the unit of volume, which for distinction may be represented by γ and γ'. The equation then becomes

$$\frac{K}{H} = \frac{3}{2} \cdot \frac{\gamma'-\gamma}{\gamma}. \quad . \quad . \quad . \quad . \quad . \quad . \quad (13)$$

If we neglect deviations which arise from an imperfect gaseous condition, and conceive all gases to be in the ideal state, then, as I have shown in my memoir "On the Moving Force of Heat*," the difference $\gamma'-\gamma$ is the same for all gases. Hence the ratio $\frac{K}{H}$ *is inversely proportional to the true specific heat of the gas calculated according to the unit of volume.*

For those simple gases which manifest no irregularities with respect to their volume, and for those compound ones which suffered no diminution of volume during the act of combination, γ, and therefore $\frac{K}{H}$ also, has the same value. For these gases we

* Poggendorff's *Annalen*, vol. lxxix. p. 394. Phil. Mag. vol. ii. p. 1.

have approximately

$$\frac{\gamma'}{\gamma}=1{\cdot}421,$$

hence

$$\frac{K}{H}=0{\cdot}6315.$$

On the contrary, for those compound gases whose volumes during combination have been diminished, γ is greater, and consequently $\frac{K}{H}$ less; and, in fact, it is less the smaller the volume of the combination compared with the sum of the volumes of its constituents, the latter being all considered gaseous.

Thus is corroborated what was before stated, that the *vis viva* of the translatory motion does not alone represent the whole quantity of heat in the gas, and that the difference is greater the greater the number of atoms of which the several molecules of the combination consist. We must conclude, therefore, that besides the translatory motion of the molecules as such, the constituents of these molecules perform other motions, whose *vis viva* also forms a part of the contained quantity of heat.

Zurich, January 5, 1857.

13

Reprinted from *Philos. Mag.* ser. 4, 17:81–91 (1859)

ON THE MEAN LENGTH OF THE PATHS DESCRIBED BY THE SEPARATE MOLECULES OF GASEOUS BODIES ON THE OCCURRENCE OF MOLECULAR MOTION; TOGETHER WITH SOME OTHER REMARKS UPON THE MECHANICAL THEORY OF HEAT

R. Clausius*

(1.) THE February Number of Poggendorff's *Annalen* contains a paper by Buijs-Ballot "On the Nature of the Motion which we call Heat and Electricity." Amongst the objections which the author there makes against the views advanced by Joule, Krönig, and myself concerning the molecular motion of gaseous bodies, the following deserves especial consideration. Attention is drawn to the circumstance that, if the molecules moved in straight lines, volumes of gases in contact would necessarily speedily mix with one another,—a result which does not actually take place. To prove that such mixture does not occur, the following facts are adduced (p. 250):—"How then does it happen that tobacco-smoke, in rooms, remains so long extended in immoveable layers?" Mention is also made of the same appearance with clouds of smoke in the open air. Further, "If sulphuretted hydrogen or chlorine be evolved in one corner of a room, entire minutes elapse before they are smelt in another corner, although the particles of gas must have had to traverse the room hundreds of times in a second." Further, "How could carbonic acid gas remain so long in an open vessel?"

These objections may, at first glance, appear to have very great weight; and I consider it therefore necessary to prove, by special considerations, that the facts adduced are perfectly reconcileable with the theory of the rectilineal motion of the molecules. Indeed, I rejoice at the discussion of this point by M. Buijs-

* Translated by Dr. F. Guthrie, from Poggendorff's *Annalen*, No. 10, 1858.

Ballot, inasmuch as it affords me a desired opportunity of completing this part of my theory (which was perhaps discussed too briefly in my paper), and to prevent thereby further misunderstandings.

(2.) It is assumed in the objections, that the molecules traverse considerable spaces in straight lines; this appears prominently in the second objection, in which it is said that a molecule must have had to traverse the room many times in a second. This assumption can, however, in nowise be considered as a necessary consequence of the views advanced by me concerning the conditions of gases. Amongst the conditions which must be satisfied if Mariotte and Gay-Lussac's law for a gas is true with perfect strictness, I have adduced the following,—"that those portions of the path of a molecule throughout which the molecular forces are of influence in sensibly altering the motion of the molecule, either in direction or velocity, must be of vanishing value compared with those portions of the path throughout which such forces may be considered as inactive." Now in actually existing gases, Mariotte and Gay-Lussac's law is not strictly, but only approximately true; and it hence follows that in them such first portions of the paths of the molecules must be small, but not vanishingly small, compared with the entire paths. Inasmuch, now, as one of the fundamental conditions upon which the whole theory rests is that the molecular forces are only effective at small distances from the molecules, a path which is very great in comparison to the sphere of action of a molecule may yet, considered absolutely, be very small.

By a few simple considerations, an approximate idea may be formed of the mean magnitude of the paths traversed by the separate molecules: I purpose endeavouring to elucidate this in what immediately follows.

(3.) For this purpose it will be advisable to prefix some remarks concerning the manner in which it is possible to view the molecular forces, and what has accordingly to be understood by the sphere of action. These remarks are not to be considered as an essential part of the subsequent development, but are merely intended to fix our ideas.

If we do not take into account the forces of chemical affinity, and only consider such molecules as are chemically indifferent to one another, I imagine that there are still two forces which are to be distinguished. I believe, namely, that when two molecules approach one another an attraction is at first exerted, which begins to be of sensible effect even at some distance, and which increases as the distance diminishes; but that, when the molecules have arrived into the immediate neighbourhood of one another, a force comes into play which seeks to drive them asunder. For the view which it is here intended to take, it is indifferent what kind of force this repulsive one is supposed to be, that is, whether, as in the case of solid elastic bodies, it only strives to separate the molecules when they are in actual contact with a force equal to that with which they are pressed together, or whether it is one which begins to act before the actual contact of the molecules. In the same manner, we need not here discuss the question as to the source of these forces, whether they are both to be ascribed to the particles of ponderable matter themselves, or whether one of them is to be referred to a more subtle substance, with which the ponderable particles of matter are furnished.

Let us now imagine two molecules moving in directions such that, if they preserved them unchanged, they would not strike one another, but pass by at some distance. Two cases may here occur. If the distance is very small, the molecules which were drawn towards one another, even from some distance, by the force of attraction, approach so closely that the repulsive force comes into play, and a rebounding of the molecules results. If the distance be somewhat greater, the paths of the molecules only suffer a certain change of direction through the attractive force, without the repulsive force being able to act. Finally, at still greater distances, the effect of the molecules upon one another may be altogether neglected.

How great the distances must be in order that the one or other case might occur, could not be determined universally, even if we possessed exact knowledge of the molecular forces; for the velocity of the molecules and the reciprocal inclination of their paths are of influence. Nevertheless, *mean* values of these distances may be obtained. We will therefore suppose that the distance ρ is given for such a mean value, which forms the boundary between the first and second case, and the meaning of which we will define with greater precision in the following manner:—If the centres of gravity of two molecules have such directions of motion that if they were to proceed in those directions in straight lines they would pass by one another at a distance greater than ρ, then the molecules only change their courses to some extent through reciprocal attraction, without the repulsive force coming into action between them. If, on the other hand, this distance is less than ρ, the latter force also comes into play, and a rebounding of the molecules takes place.

If, now, the latter case alone be considered as one of *impact*, and we do not concern ourselves with the changes of direction which the force of attraction effects at greater distances, we may, for what we have here to consider, represent a sphere of radius

ρ, described around a molecule and having its centre of gravity for a centre, by the term *sphere of action* of the molecule.

I again call attention to the fact that the special hypotheses here made, concerning the nature of the molecular forces, are not to be viewed as a necessary condition for the developments which follow; their only purpose is to facilitate the comprehension by giving something definite to the imagination. It is of no import how we consider the forces by reason of which the molecules change the directions of their motions; if we but admit that their effects are only sensible at very small distances, we may assume some distance as limiting value for the purpose of being able to neglect the actions from greater distances, and only regard those for smaller ones. A sphere described at this distance may be called a sphere of action.

(4.) If, now, in a given space, we imagine a great number of molecules moving irregularly about amongst one another, and if we select one of them to watch, such a one would ever and anon impinge upon one of the other molecules, and bound off from it. We have now, therefore, to solve the question as to how great is the mean length of the path between two such impacts; or more exactly expressed, *how far on an average can the molecule move, before its centre of gravity comes into the sphere of action of another molecule.*

We will not discuss this question, however, immediately in the form just given: we will propose instead a somewhat simpler one, which is related to the other in such a manner that the solution of the one may be derived from that of the other.

If we assume that not all the molecules present in the space are in motion, but that the one chosen for observation is the only one which moves, and all the rest remain fixed in position, the moving molecule in these circumstances also would strike here and there upon one of the others, and the number of blows which it suffers in this case during one unit of time may be compared with the numbers which it would experience in event of universal movement. On considering the matter more attentively, we are soon convinced that the number of blows amongst moving molecules must be greater than amongst stationary ones, or, which comes to the same thing, that the mean length of the paths which the molecule watched passes over between two consecutive impacts, must be less in the first case than in the second. The relation between the lengths of the two paths may be definitely found as soon as the velocity of the remaining molecules, in comparison with that of the one watched, is known. For our investigations, that case only is of special interest where the velocities of all the molecules are *on an average* equally great. In this case, if we only consider the *mean* velocities, we may more simply assume that all molecules move at the same rate; and for this case we obtain the following result:—*The mean lengths of path for the two cases* (1) *where the remaining molecules move with the same velocity as the one watched, and* (2) *where they are at rest, bear the proportion to one another of* $\frac{3}{4}$ *to* 1.

It would not be difficult to prove the correctness of this relation: it is, however, unnecessary for us to devote our time to it; for, in our consideration of the mean path, it is not the question to determine exactly its numerical value, but merely to obtain an approximate notion of its magnitude; and hence the exact knowledge of this relation is not necessary. It is even sufficient for our purpose if we may assume as certain that the mean path among moving molecules *cannot be greater* than among stationary ones; this will certainly be at once admitted. Under this hypothesis, we will confine the discussion of the question to that case *where the molecule watched alone moves, while all the others remain at rest.*

Moreover, without affecting the question in anything, we may suppose a mere moving point in place of the moving molecule; for it is in fact only the centre of gravity of the molecule which has to be considered.

(5.) Suppose, then, there is a space containing a great number of molecules, and that these are not regularly arranged, the only condition being that the density is the same throughout, *i.e.* in equal parts of the space there are the same numbers of molecules. The determination of the density may be performed conveniently for our investigation by knowing how far apart two neighbouring molecules would be separated from one another if the molecules were arranged cubically, that is, so arranged that the whole space might be supposed divided into a number of equal very small cubic spaces, in whose corners the centres of the molecules were situated. We shall denote this distance, that is, the side of one of these little cubes, by λ, and shall call it *the mean distance of the neighbouring molecules.*

If, now, a point moves through this space in a straight line, let us suppose the space to be divided into parallel layers perpendicular to the motion of the point, and let us determine *how great is the probability that the point will pass freely through a layer of the thickness* x *without encountering the sphere of action of a molecule.*

Let us first take a layer of the thickness 1, and let us denote by the fraction of unity a the probability of the point passing through this layer without meeting with any sphere of action: then the corresponding probability for a thickness 2 is a^2; for if such a layer be supposed divided into two layers of the thickness 1, the probability of the points passing free through the first layer, and thereby arriving at the second, must be multiplied by the probability of its passing through the latter one. Similarly, for a layer of the thickness 3, we have a^3, &c., and for a layer of any thickness x we may accordingly write a^x. Let us transform this expression by putting $e^{-\alpha}$ for a, in which e is the base of the natural logarithms, and $-\alpha = \log_e a$, which logarithm must be negative, because a is less than 1. If now we denote the probability of the free passage through a layer of the thickness x by W, we have the equation

$$W = e^{-\alpha x}, \quad . \quad . \quad . \quad . \quad . \quad . \quad . \quad . \quad (1)$$

and we have only to determine here the constant α.

Again, let us consider a layer of such thinness that the higher powers of the thickness may be neglected in comparison with the first. Calling this thickness δ and the corresponding probability W_δ, the former equation becomes

$$W_\delta = e^{-\alpha\delta} = 1 - \alpha\delta \quad . \quad . \quad . \quad . \quad . \quad . \quad (2)$$

The probability in this case may also be determined from special considerations. Let us direct our attention to any plane in the layer parallel to one of the bounding planes of the layer, and let us suppose all the molecules whose centres lie in the layer to be so moved perpendicular to the layer that their centres all fall upon this plane; we have now only to inquire how great the probability is that the point, in its passage through this plane, meets with no sphere of action; such probability may be simply represented by the proportion of two superficial areas. Of the entire part of the plane which falls within the given space, a certain portion is covered by the great circles of the spheres of action whose centres fall upon it, while the remaining portion is free for the passage; and the probability of the uninterrupted passage is therefore expressed *by the relation of the free portion of the plane to the whole plane.*

From the manner in which the density was determined at the beginning of this article, it follows that in a layer of thickness λ, so many molecules must be contained, that, if they be supposed brought into one and the same plane parallel to the bounding plane, and to be arranged still quadratically in this plane, then the side of the small square in whose corners would be situated the centres of the molecules would be equal to λ. Hence it follows, that the part of the plane which would be covered by the great circles of the spheres of action, would be related to the remainder of the plane as a great circle would to a square of side λ, so that, accordingly, the covered superficial area would be expressed by the fraction

$$\frac{\pi\rho^2}{\lambda^2}$$

of the entire superficial area. In order to ascertain the corresponding magnitude for a layer of the thickness δ, we have only to multiply the previous fraction by $\frac{\delta}{\lambda}$, that is,

$$\frac{\pi\rho^2}{\lambda^3}\delta;$$

and if this magnitude be subtracted from 1, the difference represents the free portion of the plane as a fraction of the whole plane.

Hence the probability that the point will pass through our plane, or, which comes to the same thing, through a layer of thickness δ, without obstruction, is determined by the equation

$$W_\delta = 1 - \frac{\pi\rho^2}{\lambda^3}\delta; \quad . \quad . \quad . \quad . \quad . \quad (3)$$

and on comparing this expression for W_δ with that given in equation (2), we find that

$$\alpha = \frac{\pi\rho^2}{\lambda^3}, \quad . \quad . \quad . \quad . \quad . \quad . \quad . \quad . \quad (4)$$

and hence the general equation (1) is transformed into

$$W = e^{-\frac{\pi\rho^2}{\lambda^3}\cdot x}. \quad . \quad . \quad . \quad . \quad . \quad . \quad . \quad . \quad (5)$$

(6.) By means of this equation we can now determine the mean value of the path which the point has to traverse before it meets with a sphere of action.

Let us suppose that a great number (N) of points are thrown through space in one direction, and let us suppose the space to be divided into very thin layers perpendicular to the direction of motion; then a small number of the points would be detained in the first layer by the spheres of action, another lot in the second, another in the third, and so on. If, now, each of these small numbers be multiplied by the length of path, the products added, and the sum obtained divided by the whole number N, the quotient will be the mean length of the path which we seek.

According to equation (5), the number of points which either reach or pass the distance x from the commencement of the motion is represented by

$$Ne^{-\frac{\pi\rho^2}{\lambda^3}x};$$

and accordingly the number which reach or pass the distance $x+dx$ is expressed by

$$Ne^{-\frac{\pi\rho^2}{\lambda^3}(x+dx)} = Ne^{-\frac{\pi\rho^2}{\lambda^3}x}\left(1 - \frac{\pi\rho^2}{\lambda^3}dx\right).$$

The difference of these two expressions, namely,

$$Ne^{-\frac{\pi\rho^2}{\lambda^3}x} \cdot \frac{\pi\rho^2}{\lambda^3} dx,$$

represents the number of those points which are detained between x and $x+dx$. The path traversed by these points may be considered as x if we neglect infinitely small differences; and hence the above expression must be multiplied by this length in order to obtain one of the products mentioned before, namely,

$$Ne^{-\frac{\pi\rho^2}{\lambda^3}x} \cdot \frac{\pi\rho^2}{\lambda^3} x dx.$$

If, now, it be desired to obtain the sum of all products of this kind which correspond to the several layers of the thickness dx, this must of course, in the case in point where the layers are infinitely thin, be effected by integration. Hence the above formula has to be integrated from $x=0$ to $x=\infty$, whence the following expression is obtained,

$$N\frac{\lambda^3}{\pi\rho^2}.$$

This expression has now only to be divided by N in order to arrive at the mean length of path required. If this be called l', the equation is

$$l' = \frac{\lambda^3}{\pi\rho^2}. \quad \ldots \ldots \ldots \quad (6)$$

In the case where not one molecule only is in motion while all the others are at rest, but where all molecules move with equal velocity, the mean length of way, as mentioned before, is less than that above considered in the proportion of $\frac{3}{4}$ to 1. Hence if we put the simple letter l for this case, we have

$$l = \frac{3}{4}\frac{\lambda^3}{\pi\rho^2}. \quad \ldots \ldots \ldots \quad (7)$$

Writing this equation in the form

$$\frac{l}{\rho} = \frac{\lambda^3}{\frac{4}{3}\pi\rho^3}, \quad \ldots \ldots \ldots \quad (7a)$$

a simple law results. It follows from the manner in which we determined the density, that the part of the given space filled by the spheres of action of the molecules is related to the whole given space as a sphere of action to a cube of the side λ, that is, as

$$\frac{4}{3}\pi\rho^3 : \lambda^3.$$

Accordingly the meaning of the previous equation may be so put:—*The mean length of path of a molecule is in the same pro portion to the radius of the sphere of action as the entire space occupied by the gas, to that portion of the space which is actually filled up by the spheres of action of the molecules.*

(7.) In order to have a definite numerical example, let us assume, in round numbers, that the spheres of action of the molecules are so small that only a thousandth of the space occupied by the gas is really filled out by the spheres of action, and that the whole remaining space be free for motion.

For this case we have

$$\frac{\lambda^3}{\frac{4}{3}\pi\rho^3}=1000,$$

whence it follows that

$$\frac{\lambda}{\rho}=16\cdot12. \quad . \quad . \quad . \quad . \quad . \quad . \quad (8)$$

On applying these values we obtain from equations (6) and (7),

$$l'=1333\rho=83\lambda, \quad . \quad . \quad . \quad . \quad . \quad (9)$$

$$l=1000\rho=62\lambda. \quad . \quad . \quad . \quad . \quad . \quad (10)$$

The first expressions in both equations show that, under the assumption made, the mean path has a considerable length in comparison to the radius of the spheres of action, and that therefore, as far as the effect of this circumstance is concerned, Mariotte and Gay-Lussac's law may be very nearly true for the gas. By a simple calculation it may be shown that the relation of 1000 to 1 completely suffices, even for those approximations found by Regnault with permanent gases. It follows that the magnitude of the spheres of action which was taken for illustration, although arbitrarily chosen, may yet be regarded as one within the bounds of possibility.

But if we now regard this same mean value of the length of path in such a manner as to compare it, not with the sizes of molecules, but with our usual units of length, we obtain totally different relations. In all physical and chemical investigations in which opportunity presents itself for drawing conclusions concerning the weight and size of the separate molecules, we are invariably led to the conclusion that, compared with all measurable magnitudes, molecules must be of extraordinarily small size. As yet, no one has been able to establish a bounding line on the other side (for smallness). Accordingly, when an ordinary unit of measure, *e. g.* a litre, is filled with gas at the ordinary atmospheric pressure, we must assume that the number of molecules present is very great, and that consequently the distances between the molecules is very small. Accordingly the values previously found for l' and l, namely, 83λ and 62λ, must only be regarded as small magnitudes.

(8.) After the above determination of the length of the *mean* path, we still have to consider how the separate paths which really occur are related to the mean path.

The first question which presents itself is, in what proportion is the number of cases in which the real path is less than the mean path, to that of the cases in which it is greater. For answering this question, use is made of (5), in which we have only to substitute the mean value l' for x in order to find what probability there is that the true path is equal to or greater than the mean one. If for l' we here make use of the expression in (6), and denote the corresponding value of W by W_1, then

$$W_1 = e^{-1} = 0{\cdot}3679. \quad . \quad . \quad . \quad . \quad . \quad (11)$$

From the above equation it follows, that out of N cases only 0·3679 N occur in which the real path is equal to, or greater than the mean one, while in the

0·6321 N

cases the true path is the smaller one.

If, further, it be required to know the number of cases in which the true path is equal to or above the double, treble, &c. of the mean one, the same process may be adopted as before. Calling the probabilities in question W_2, W_3, &c., we have

$$\left.\begin{aligned} W_2 &= e^{-2} \\ W_3 &= e^{-} \\ &\text{\&c.} \end{aligned}\right\} \quad . \quad . \quad . \quad . \quad . \quad . \quad (12)$$

These numbers evidently diminish very rapidly, since, for instance, $e^{-10} = 0{\cdot}000045$; and we gather from this that, although in isolated cases a molecule may traverse a path considerably longer than the mean one, such cases are comparatively rare, and that in the majority of cases the actual path is smaller or very little larger than the small mean value found above.

(9.) If, now, these results be applied to the externally recognizable behaviours of a gas, in which it is presumed that no other motion common to the whole mass besides the molecular one is present, it is easy to convince oneself that the theory which explains the expansive force of gases does not lead to the conclusion that two quantities of gas bounding one another must mix with one another quickly and violently, but that only a comparatively small number of atoms can arrive quickly at a great distance, while the chief quantities only gradually mix at the surface of their contact.

From this it is clear why clouds of smoke only slowly lose their form on quiet days. Even when the air is in motion, provided such motion consists of a uniform one of the entire current, a cloud of smoke may be carried off without quickly losing its form. Both the other facts adduced by Buijs-Ballot also admit of simple explanation. The remark made by him attached to one of his objections, that the molecules of gas in a room must traverse the room many hundred times in one second, is completely foreign to the theory. Perhaps it might be said of a remark which occurs in the mathematical development attached to my previous paper, that it afforded reason for such an idea. I assumed there, namely, that the gas was in a very flat vessel, and I then assumed that the molecules of gas without disturbing one another, sped backwards and forwards between the two great parallel sides. Nevertheless, this assumption was there introduced with the following words: "In estimating the pressure, *instead of regarding the motion as it really occurs,* we may introduce certain simplifications." I believe I thereby called sufficient attention to the fact that this assumption should not serve to furnish an image of the real process, but only to simplify the calculation there intended, the result of which could not be thereby changed.

14 A

Reprinted from *Philos. Mag.* ser. 4, 19:19–32 (1860)

ILLUSTRATIONS OF THE DYNAMICAL THEORY OF GASES. PART I. ON THE MOTIONS AND COLLISIONS OF PERFECTLY ELASTIC SPHERES

J. C. Maxwell*

SO many of the properties of matter, especially when in the gaseous form, can be deduced from the hypothesis that their minute parts are in rapid motion, the velocity increasing with the temperature, that the precise nature of this motion becomes a subject of rational curiosity. Daniel Bernouilli, Herapath, Joule, Krönig, Clausius, &c. have shown that the relations between pressure, temperature, and density in a perfect gas can be explained by supposing the particles to move with uniform velocity in straight lines, striking against the sides of the containing vessel and thus producing pressure. It is not necessary to suppose each particle to travel to any great distance in the same straight line; for the effect in producing pressure will be the same if the particles strike against each other; so that the straight line described may be very short. M. Clausius has determined the mean length of path in terms of the average distance

* Communicated by the Author, having been read at the Meeting of the British Association at Aberdeen, September 21, 1859.

of the particles, and the distance between the centres of two particles when collision takes place. We have at present no means of ascertaining either of these distances; but certain phænomena, such as the internal friction of gases, the conduction of heat through a gas, and the diffusion of one gas through another, seem to indicate the possibility of determining accurately the mean length of path which a particle describes between two successive collisions. In order to lay the foundation of such investigations on strict mechanical principles, I shall demonstrate the laws of motion of an indefinite number of small, hard, and perfectly elastic spheres acting on one another only during impact.

If the properties of such a system of bodies are found to correspond to those of gases, an important physical analogy will be established, which may lead to more accurate knowledge of the properties of matter. If experiments on gases are inconsistent with the hypothesis of these propositions, then our theory, though consistent with itself, is proved to be incapable of explaining the phænomena of gases. In either case it is necessary to follow out the consequences of the hypothesis.

Instead of saying that the particles are hard, spherical, and elastic, we may if we please say that the particles are centres of force, of which the action is insensible except at a certain small distance, when it suddenly appears as a repulsive force of very great intensity. It is evident that either assumption will lead to the same results. For the sake of avoiding the repetition of a long phrase about these repulsive forces, I shall proceed upon the assumption of perfectly elastic spherical bodies. If we suppose those aggregate molecules which move together to have a bounding surface which is not spherical, then the rotatory motion of the system will store up a certain proportion of the whole *vis viva*, as has been shown by Clausius, and in this way we may account for the value of the specific heat being greater than on the more simple hypothesis.

On the Motion and Collision of Perfectly Elastic Spheres.

Prop. I. Two spheres moving in opposite directions with velocities inversely as their masses strike one another; to determine their motions after impact.

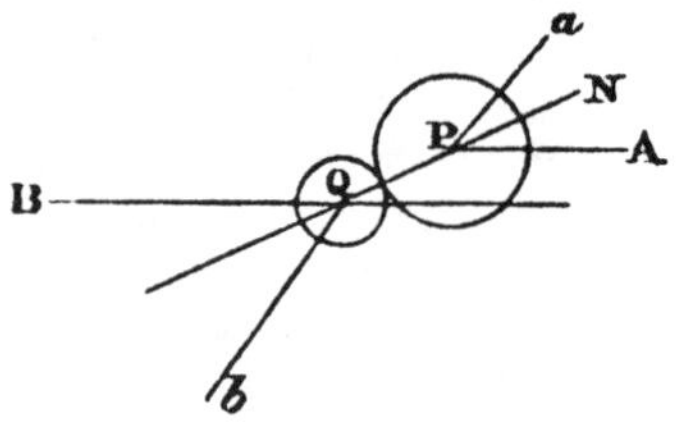

Let P and Q be the position of the centres at impact; A P, B Q the directions and magnitudes of the velocities before impact; P *a*, Q *b* the same after impact; then, resolving the velocities parallel and perpendicular to P Q the line of centres, we find that the velocities parallel to the line of centres are exactly reversed, while those perpendicular to that line are unchanged. Compounding these velocities again, we find that the velocity of each ball is the same before and after impact, and that the directions before and after impact lie in the same plane with the line of centres, and make equal angles with it.

Prop. II. To find the probability of the direction of the velocity after impact lying between given limits.

In order that a collision may take place, the line of motion of one of the balls must pass the centre of the other at a distance less than the sum of their radii; that is, it must pass through a circle whose centre is that of the other ball, and radius (s) the sum of the radii of the balls. Within this circle every position is equally probable, and therefore the probability of the distance from the centre being between r and $r+dr$ is

$$\frac{2r\,dr}{s^2}.$$

Now let ϕ be the angle APa between the original direction and the direction after impact, then $\mathrm{APN}=\frac{1}{2}\phi$, and $r=s\sin\frac{1}{2}\phi$, and the probability becomes

$$\tfrac{1}{2}\sin\phi\,d\phi.$$

The area of a spherical zone between the angles of polar distance ϕ and $\phi+d\phi$ is

$$2\pi\sin\phi\,d\phi;$$

therefore if ω be any small area on the surface of a sphere, radius unity, the probability of the direction of rebound passing through this area is

$$\frac{\omega}{4\pi};$$

so that the probability is independent of ϕ, that is, all directions of rebound are equally likely.

Prop. III. Given the direction and magnitude of the velocities of two spheres before impact, and the line of centres at impact; to find the velocities after impact.

Let O A, O B represent the velocities before impact, so that if there had been no action between the bodies they would have been at A and B at the end of a second. Join A B, and let G be their centre of gravity, the position of which is not affected by their mutual action. Draw G N parallel to the line of centres at impact (not necessarily in the plane A O B). Draw aGb in the

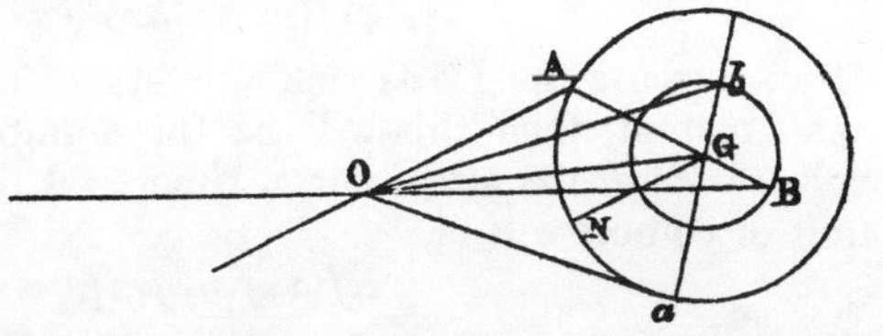

plane AGN, making N G a=N G A, and G a=G A and G b=G B; then by Prop. I. G a and G b will be the velocities relative to G; and compounding these with O G, we have O a and O b for the true velocities after impact.

By Prop. II. all directions of the line a G b are equally probable. It appears therefore that the velocity after impact is compounded of the velocity of the centre of gravity, and of a velocity equal to the velocity of the sphere relative to the centre of gravity, which may with equal probability be in any direction whatever.

If a great many equal spherical particles were in motion in a perfectly elastic vessel, collisions would take place among the particles, and their velocities would be altered at every collision; so that after a certain time the *vis viva* will be divided among the particles according to some regular law, the average number of particles whose velocity lies between certain limits being ascertainable, though the velocity of each particle changes at every collision.

Prop. IV. To find the average number of particles whose velocities lie between given limits, after a great number of collisions among a great number of equal particles.

Let N be the whole number of particles. Let x, y, z be the components of the velocity of each particle in three rectangular directions, and let the number of particles for which x lies between x and $x+dx$ be $\mathrm{N}f(x)dx$, where $f(x)$ is a function of x to be determined.

The number of particles for which y lies between y and $y+dy$ will be $\mathrm{N}f(y)dy$; and the number for which z lies between z and $z+dz$ will be $\mathrm{N}f(z)dz$, where f always stands for the same function.

Now the existence of the velocity x does not in any way affect that of the velocities y or z, since these are all at right angles to each other and independent, so that the number of particles whose velocity lies between x and $x+dx$, and also between y and $y+dy$, and also between z and $z+dz$, is

$$\mathrm{N}f(x)\,f(y)\,f(z)dx\,dy\,dz.$$

If we suppose the N particles to start from the origin at the same instant, then this will be the number in the element of volume $(dx\,dy\,dz)$ after unit of time, and the number referred to unit of volume will be

$$\mathrm{N}f(x)\,f(y)\,f(z).$$

But the directions of the coordinates are perfectly arbitrary, and therefore this number must depend on the distance from the origin alone, that is

$$f(x)\,f(y)\,f(z)=\phi(x^2+y^2+z^2).$$

Solving this functional equation, we find

$$f(x)=Ce^{Ax^2}, \quad \phi(r^2)=C^3e^{Ar^2}.$$

If we make A positive, the number of particles will increase with the velocity, and we should find the whole number of particles infinite. We therefore make A negative and equal to $-\frac{1}{\alpha^2}$, so that the number between x and $x+dx$ is

$$NCe^{-\frac{x^2}{\alpha^2}}dx.$$

Integrating from $x=-\infty$ to $x=+\infty$, we find the whole number of particles,

$$NC\sqrt{\pi}\alpha=N, \quad \therefore C=\frac{1}{\alpha\sqrt{\pi}},$$

$f(x)$ is therefore

$$\frac{1}{\alpha\sqrt{\pi}}e^{-\frac{x^2}{\alpha^2}}.$$

Whence we may draw the following conclusions:—

1st. The number of particles whose velocity, resolved in a certain direction, lies between x and $x+dx$ is

$$N\frac{1}{\alpha\sqrt{\pi}}e^{-\frac{x^2}{\alpha^2}}dx. \quad \ldots \ldots \quad (1)$$

2nd. The number whose actual velocity lies between v and $v+dv$ is

$$N\frac{4}{\alpha^3\sqrt{\pi}}v^2e^{-\frac{v^2}{\alpha^2}}dv. \quad \ldots \ldots \quad (2)$$

3rd. To find the mean value of v, add the velocities of all the particles together and divide by the number of particles; the result is

$$\text{mean velocity} = \frac{2\alpha}{\sqrt{\pi}}. \quad \ldots \ldots \quad (3)$$

4th. To find the mean value of v^2, add all the values together and divide by N,

$$\text{mean value of } v^2=\tfrac{3}{2}\alpha^2. \quad \ldots \ldots \quad (4)$$

This is greater than the square of the mean velocity, as it ought to be.

It appears from this proposition that the velocities are distributed among the particles according to the same law as the errors are distributed among the observations in the theory of the "method of least squares." The velocities range from 0 to ∞, but the number of those having great velocities is comparatively small. In addition to these velocities, which are in all directions equally, there may be a general motion of translation of the entire system of particles which must be compounded with the motion of the particles relatively to one another. We may call the one the motion of translation, and the other the motion of agitation.

Prop. V. Two systems of particles move each according to the law stated in Prop. IV.; to find the number of pairs of particles, one of each system, whose relative velocity lies between given limits.

Let there be N particles of the first system, and N′ of the second, then NN′ is the whole number of such pairs. Let us consider the velocities in the direction of x only; then by Prop. IV. the number of the first kind, whose velocities are between x and $x+dx$, is

$$N\frac{1}{\alpha\sqrt{\pi}}\epsilon^{-\frac{x^2}{\alpha^2}}dx,$$

The number of the second kind, whose velocity is between $x+y$ and $x+y+dy$, is

$$N'\frac{1}{\beta\sqrt{\pi}}\epsilon^{-\frac{(x+y)^2}{\beta^2}}dy,$$

where β is the value of α for the second system.

The number of pairs which fulfil both conditions is

$$NN'\frac{1}{\alpha\beta\pi}\epsilon^{-\left(\frac{x^2}{\alpha^2}+\frac{(x+y)^2}{\beta^2}\right)}dx\,dy.$$

Now x may have any value from $-\infty$ to $+\infty$ consistently with the difference of velocities being between y and $y+dy$; therefore integrating between these limits, we find

$$NN'\frac{1}{\sqrt{\alpha^2+\beta^2}\sqrt{\pi}}\epsilon^{-\frac{y^2}{\alpha^2+\beta^2}}dy \quad . \quad . \quad . \quad . \quad (5)$$

for the whole number of pairs whose difference of velocity lies between y and $y+dy$.

This expression, which is of the same form with (1) if we put NN′ for N, $\alpha^2+\beta^2$ for α^2, and y for x, shows that the distribution of relative velocities is regulated by the same law as that of the velocities themselves, and that the mean relative velocity is the square root of the sum of the squares of the mean velocities of the two systems.

Since the direction of motion of every particle in one of the systems may be reversed without changing the distribution of velocities, it follows that the velocities compounded of the velocities of two particles, one in each system, are distributed according to the same formula (5) as the relative velocities.

Prop. VI. Two systems of particles move in the same vessel;

to prove that the mean *vis viva* of each particle will become the same in the two systems.

Let P be the mass of each particle of the first system, Q that of each particle of the second. Let p, q be the mean velocities in the two systems before impact, and let p', q' be the mean velocities after one impact. Let $AO=p$ and $OB=q$, and let AOB be a right angle; then, by Prop. V., AB will be the mean relative velocity, OG will be the mean velocity of centre of gravity; and drawing aGb at right angles to OG, and making $aG=AG$ and $bG=BG$, then Oa will be the mean velocity of P after impact, compounded of OG and Ga, and Ob will be that of Q after impact.

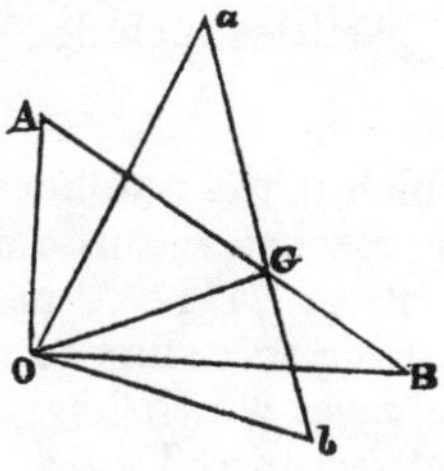

Now

$$AB=\sqrt{p^2+q^2},\quad AG=\frac{Q}{P+Q}\sqrt{p^2+q^2},\quad BG=\frac{P}{P+Q}\sqrt{p^2+q^2},$$

$$OG=\frac{\sqrt{P^2p^2+Q^2q^2}}{P+Q},$$

therefore

$$p'=Oa=\frac{\sqrt{Q^2(p^2+q^2)+P^2p^2+Q^2q^2}}{P+Q}$$

and

$$q'=Ob=\frac{\sqrt{P^2(p^2+q^2)+P^2p^2+Q^2q^2}}{P+Q},$$

and

$$Pp'^2-Qq'^2=\left(\frac{P-Q}{P+Q}\right)^2(Pp^2-Qq^2). \quad . \quad . \quad . \quad . \quad (6)$$

It appears therefore that the quantity Pp^2-Qq^2 is diminished at every impact in the same ratio, so that after many impacts it will vanish, and then

$$Pp^2=Qq^2.$$

Now the mean *vis viva* is $\frac{3}{2}P\alpha^2=\frac{3\pi}{8}Pp^2$ for P, and $\frac{3\pi}{8}Qq^2$ for Q; and it is manifest that these quantities will be equal when $Pp^2=Qq^2$.

If any number of different kinds of particles, having masses P, Q, R, and velocities p, q, r respectively, move in the same vessel, then after many impacts

$$Pp^2=Qq^2=Rr^2, \text{ \&c.} \quad . \quad . \quad . \quad . \quad . \quad (7)$$

Prop. VII. A particle moves with velocity r relatively to a number of particles of which there are N in unit of volume; to find the number of these which it approaches within a distance s in unit of time.

If we describe a tubular surface of which the axis is the path of the particle, and the radius the distance s, the content of this surface generated in unit of time will be $\pi r s^2$, and the number of particles included in it will be

$$N\pi r s^2, \quad . \quad . \quad . \quad . \quad . \quad . \quad . \quad . \quad (8)$$

which is the number of particles to which the moving particle approaches within a distance s.

Prop. VIII. A particle moves with velocity v in a system moving according to the law of Prop. IV.; to find the number of particles which have a velocity relative to the moving particle between r and $r+dr$.

Let u be the actual velocity of a particle of the system, v that of the original particle, and r their relative velocity, and θ the angle between v and r, then

$$u^2=v^2+r^2-2vr\cos\theta.$$

If we suppose, as in Prop. IV., all the particles to start from the origin at once, then after unit of time the "density" or number of particles to unit of volume at distance u will be

$$N\frac{1}{\alpha^3\pi^{\frac{3}{2}}}e^{-\frac{u^2}{\alpha^2}}.$$

From this we have to deduce the number of particles in a shell whose centre is at distance v, radius $= r$, and thickness $=dr$,

$$N\frac{1}{\alpha\sqrt{\pi}}\frac{r}{v}\left\{e^{-\frac{(r-v)^2}{\alpha^2}}-e^{-\frac{(r+v)^2}{\alpha^2}}\right\}dr, \quad . \quad . \quad . \quad (9)$$

which is the number required.

Cor. It is evident that if we integrate this expression from $r=0$ to $r=\infty$, we ought to get the whole number of particles $=N$, whence the following mathematical result,

$$\int_0^\infty dx\,.\,x\left(e^{-\frac{(x-a)^2}{\alpha^2}}-e^{-\frac{(x+a)^2}{\alpha^2}}\right)=\sqrt{\pi}a\alpha. \quad . \quad . \quad (10)$$

Prop. IX. Two sets of particles move as in Prop. V.; to find the number of pairs which approach within a distance s in unit of time.

The number of the second kind which have a velocity between v and $v+dv$ is

$$N'\frac{4}{\beta^3\sqrt{\pi}}v^2e^{-\frac{v^2}{\beta^2}}dv=n'.$$

The number of the first kind whose velocity relative to these is between r and $r+dr$ is

$$N\frac{1}{\alpha\sqrt{\pi}}\frac{r}{v}\left(e^{-\frac{(r-v)^2}{\alpha^2}}-e^{-\frac{(r+v)^2}{\alpha^2}}\right)dr=n,$$

and the number of pairs which approach within distance s in unit of time is

$$n\,n'\pi r s^2,$$

$$=NN'\frac{4}{\alpha\beta^3}s^2r^2v\,e^{-\frac{v^2}{\beta}}\left\{e^{-\frac{(v-r)^2}{\alpha^2}}-e^{-\frac{(v+r)^2}{\alpha^2}}\right\}dr\,dv.$$

By the last proposition we are able to integrate with respect to v, and get

$$NN'\frac{4\sqrt{\pi}}{(\alpha^2+\beta^2)^{\frac{3}{2}}}s^2r^3e^{-\frac{r^2}{\alpha^2+\beta^2}}dr.$$

Integrating this again from $r=0$ to $r=\infty$,

$$2NN'\sqrt{\pi}\sqrt{\alpha^2+\beta^2}\,s^2 \quad . \quad . \quad . \quad . \quad . \quad (11)$$

is the number of collisions in unit of time which take place in unit of volume between particles of different kinds, s being the distance of centres at collision. The number of collisions between two particles of the first kind, s_1 being the striking distance, is

$$2N^2\sqrt{\pi}\sqrt{2\alpha^2}\,s_1^{\,2};$$

and for the second system it is

$$2N'^2\sqrt{\pi}\sqrt{2\beta^2}\,s_2^{\,2}.$$

The mean velocities in the two systems are $\frac{2\alpha}{\sqrt{\pi}}$ and $\frac{2\beta}{\sqrt{\pi}}$; so that if l_1 and l_2 be the mean distances travelled by particles of the first and second systems between each collision, then

$$\frac{1}{l_1}=\pi N_1\sqrt{2}\,s_1^{\,2}+\pi N_2\frac{\sqrt{\alpha^2+\beta^2}}{\alpha}s^2,$$

$$\frac{1}{l_2}=\pi N_1\frac{\sqrt{\alpha^2+\beta^2}}{\beta}s^2+\pi N_2\sqrt{2}\,s_2^{\,2}.$$

Prop. X. To find the probability of a particle reaching a given distance before striking any other.

Let us suppose that the probability of a particle being stopped while passing through a distance dx, is αdx; that is, if N particles arrived at a distance x, $N\alpha dx$ of them would be stopped before getting to a distance $x+dx$. Putting this mathematically,

$$\frac{dN}{dx}=-N\alpha, \text{ or } N=Ce^{-\alpha x}.$$

Putting $N=1$ when $x=0$, we find $e^{-\alpha x}$ for the probability of a particle not striking another before it reaches a distance x.

The *mean distance* travelled by each particle before striking is $\frac{1}{\alpha}=l$. The probability of a particle reaching a distance $=nl$ without being struck is e^{-n}. (See a paper by M. Clausius, Philosophical Magazine, February 1859.)

If all the particles are at rest but one, then the value of α is

$$\alpha=\pi s^2 N,$$

where s is the distance between the centres at collision, and N is the number of particles in unit of volume. If v be the velocity of the moving particle relatively to the rest, then the number of collisions in unit of time will be

$$v\pi s^2 N;$$

and if v_1 be the actual velocity, then the number will be $v_1\alpha$; therefore

$$\alpha=\frac{v}{v_1}\pi s^2 N,$$

where v_1 is the actual velocity of the striking particle, and v its velocity relatively to those it strikes. If v_2 be the actual velocity of the other particles, then $v=\sqrt{v_1^2+v_2^2}$. If $v_1=v_2$, then $v=\sqrt{2}v_1$, and

$$\alpha=\sqrt{2}\pi s^2 N.$$

Note.—M. Clausius makes $\alpha=\frac{4}{3}\pi s^2 N$.

Prop. XI. In a mixture of particles of two different kinds, to find the mean path of each particle.

Let there be N_1 of the first, and N_2 of the second in unit of volume. Let s_1 be the distance of centres for a collision between two particles of the first set, s_2 for the second set, and s' for collision between one of each kind. Let v_1 and v_2 be the coefficients of velocity, M_1 M_2 the mass of each particle.

The probability of a particle M_1 not being struck till after reaching a distance x_1 by another particle of the same kind is

$$e^{-\sqrt{2}\pi s_1^2 N_1 x}.$$

The probability of not being struck by a particle of the other kind in the same distance is

$$e^{-\sqrt{1+\frac{v_2^2}{v_1^2}}\pi s'^2 N_2 x}.$$

Therefore the probability of not being struck by any particle before reaching a distance x is

$$e^{-\pi\left(\sqrt{2}s_1^2 N_1+\sqrt{1+\frac{v_2^2}{v_1^2}}s'^2 N_2\right)x};$$

and if l_1 be the *mean distance* for a particle of the first kind,

$$\frac{1}{l_1} = \sqrt{2}\pi s_1^2 N_1 + \pi\sqrt{1+\frac{v_2^2}{v_1^2}}s'^2 N_2. \quad . \quad . \quad (12)$$

Similarly, if l_2 be the mean distance for a particle of the second kind,

$$\frac{1}{l_2} = \sqrt{2}\pi s_2^2 N_2 + \pi\sqrt{1+\frac{v_1^2}{v_2^2}}s'^2 N_1. \quad . \quad . \quad (13)$$

The mean density of the particles of the first kind is $N_1 M_1 = \rho_1$, and that of the second $N_2 M_2 = \rho_2$. If we put

$$A = \sqrt{2}\frac{\pi s_1^2}{M_1}, \quad B = \pi\sqrt{1+\frac{v_2^2}{v_1^2}}\frac{s'^2}{M_2}, \quad C = \pi\sqrt{1+\frac{v_1^2}{v_2^2}}\frac{s'^2}{M_1},$$

$$D = \sqrt{2}\frac{\pi s_2}{M_2}, \quad . \quad . \quad . \quad . \quad . \quad . \quad . \quad . \quad . \quad . \quad (14)$$

$$\frac{1}{l_1} = A\rho_1 + B\rho_2, \quad \frac{1}{l_2} = C\rho_1 + D\rho^2, \quad . \quad . \quad (15)$$

and

$$\frac{B}{C} = \frac{M_1 v_2}{M_2 v_1} = \frac{v_2^3}{v_1^3}. \quad . \quad . \quad . \quad . \quad . \quad . \quad . \quad . \quad (16)$$

Prop. XII. To find the pressure on unit of area of the side of the vessel due to the impact of the particles upon it.

Let N = number of particles in unit of volume;
M = mass of each particle;
v = velocity of each particle,
l = mean path of each particle;

then the number of particles in unit of area of a stratum dz thick is

$$N dz. \quad . \quad . \quad . \quad . \quad . \quad . \quad . \quad . \quad . \quad (17)$$

The number of collisions of these particles in unit of time is

$$N dz \frac{v}{l}. \quad . \quad . \quad . \quad . \quad . \quad . \quad . \quad . \quad . \quad (18)$$

The number of particles, which after collision reach a distance between nl and $(n+dn)l$, is

$$N\frac{v}{l}e^{-n}\, dz\, dn. \quad . \quad . \quad . \quad . \quad . \quad . \quad . \quad (19)$$

The proportion of these which strike on unit of area at distance z is

$$\frac{nl-z}{2nl}; \quad . \quad . \quad . \quad . \quad . \quad . \quad . \quad . \quad . \quad (20)$$

the mean velocity of these in the direction of z is

$$v\frac{nl+z}{2nl}. \quad . \quad . \quad . \quad . \quad . \quad . \quad . \quad . \quad (21)$$

Multiplying together (19), (20), and (21), and M, we find the momentum at impact

$$MN\frac{v^2}{4n^2l^3}(n^2l^2-z^2)e^{-n}\,dz\,dn.$$

Integrating with respect to z from 0 to nl, we get

$$\tfrac{1}{6}MNv^2\,ne^{-n}\,dn.$$

Integrating with respect to n from 0 to ∞, we get

$$\tfrac{1}{6}MNv^2$$

for the momentum in the direction of z of the striking particles; the momentum of the particles after impact is the same, but in the opposite direction; so that the whole pressure on unit of area is twice this quantity, or

$$p=\tfrac{1}{3}MNv^2. \quad \ldots\ldots\ldots \quad (22)$$

This value of p is independent of l the length of path. In applying this result to the theory of gases, we put $MN=\rho$, and $v^2=3k$, and then

$$p=k\rho,$$

which is Boyle and Mariotte's law. By (4) we have

$$v^2=\tfrac{3}{2}\alpha^2, \quad \therefore\ \alpha^2=2k. \quad \ldots\ldots \quad (23)$$

We have seen that, on the hypothesis of elastic particles moving in straight lines, the pressure of a gas can be explained by the assumption that the square of the velocity is proportional directly to the absolute temperature, and inversely to the specific gravity of the gas at constant temperature, so that at the same pressure and temperature the value of NMv^2 is the same for all gases. But we found in Prop. VI. that when two sets of particles communicate agitation to one another, the value of Mv^2 is the same in each. From this it appears that N, the number of particles in unit of volume, is the same for all gases at the same pressure and temperature. This result agrees with the chemical law, that equal volumes of gases are chemically equivalent.

We have next to determine the value of l, the mean length of the path of a particle between consecutive collisions. The most direct method of doing this depends upon the fact, that when different strata of a gas slide upon one another with different velocities, they act upon one another with a tangential force tending to prevent this sliding, and similar in its results to the friction between two solid surfaces sliding over each other in the same way. The explanation of gaseous friction, according to our hypothesis, is, that particles having the mean velocity of translation belonging to one layer of the gas, pass out of it into another layer having a different velocity of translation; and by striking against the particles of the second layer, exert upon it a tangential force which constitutes the internal friction of the gas. The whole friction between two portions of gas separated by a plane surface, depends upon the total action between all the layers on the one side of that surface upon all the layers on the other side.

Prop. XIII. To find the internal friction in a system of moving particles.

Let the system be divided into layers parallel to the plane of xy, and let the motion of translation of each layer be u in the direction of x, and let $u = \mathrm{A} + \mathrm{B}z$. We have to consider the mutual action between the layers on the positive and negative sides of the plane xy. Let us first determine the action between two layers dz and dz', at distances z and $-z'$ on opposite sides of this plane, each unit of area. The number of particles which, starting from dz in unit of time, reach a distance between nl and $(n+dn)l$ is by (19),

$$\mathrm{N}\frac{v}{l}e^{-n}\,dz\,dn.$$

The number of these which have the ends of their paths in the layer dz' is

$$\mathrm{N}\frac{v}{2nl^2}e^{-n}\,dz\,dz'\,dn.$$

The mean velocity in the direction of x which each of these has before impact is $\mathrm{A}+\mathrm{B}z$, and after impact $\mathrm{A}+\mathrm{B}z'$; and its mass is M, so that a mean momentum $=\mathrm{MB}(z-z')$ is communicated by each particle. The whole action due to these collisions is therefore

$$\mathrm{NMB}\frac{v}{2nl^2}(z-z')e^{-n}\,dz\,dz'\,dn.$$

We must first integrate with respect to z' between $z'=0$ and $z'=z-nl$; this gives

$$\tfrac{1}{2}\mathrm{NMB}\frac{v}{2nl^2}(n^2l^2-z^2)e^{-n}\,dz\,dn$$

for the action between the layer dz and all the layers below the plane xy. Then integrate from $z=0$ to $z=nl$,

$$\tfrac{1}{6}\mathrm{MNB}lvn^2e^{-n}\,dn.$$

Integrate from $n=0$ to $n=\infty$, and we find the whole friction between unit of area above and below the plane to be

$$\mathrm{F}=\tfrac{1}{3}\mathrm{MN}lv\mathrm{B}=\tfrac{1}{3}\rho lv\frac{du}{dz}=\mu\frac{du}{dz},$$

where μ is the ordinary coefficient of internal friction,

$$\mu=\tfrac{1}{3}\rho lv=\frac{1}{3\sqrt{2}}\frac{\mathrm{M}v}{\pi s^2}, \quad . \quad . \quad . \quad . \quad . \quad (24)$$

where ρ is the density, l the mean length of path of a particle, and v the mean velocity $v = \frac{2\alpha}{\sqrt{\pi}} = 2\sqrt{\frac{2k}{\pi}}$,

$$l = \tfrac{3}{2}\frac{\mu}{\rho}\sqrt{\frac{\pi}{2k}}. \quad \ldots \ldots \quad (25)$$

Now Professor Stokes finds by experiments on air,

$$\sqrt{\frac{\mu}{\rho}} = \cdot 116.$$

If we suppose $\sqrt{k} = 930$ feet per second for air at 60°, and therefore the mean velocity $v = 1505$ feet per second, then the value of l, the mean distance travelled over by a particle between consecutive collisions, $= \frac{1}{447000}$th of an inch, and each particle makes 8,077,200,000 collisions per second.

A remarkable result here presented to us in equation (24), is that if this explanation of gaseous friction be true, the coefficient of friction is independent of the density. Such a consequence of a mathematical theory is very startling, and the only experiment I have met with on the subject does not seem to confirm it. We must next compare our theory with what is known of the diffusion of gases, and the conduction of heat through a gas.

14B

Reprinted from pages 21–23 and 31–37 of *Philos. Mag.* ser. 4, **20**:21–37 (1860)

ILLUSTRATIONS OF THE DYNAMICAL THEORY OF GASES. PART II. ON THE PROCESS OF DIFFUSION OF TWO OR MORE KINDS OF MOVING PARTICLES AMONG ONE ANOTHER

J. C. Maxwell

WE have shown, in the first part of this paper, that the motions of a system of many small elastic particles are of two kinds: one, a general motion of translation of the whole system, which may be called the motion in mass; and the other a motion of agitation, or molecular motion, in virtue of which velocities in all directions are distributed among the particles according to a certain law. In the cases we are considering, the collisions are so frequent that the law of distribution of the molecular velocities, if disturbed in any way, will be re-established in an inappreciably short time; so that the motion will always con-

sist of this definite motion of agitation, combined with the general motion of translation.

When two gases are in communication, streams of the two gases might run freely in opposite directions, if it were not for the collisions which take place between the particles. The rate at which they actually interpenetrate each other must be investigated. The diffusion is due partly to the spreading of the particles by the molecular agitation, and partly to the actual motion of the two opposite currents in mass, produced by the pressure behind, and resisted by the collisions of the opposite stream. When the densities are equal, the diffusions due to these two causes respectively are as 2 to 3.

Prop. XIV. *In a system of particles whose density, velocity, &c. are functions of* x, *to find the quantity of matter transferred across the plane of* yz, *due to the motion of agitation alone.*

If the number of particles, their velocity, or their length of path is greater on one side of this plane than on the other, then more particles will cross the plane in one direction than in the other; and there will be a transference of matter across the plane, the amount of which may be calculated.

Let there be taken a stratum whose thickness is dx, and area unity, at a distance x from the origin. The number of collisions taking place in this stratum in unit of time will be

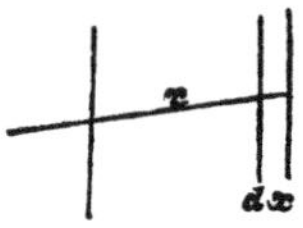

$$N \frac{v}{l} dx.$$

The proportion of these which reach a distance between nl and $(n+dn)l$ before they strike another particle is

$$e^{-n} dn.$$

The proportion of these which pass through the plane yz is

$$\frac{nl+x}{2nl} \text{ when } x \text{ is between } -nl \text{ and } 0,$$

and

$$-\frac{nl-x}{2nl} \text{ when } x \text{ is between } 0 \text{ and } +nl;$$

the sign being negative in the latter case, because the particles cross the plane in the negative direction. The mass of each particle is M; so that the quantity of matter which is projected from the stratum dx, crosses the plane yz in a positive direction, and strikes other particles at distances between nl and $(n+dn)l$ is

$$\frac{MNv(x \mp nl)}{2nl^2} dx\, e^{-n} dn, \quad . \quad . \quad . \quad . \quad . \quad (26)$$

where x must be between $\pm nl$, and the upper or lower sign is to be taken according as x is positive or negative.

In integrating this expression, we must remember that N, v, and l are functions of x, not vanishing with x, and of which the variations are very small between the limits $x = -nl$ and $x = +nl$.

As we may have occasion to perform similar integrations, we may state here, to save trouble, that if U and r are functions of x not vanishing with x, whose variations are very small between the limits $x = +r$ and $x = -r$,

$$\int_{-r}^{+r} \pm \mathrm{U} x^m dx = \frac{2}{m+2} \frac{d}{dx} (\mathrm{U} r^{m+2}). \quad . \quad . \quad . \quad (27)$$

When m is an odd number, the upper sign only is to be considered; when m is even or zero, the upper sign is to be taken with positive values of x, and the lower with negative values. Applying this to the case before us,

$$\int_{-nl}^{+nl} \frac{\mathrm{MN} v x \, dx}{2nl^2} = \tfrac{1}{3} \frac{d}{dx} (\mathrm{MN} v n^2 l),$$

$$\int_{-nl}^{+nl} \mp \frac{\mathrm{MN} v}{2l} dx = -\tfrac{1}{2} \frac{d}{dx} (\mathrm{MN} v n^2 l).$$

We have now to integrate

$$\int_0^\infty -\tfrac{1}{6} \frac{d}{dx} (\mathrm{MN} v l) n^2 e^{-n} dn,$$

n being taken from 0 to ∞. We thus find for the quantity of matter transferred across unit of area by the motion of agitation in unit of time,

$$q = -\tfrac{1}{3} \frac{d}{dx} (\rho v l), \quad . \quad . \quad . \quad . \quad . \quad . \quad (28)$$

where $\rho = \mathrm{MN}$ is the density, v the mean velocity of agitation, and l the mean length of path.

Prop. XV. The quantity transferred, in consequence of a mean motion of translation V, would obviously be

$$\mathrm{Q} = \mathrm{V} \rho. \quad . \quad . \quad . \quad . \quad . \quad . \quad . \quad . \quad . \quad . \quad (29)$$

[*Editor's Note:* Material has been omitted at this point.]

Prop. XXI. *To find the amount of energy which crosses unit of area in unit of time when the velocity of agitation is greater on one side of the area than on the other.*

The energy of a single particle is composed of two parts,—the *vis viva* of the centre of gravity, and the *vis viva* of the various motions of rotation round that centre, or, if the particle be capable of internal motions, the *vis viva* of these. We shall suppose that the whole *vis viva* bears a constant proportion to that due to the motion of the centre of gravity, or

$$E=\tfrac{1}{2}\beta Mv^2,$$

where β is a coefficient, the experimental value of which is 1·634. Substituting E for M in Prop. XIV., we get for the transference of energy across unit of area in unit of time,

$$Jgq=-\tfrac{1}{3}\frac{d}{dx}(\tfrac{1}{2}\beta Mv^2Nvl),$$

where J is the mechanical equivalent of heat in foot-pounds, and q is the transfer of heat in thermal units.

Now $MN=\rho$, and $l=\frac{1}{A\rho}$, so that $MNl=\frac{1}{A}$;

$$\therefore\ Jgq=-\tfrac{1}{2}\frac{\beta v^2}{A}\frac{dv}{dx}. \quad . \quad . \quad . \quad . \quad . \quad . \quad (59)$$

Also, if T is the absolute temperature,

$$\frac{1}{T}\frac{dT}{dx}=\frac{2}{v}\frac{dv}{dx};$$

$$\therefore\ Jgq=-\tfrac{5}{4}\beta plv\frac{1}{T}\frac{dT}{dx}, \quad . \quad . \quad . \quad . \quad . \quad (60)$$

where p must be measured in dynamical units of force.

Let $J=772$ foot-pounds, $p=2116$ pounds to square foot, $l=\frac{1}{400000}$ inch, $v=1505$ feet per second, $T=522$ or 62° Fahrenheit; then

$$q=\frac{T'-T}{40000x}, \quad . \quad . \quad . \quad . \quad . \quad . \quad . \quad . \quad . \quad (61)$$

where q is the flow of heat in thermal units per square foot of area; and T′, and T are the temperatures at the two sides of a stratum of air x *inches* thick.

In Prof. Rankine's work on the Steam-engine, p. 259, values of the *thermal resistance*, or the reciprocal of the *conductivity*, are given for various substances as computed from a Table of conductivities deduced by M. Peclet from experiments by M. Despretz :—

	Resistance.
Gold, Platinum, Silver . .	0·0036
Copper	0·0040
Iron	0·0096
Lead	0·0198
Brick	0·3306
Air by our calculation . . .	40000

It appears, therefore, that the resistance of a stratum of air to the conduction of heat is about 10,000,000 times greater than that of a stratum of copper of equal thickness. It would be almost impossible to establish the value of the conductivity of a gas by direct experiment, as the heat radiated from the sides of the vessel would be far greater than the heat conducted through the air, even if currents could be entirely prevented.

PART III. *On the Collision of Perfectly Elastic Bodies of any Form.*

When two perfectly smooth spheres strike each other, the force which acts between them always passes through their centres of gravity; and therefore their motions of rotation, if they have any, are not affected by the collision, and do not enter into our calculations. But, when the bodies are not spherical, the force of compact will not, in general, be in the line joining their centres of gravity; and therefore the force of impact will depend both on the motion of the centres and the motions of rotation before impact, and it will affect both these motions after impact.

In this way the velocities of the centres and the velocities of rotation will act and react on each other, so that finally there will be some relation established between them; and since the rotations of the particles about their three axes are quantities related to each other in the same way as the three velocities of their centres, the reasoning of Prop. IV. will apply to rotation as well as velocity, and both will be distributed according to the law

$$\frac{dN}{dx}=N\frac{1}{\alpha\sqrt{\pi}}e^{-\frac{x^2}{\alpha^2}}.$$

Also, by Prop. V., if x be the average velocity of one set of particles, and y that of another, then the average value of the sum or difference of the velocities is

$$\sqrt{x^2+y^2};$$

from which it is easy to see that, if in each individual case

$$u=ax+by+cz,$$

where x, y, z are independent quantities distributed according to the law above stated, then the *average values* of these quantities will be connected by the equation

$$u^2=a^2x^2+b^2y^2+c^2z^2.$$

Prop. XXII. *Two perfectly elastic bodies of any form strike each other: given their motions before impact, and the line of impact, to find their motions after impact.*

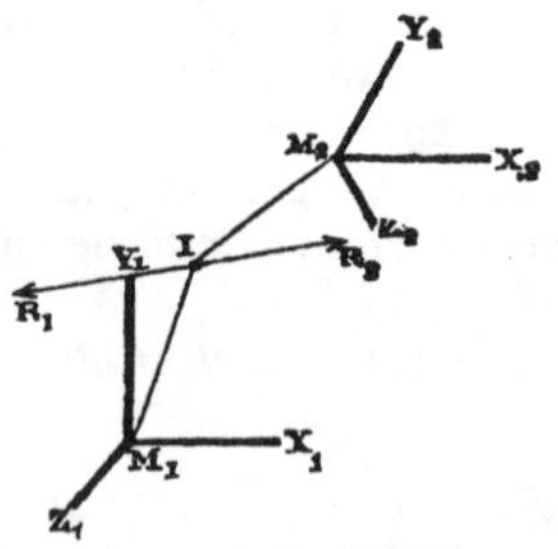

Let M_1 and M_2 be the centres of gravity of the two bodies. $M_1 X_1$, $M_1 Y_1$, and $M_1 Z_1$ the principal axes of the first; and $M_2 X_2$, M_2, Y_2, and $M_2 Z_2$ those of the second. Let I be the point of impact, and $R_1 I R_2$ the line of impact.

Let the coordinates of I with respect to M_1 be $x_1 y_1 z_1$, and with respect to M_2 let them be $x_2 y_2 z_2$.

Let the direction-cosines of the line of impact $R_1 I R_2$ be

$$l_1 m_1 n_1 \text{ with respect to } M_1, \text{ and } l_2 m_2 n_2 \text{ with respect to } M_2.$$

Let M_1 and M_2 be the masses, and $A_1 B_1 C_1$ and $A_2 B_2 C_2$ the moments of inertia of the bodies about their principal axes.

Let the velocities of the centres of gravity, resolved in the direction of the principal axes of each body, be

$$U_1 \; V_1 \; W_1 \text{ and } U_2 \; V_2 \; W_2 \text{ before impact,}$$

and

$$U'_1 \; V'_1 \; W'_1 \text{ and } U'_2 \; V'_2 \; W'_2 \text{ after impact.}$$

Let the angular velocities round the same axes be

$$p_1 \; q_1 \; r_1 \text{ and } p_2 \; q_2 \; r_2 \text{ before impact,}$$

and

$$p'_1 \; q'_1 \; r'_1 \text{ and } p'_2 \; q'_2 \; r'_2 \text{ after impact.}$$

Let R be the impulsive force between the bodies, measured by the momentum it produces in each.

Then, for the velocities of the centres of gravity, we have the following equations:

$$U'_1 = U_1 + \frac{Rl_1}{M_1}, \quad U'_2 = U_2 - \frac{Rl_2}{M_2}, \quad \ldots \quad (62)$$

with two other pairs of equations in V and W. The equations for the angular velocities are

$$p'_1 = p_1 + \frac{R}{A_1}(y_1 n_1 - z_1 m_1), \quad p'_2 = p_2 - \frac{R}{A_2}(y_2 n_2 - z_2 m_2), \quad (63)$$

with two other pairs of equations for q and r.

The condition of perfect elasticity is that the whole *vis viva* shall be the same after impact as before, which gives the equation

$$M_1(U'^2_1 - U^2_1) + M_2(U'^2_2 - U^2_2) + A_1(p'^2_1 - p^2_1) + A_2(p'^2_2 - p^2_2) + \&c. = 0. \quad (64)$$

The terms relating to the axis of x are here given; those relating to y and z may be easily written down.

Substituting the values of these terms, as given by equations

(62) and (63), and dividing by R, we find

$$l_1(U'_1+U_1)-l_2(U'_2+U_2)+(y_1n_1-z_1m_1)(p'_1+p_1)$$
$$-(y_2n_2-z_2m_2)(p'_2+p_2)+\&c.=0. \quad . \quad . \quad . \quad . \quad (65)$$

Now if v_1 be the velocity of the striking-point of the first body before impact, resolved along the line of impact,

$$v_1=l_1U_1+(y_1n_1-z_1m_1)p_1+\&c.;$$

and if we put v_2 for the velocity of the other striking-point resolved along the same line, and v'_1 and v'_2 the same quantities after impact, we may write equation (65),

$$v_1+v'_1-v_2-v'_2=0, \quad . \quad . \quad . \quad . \quad . \quad (66)$$

or

$$v_1-v_2=v'_2-v'_1, \quad . \quad . \quad . \quad . \quad . \quad . \quad . \quad . \quad (67)$$

which shows that the velocity of separation of the striking-points resolved in the line of impact is equal to that of approach.

Substituting the values of the accented quantities in equation (65) by means of equations (63) and (64), and transposing terms in R, we find

$$2\{U_1l_1-U_2l_2+p_1(y_1n_1-z_1m_1)-p_2(y_2n_2-z_2m_2)\}+\&c.$$
$$=-R\left\{\frac{l_1^2}{M_1}+\frac{l_2^2}{M_2}+\frac{(y_1n_1-z_1m_1)^2}{A_1}+\frac{(y_2n_2-z_2m_2)^2}{A_2}+\&c., \right. \quad (68)$$

the other terms being related to y and z as these are to x. From this equation we may find the value of R; and by substituting this in equations (63), (64), we may obtain the values of all the velocities after impact.

We may, for example, find the value of U'_1 from the equation

$$\left.\begin{array}{l} U'_1\left\{\dfrac{l_1^2}{M_1}+\dfrac{l_2^2}{M_2}+\dfrac{(y_1n_1-z_1m_1)^2}{A_1}+\dfrac{(y_2n_2-z_2m_2)^2}{A_2}+\&c.\right\}\dfrac{M_1}{l_1} \\ =U_1\left\{-\dfrac{l_1^2}{M_1}+\dfrac{l_2^2}{M_2}+\dfrac{(y_1n_1-z_1m_1)^2}{A_1}+\dfrac{(y_2n_2-z_2m_2)^2}{A_2}+\&c.\right\}\dfrac{M_1}{l_1} \\ +2U_2l_2-2p_1(y_1n_1-z_1m_1)+2p_2(y_2n_2-z_2m_2)-\&c. \end{array}\right\}. \quad (69)$$

Prop. XXIII. *To find the relations between the average velocities of translation and rotation after many collisions among many bodies.*

Taking equation (69), which applies to an individual collision, we see that U'_1 is expressed as a linear function of U_1, U_2, p_1, p_2, &c., all of which are quantities of which the values are distributed among the different particles according to the law of Prop. IV. It follows from Prop. V., that if we square every term of the equation, we shall have a new equation between the *average values* of the different quantities. It is plain that, as soon as the required relations have been established, they will remain the same after collision, so that we may put $U_1'^2 = U_1^2$ in the equation of averages. The equation between the average values may then be written

$$(M_1U_1^2 - M_2U_2^2)\frac{l_2^2}{M_2} + (M_1U_1^2 - A_1p_1^2)\frac{(y_1n_1 - z_1m_1)^2}{A_1}$$
$$+ (M_1U_1^2 - A_2p_2^2)\frac{(y_2n_2 - z_2m_2)^2}{A_2} + \&c. = 0.$$

Now since there are collisions in every possible way, so that the values of l, m, n, &c. and x, y, z, &c. are infinitely varied, this equation cannot subsist unless

$$M_1U_1^2 = M_2U_2^2 = A_1p_1^2 = A_2p_2^2 = \&c.$$

The final state, therefore, of any number of systems of moving particles of any form is that in which the average *vis viva* of translation along each of the three axes is the same in all the systems, and equal to the average *vis viva* of rotation about each of the three principal axes of each particle.

Adding the *vires vivæ* with respect to the other axes, we find that the whole *vis viva* of translation is equal to that of rotation in each system of particles, and is also the same for different systems, as was proved in Prop. VI.

This result (which is true, however nearly the bodies approach the spherical form, provided the motion of rotation is at all affected by the collisions) seems decisive against the unqualified acceptation of the hypothesis that gases are such systems of hard elastic particles. For the ascertained fact that γ, the ratio of the specific heat at constant pressure to that at constant volume, is equal to 1·408, requires that the ratio of the whole *vis viva* to the *vis viva* of translation should be

$$\beta = \frac{2}{3(\gamma - 1)} = 1{\cdot}634;$$

whereas, according to our hypothesis, $\beta = 2$.

We have now followed the mathematical theory of the collisions of hard elastic particles through various cases, in which there seems to be an analogy with the phænomena of gases. We have deduced, as others have done already, the relations of pressure, temperature, and density of a single gas. We have also proved that when two different gases act freely on each other (that is, when at the same temperature), the mass of the single particles of each is inversely proportional to the square of the molecular velocity; and therefore, at equal temperature and pressure, *the number of particles in unit of volume is the same.*

We then offered an explanation of the internal friction of

gases, and deduced from experiments a value of the mean length of path of a particle between successive collisions.

We have applied the theory to the law of diffusion of gases, and, from an experiment on olefiant gas, we have deduced a value of the length of path not very different from that deduced from experiments on friction.

Using this value of the length of path between collisions, we found that the resistance of air to the conduction of heat is 10,000,000 that of copper, a result in accordance with experience.

Finally, by establishing a necessary relation between the motions of translation and rotation of all particles not spherical, we proved that a system of such particles could not possibly satisfy the known relation between the two specific heats of all gases.

Part IV
THE QUANTUM THEORY

Editor's Comments on Papers 15 Through 18

15 PLANCK
On the Theory of the Law of the Distribution of Energy in the Normal Spectrum

16 EINSTEIN
On a Hueristic Viewpoint Concerning the Production and Transformation of Light

17 COMPTON
A Quantum Theory of the Scattering of X-Rays by Light Elements

18 BOTHE and GEIGER
Experimental Test of the Theory of Bohr, Kramers, and Slater

By the close of the nineteenth century, the foundations of molecular gas dynamics had been well established. The ground-breaking work of Joule, Krönig, Clausius, Maxwell, and Boltzmann had been followed up and refined by many others, notably Kirchhoff, Lorentz, Tait, Kelvin, Van der Waals, Gibbs, and Rayleigh. Important physical quantities such as specific heats, viscosity, thermal conductivity, and diffusion coefficients had been measured, and satisfactory theoretical understanding of the experimental results had been obtained for gases under standard conditions of pressure and temperature. All the theoretical work presumed a basic model in which atoms and molecules behave like elastic spheres when colliding and obey the Newtonian laws of mechanics. Calculation of molecular dimensions led to plausible results. However, as Sir James Jeans pointed out in quoting from Lord Kelvin's address to the Royal Institution in 1900, "The beauty and clearness of the dynamical theory which asserts heat and light to be modes of motion, is at present obscured by two clouds" (Jeans 1954). The first of these clouds concerned the composition of the so-called aether of space. This matter does not bear directly on the development of atomic theory at this point and will be dealt with later in Part VII in connection with energy and relativity. The second of the clouds had to be taken seriously by those who were seeking a more complete understanding of thermal equilibrium and the ways in which the atomic constituents of a system in equilibrium share energy. The question essentially reduces to how a gas composed of elastic spheres can sustain its translational energy when it is continually being subjected to processes that transfer some of this energy into internal vibrations and rotations and dissipate the remainder ultimately to the surrounding ether. The resolution of this question was to require a radical revision in the laws of physics governing energy exchanges between systems of molecules and their surroundings. The person most responsible for the creation of this new branch, called "quantum physics," was Max Planck (1858–1947).

Max Planck was born in Kiel, Germany, the son of Johann von Planck, a professor of civil law at the University of Kiel. The family moved to Munich when Planck was nine years old, and he received his university education at Munich, taking his doctorate there in physics in 1879. Following service as a professor at Kiel from 1885 to 1888, Planck was appointed Kirchhoff's successor at the University of Berlin. He remained there until his retirement in 1926. His latter retirement years were marred by misfortunes during World War II, which saw many of his personal possessions destroyed during the air raids on Berlin. Prior to his death in Göttingen in 1947, he was honored by the founding of the Max Planck Association for the Advancement of Science and the establishment of research institutes in his name at various places in West Germany. Planck received the Nobel Prize for physics in 1918 in recognition of his development of and contributions to quantum theory.

Planck was led to the quantum theory by the realization that the key to an understanding of the energy transfer from matter to its surroundings lay in the explanation of the spectral distribution of energy radiated by the so-called ideal cavity radiator when in thermal equilibrium at some specified temperature. The background factors leading to an awareness of the importance of cavity radiation represented the convergence of two important growing fields of physics. One was electromagnetism, whose laws had been codified by Coulomb, Ampere, Faraday, Lenz, and Maxwell and which had culminated in the striking experimental confirmation by Heinrich Hertz during the late 1800s of the generation of electromagnetic waves by oscillating electrical charges, whose existence had been predicted by Maxwell in his electromagnetic theory. The other field was the kinetic theory, statistical mechanics, and thermodynamics already alluded to. Experimental observations of cavity radiation showed that the rate of energy radiated (electromagnetic radiation does represent the propagation of energy) per wave-length interval per unit surface area is represented by a bell-shaped curve with the maximum rate occurring at a wavelength depending on the temperature. The radiation itself originates from the Hertzian resonators in the radiating matter, producing standing electromagnetic waves in the cavity. Because of the vast number of resonators of differing frequencies, these standing electromagnetic waves possess a wide range of wavelengths and energies when in thermal equilibrium in the cavity. A logical approach to finding the spectral distribution was to determine the average energy per mode of oscillation and multiply that by the density of modes in a given wavelength interval. The original calculations assumed that a *continuum* of energy states is available for these standing waves and yields the result that just as for the molecules in a gas in thermal equilibrium, the average energy per degree of freedom should be kT, where k is the Boltzmann constant, and T the absolute temperature. However, the use of this value predicts a spectral distribution that has no maximum and becomes infinitely large as the wavelength approaches zero, in direct contradiction to the experimental data. In 1900 Planck resolved the difficulty by showing that to agree with observations, each oscillator has to be assumed to radiate energy in discrete or *quantized* amounts proportional to its frequency. Fitting the experimental data to the new theory established the numerical value of the proportionality constant at about 10^{-26} erg sec. Thus energy discreteness was first introduced into the radiation process, a fresh conception that was to have far-reaching consequences on the description and understanding of all energy-transfer processes involving atomic scale particles. Planck first presented his analysis in a paper read before the Berlin Physical Society on October 19, 1900. He refined and clarified his results in a paper presented to a meeting of the German Physical Society on December 14, 1900. This paper, "On the Theory of the Law of the Distribution of Energy in the Normal Spectrum," is reproduced here in translation as Paper 15.

From the standpoint of energy, special significance attaches to Planck's discovery of the preceding proportionality constant, which he called h. The constant has the physical dimensions of action, i.e., energy multiplied by time. The action associated with the motion of a particle represents either the integral of its momentum over space or of its energy over time. In the foundations of mechanics, momentum and displacement on the one hand and energy and time on the other are respectively pairs of conjugate variables. According to classical laws, motion occurs along paths such that the action has a stationary value, and presumably the change in action between possible paths as the conjugate coordinates change is continuous. The fact that action is now restricted in Planck's theory to discrete or quantized values means that dynamic processes on the atomic scale have discontinuous characteristics. This fact has had profound implications in our understanding of the structural stability of atoms and their constituents. The subsequent course of much twentieth-century physics has turned out to be influenced in a major way by quantization.

The full exploitation of Planck's discovery was by no means immediate, though one important application and extension did occur rather soon, i.e., the clarification of the hitherto puzzling photoelectric effect. In 1905 Albert Einstein (1879–1955), while working in Berne as an examiner in the Swiss Patent Office, realized that this effect wherein electrons are released by light irradiation at the surface of a metal only when the light frequency is above a certain threshold value could be explained if Planck's original idea on the quantization of radiant energy emitted by resonators were generalized to include quantization of absorption of radiation by matter. An electron has to receive a discrete amount of energy equal to or greater than the binding energy of the surface in order to be released. Einstein found that the maximum kinetic energy of the electron that escapes from the surface of the metal is equal to the product hf, where f is the frequency of the light, and h is Planck's constant of action, minus the binding energy. Einstein presented these ideas in a paper entitled in English translation "On a Heuristic Viewpoint Concerning the Production and Transformation of Light," which is reproduced here as Paper 16. In this paper Einstein also proposed that the energy of electromagnetic radiation, rather than being distributed uniformly throughout space in continuous fashion, is distributed discontinuously and that in certain experimental situations, as in the photoelectric effect, the discreteness properties govern. This proposal led to the emergence of the dual wave-corpuscular theory of light, in accordance with which light displays particle-like behavior under certain conditions with the radiant energy being quantized in quanta of magnitude hf, whereas under other conditions it displays continuous wave-like properties, as for example in diffraction experiments.

The experimental confirmation of the mechanical particle behavior of light quanta when interacting with matter was not to be made for another twenty years. In 1923 the American psysicist, Arthur Holly Compton (1892–1962) showed that when radiation in the form of x-rays, possessing wavelengths of the order of 10^{-8} cm, is scattered from metal foil targets, the scattering pattern is different from that predicted by classical electromagnetic theory. The wavelength of the scattered rays increases, and the increase depends on the scattering angle. By way of explanation of his experimental results, Compton proposed that an x-ray quantum collides with a single electron in the foil and gives up energy and momentum just as in any mechanical collision between a moving particle and a stationary one. By using the equations for the conservation of momentum and energy expressed in the necessary relativistic form, Compton was able to derive a formula for the wavelength change with scattering angle that agrees with experiment. It should be noted that an important part of Compton's hypothesis was the use of the vector law of momentum conservation, which confers a definite directional character on both the incident and scattered radiation and thus imputes directional motion to the individual quanta. Paper 17 contains Compton's description of his theory and experiment.

Arthur H. Compton was born in Wooster, Ohio, where his father was professor of philosophy and dean at the College of Wooster. He earned his undergraduate degree at Wooster and received the Ph.D. from Princeton in 1916. Following appointments at Minnesota and Westinghouse Research Laboratories, Compton spent a year at the Cavendish Laboratory in Cambridge. While there, he became interested in gamma-ray-scattering experiments. Upon his return to the United States in 1920 to take an appointment as professor of physics at Washington University in Saint Louis, he initiated the experimental and theoretical efforts that culminated in the 1923 paper. In 1927 he shared the Nobel Prize in physics with C. T. R. Wilson, with Compton's recognition coming from the x-ray-scattering work. In 1923 Compton succeeded R. A. Millikan in the physics department at the University of Chicago. Following a distinguished research career, which included important pioneering studies on cosmic rays, he became chancellor of Washington University at the close of World War II.

There still remained a fundamental question concerning the applicability of the principles of conservation of energy and momentum to individual interactions between light quanta and material particles. In 1924 Bohr, Kramers, and Slater advanced the hypothesis that in such an interaction, these conservation principles hold only on the average for a large number of scattering events (see Paper 18 for reference). In 1925 two German physicists, Walther W. G. Bothe (1891–1957) and Hans W. Geiger (1882–1945), confirmed experimentally that the Compton effect is associated with a recoil electron's being produced in coincidence with a scattered x-ray with a coincidence rate about one million times greater than would be expected by chance. They concluded that the conservation principles do apply to *individual* atomic interactions and that Compton was correct in his interpretation (Paper 18).

Walther Bothe was born in Oranienburg, Germany, and was educated at the University of Berlin where he was a student of Planck. Following the receipt of his doctorate in 1914, he held teaching appointments at the Universities of Berlin and Giessen. In 1934 he was appointed head of the Max Planck Institute at Heidelberg and held that position the rest of his life. In addition to the Compton scattering-coincidence experiment, his work included many fundamental studies of cosmic rays and light nuclei transmutations. For his work on cosmic rays, he was awarded a share of the Nobel Prize in 1954.

Hans Geiger was born in Neustadt an der Haardt, the son of a professor of philology at Erlangen. He was educated at Munich and Erlangen and took his doctorate at the latter institution in 1906. During the period of 1906–1912, he spent considerable time at the University of Manchester in England working with Rutherford and Marsden on the perfection of nuclear counting techniques. Following World War I he went to the Physikalisch-Technische Reichsanstalt in Berlin where he and Bothe performed the coincidence experiment on Compton scattering. In 1925 Geiger became professor of physics at the University of Kiel and in collaboration with Müller, developed the nuclear counter whose name is now virtually a household word. He subsequently held chairs in physics at the University of Tübingen and the Berlin Technische Hochschule, carrying out research in nuclear physics.

REFERENCE

Jeans, James H., 1954. *The Dynamical Theory of Gases,* 4th ed.: Dover, pp. 11–13.

15

ON THE THEORY OF THE LAW OF THE DISTRIBUTION OF ENERGY IN THE NORMAL SPECTRUM

M. Planck

This article was translated expressly for this Benchmark volume by R. Bruce Lindsay, Brown University, from "Zur Theorie des Gesetzes der Energieverteilung im Normal Spectrum" Verh. Dtsch. Phys. Ges. **2**:237–245 (1900), *by permission of the publisher.*

When several weeks ago I had the honor to draw your attention to a new formula that appears appropriate to express the law for the distribution of radiant energy in all parts of the normal spectrum,[1] I founded my opinion of the utility of the formula not merely on the apparently good agreement between the few calculated numbers, which I could communicate to you at that time, and the available experimental data, but principally on the simple structure of the formula. I emphasized in particular that the formula provides a very simple logarithmic expression for the dependence of the entropy of an irradiated monochromatic oscillating resonator on its energy of oscillation. This expression seems to hold promise of a general significance better than any other radiation formula hitherto proposed, with the exception of that of Wien, which, however, is not in agreement with the experimental facts.

Entropy implies disorder, and I thought I had to discern this disorder in the irregularity with which the resonator oscillations change their amplitudes and phases even in a stationary radiation field, in so far as we consider time intervals that are large compared with the period of an oscillation but small compared with the time of a measurement. The constant energy of the stationary oscillating resonator is therefore to be considered only as a time average or, what comes to the same thing, the momentary average of the energies of a large number of similarly constructed resonators, which in the stationary radiation field are far enough apart from each other so that they do not directly influence each other. Since in this way the entropy of a resonator is controlled by the simultaneous energy distribution over many resonators, I conjectured that this quantity would have to be calculated through the introduction of probability considerations whose significance for the second law of thermodynamics was first discovered by Ludwig Boltzmann.[2] This conjecture has been confirmed. I have found it possible to deduce an expression for the entropy of a monochromatically oscillating resonator and hence to find the distribution of the energy in a stationary radiation state, that is, in the normal spec-

1. M. Planck, *Verh. Dtsch. Phys. Ges.*, **2**, 202, 1900.
2. *Sitzungsber. k. Akad. Wiss. zu Wien (II)*, **76**, 373, 1877.

trum. For this it was necessary for me only to provide a somewhat more profound interpretation than previously of the hypothesis of "natural radiation," which I had already introduced into electromagnetic theory. In addition, other relations have emerged that appear to me to have significant future promise in other domains of physics and chemistry.

It is not my purpose today to carry out systematically in all its details the deduction that has its basis in the laws of electromagnetic radiation, thermodynamics, and the theory of probability; rather I propose to lay before you as clearly as possible the essential kernel of the whole theory. This can be best accomplished by describing to you a new and elementary procedure by means of which, without the knowledge of a spectrum formula or any other theory but with the help of a single natural constant, we can evaluate the distribution of a given amount of energy among the individual colors of the normal spectrum, and then by means of a second natural constant evaluate the temperature of this energy radiation. In the procedure in question there will appear much that is arbitrary and intricate, but I shall not here stress the proof of its necessity and its easy applicability in practice but rather emphasize only the clarity of the given prescription for the solution of the problem.

Suppose we have a medium enclosed by diathermic, reflecting walls. In this medium in which the velocity of light is c there are assumed to be at appropriate distances from each other a large number of linear monochromatic resonators. We assume that there are N with frequency ν, N' with frequency ν', N'' with frequency ν'', etc., when all Ns are very large compared with unity. The system is assumed to contain a given quantity of energy, namely, total energy E_t in ergs, which appears partly in the medium as propagating radiation energy and partly in the resonators as vibrational energy. The question is, How is this energy distributed in a stationary state among the vibrations of the resonators and among the different colors of the radiation in the medium; also what temperature is possessed by the whole system?

In seeking an answer this question, we consider at first only the oscillations of the resonators, and, for the sake of argument, assign them certain arbitrary energy values. Thus we assign energy E to the N resonators with frequency ν, energy E' to the N' resonators with frequency ν', etc. Then the sum

$$E + E' + E'' + \cdots = E_0$$

must obviously be less than E_t. The difference, that is, $E_t - E_0$, must then be in the radiation in the medium. We must still consider the distribution of the energy among the individual resonators, e.g., the distribution of energy E among the N resonators with frequency ν. If E were to be looked upon as a quantity divisible into infinitely many parts, then there would be infinitely many ways of distributing it. However, and this is the essential point of the whole calculation, we shall consider that E is made up of a definite number of finite, equal parts and in doing this, utilize the constant of nature $h = 6.55 \times 10^{-27}$ erg s. This constant multiplied by the frequency ν of the resonators in question provides the element of energy ϵ in ergs. By division of E by ϵ, we get the number P of energy elements that are to be distributed among the N resonators. If the quotient so obtained is not integral, we take for its value a neighboring integer.

Now it is obvious that the distribution of the P energy elements among the N

resonators can take place only in a finite and precisely determined number of ways. Here we follow Boltzmann's terminology and call each possible kind of distribution a "complexion." If we designate the resonators with the indexes 1, 2, 3, . . . and place under each index the corresponding number of energy elements, we can represent each distribution symbolically as in the following example

1	2	3	4	5	6	7	8	9	10
7	38	11	0	9	2	20	4	4	5

In this special example, $N = 10$ and $P = 100$. The number of all possible complexions is obviously equal to the number of all possible sets of numbers we can get for the lower row in the figure. From the algebraic theory of combinations, the number of possible complexions becomes

$$\frac{N\,(N+1)\,(N+2)\cdots(N+P-1)}{1\cdot 2\cdots P} = \frac{(N+P-1)!}{(N-1)!\,P!}$$

To avoid misunderstanding, it must be remarked that two complexions are to be considered as different if the collection of numbers is the same, but they are arranged in a different order. The preceding formula can be approximated sufficiently by

$$\frac{(N+P)^{N+P}}{N^N P^P}$$

We carry out the same calculation for the resonators of the other frequencies and obtain the appropriate number of complexions for each set of resonators. The product of all the numbers so obtained then gives the totality $\mathfrak{R}$ of complexions for the whole system of resonators. Among all the possible distributions of total energy among the resonators corresponding to constant total energy, there will be one for which the number of possible complexions $\mathfrak{R}_0$ is greater than that for all other distributions. We seek the corresponding energy distribution, which is precisely the one that the resonators assume in the stationary radiation field if they as a whole possess the energy E_0. By division of E by ν, E' by ν', etc., we then obtain the stationary values of the energy of the individual resonators $U_\nu, U'_{\nu'}, U''_{\nu''}, \ldots$. From these we can compute the spatial density of the radiant energy in the medium corresponding to the energy interval from ν to $\nu + d\nu$ in the form

$$u_\nu d_\nu = \frac{8\pi\nu^3}{c^3} \cdot U_\nu d\nu$$

from which the energy in the medium is also determined.

Of all the quantities introduced, only E_0 appears to be chosen arbitrarily. However, we easily see how E_0 can be calculated from the total energy E_t, for if the chosen value of E_0 should happen to lead to too large a value for E_t, it must be appropriately decreased, and vice versa.

After the stationary energy distribution has been evaluated with the help of the constant h, the corresponding temperature θ in degrees Celsius can be found through the use of a second constant of nature $k = 1.346 \times 10^{-16}$ erg/°C by means of the equation

$$\frac{1}{\theta} = k \frac{d \log \mathfrak{R}_0}{dE_0}$$

The product $k \log \mathfrak{R}_0$ is the entropy S_0 of the resonator system. It is the sum of the entropies of all the individual resonators.

It would be rather tedious to carry out the calculations just indicated, though it would be not without interest to test in a simple case the approximation to the truth arrived at in this way. A more general type of derivation as indicated in my earlier paper[3] shows that the desired normal energy distribution has the form

$$u_\nu d\nu = \frac{8\pi h\nu^3}{c^3} \cdot \frac{d\nu}{e^{h\nu/k\theta} - 1}$$

which corresponds exactly to the formula I previously obtained

$$E_\lambda d\lambda = \frac{c_1 \lambda^{-5}}{e^{c_2/\lambda\theta} - 1} \cdot d\lambda$$

The formal differences are brought about by the difference in the definitions of u_ν and E_λ. The upper formula is indeed somewhat more general since it holds for a wholly arbitrary diathermic medium in which the velocity of light is c. The numerical values of h and k, which I have mentioned have been calculated from this formula using the measurements of F. Kurlbaum[4]

$$S_{100} - S_0 = 0.0731 \text{ watt/cm}^2$$

and O. Lummer and E. Pringsheim[5]

$$\lambda_m \cdot \theta = 2940\ \mu°$$

I now turn to a few remarks on the question of the necessity of the deduction just discussed. That the energy element ϵ, assumed for the resonators, must be proportional to the frequency ν follows directly from Wien's displacement law. The relation between u and U is one of the fundamental equations of electromagnetic radiation theory. For the rest, the whole deduction is based on the single law that the entropy of a system of resonators with given energy is proportional to the logarithm of the total number of possible complexions corresponding to this energy. We can decompose this theorem into two others: (1) The entropy of the

3. M. Planck, *Verh. Dtsch. Phys. Ges.*, **2**, 202, 1900.
4. *Wiederherstellungschir. Ann.*, **65**, 759, 1888.
5. *Verh. Dtsch. Phys. Ges.*, **2**, 176, 1900.

system in a given state is proportional to the logarithm of the probability of this state, and (2) the probability of a given state is proportional to the number of complexions corresponding to it. In other words, each complexion is just as probable as any other. As applied to radiation theory the first theorem results from the definition of the probability of a state, in so far as there is no other *a priori* way of defining the probability of a state in a radiation field save by the determination of the entropy. Herein lies one of the differences with kinetic theory. The second theorem contains the kernel of the whole theory being presented here. In the last analysis, its verification must depend wholly on experiment. It can indeed be looked on as a more precise statement of my earlier hypothesis on the nature of natural radiation, i.e., that the energy of such radiation distributes itself in completely random fashion over the partial modes of oscillation in the radiation.[6] I intend to communicate shortly in another place all the detailed calculations bearing on the considerations pointed out in this paper and at the same time shall provide a survey of the previous development of the theory.

In conclusion, I might point out an important consequence of the theory described here, which at the same time provides a further test of its applicability. Boltzmann[7] has shown that the entropy of a monatomic gas in equilibrium is equal to $\omega R \log \mathscr{P}_0$, where $\mathscr{P}_0$ is the number of possible complexions for the most probable velocity distribution (the "permutability"), R is the gas constant per mole (8.31×10^7 for oxygen 16), and ω is the ratio of the molecular mass to the mass of a gram molecule. If radiating resonators are present in the gas, according to the theory developed here, the entropy of the whole system is proportional to the logarithm of the number of all possible complexions, velocities, and radiation taken together. Since, however, according to the electromagnetic theory of radiation, the velocities of the atoms are completely independent of the distribution of radiant energy, the total number of complexions is simply equal to the product of the number corresponding to the velocities and the radiation separately. Hence the total entropy (if f denotes a proportionality factor) becomes

$$f \log (\mathscr{P}_0 \mathscr{P}_0) = f \log \mathscr{P}_0 + f \log \mathscr{R}_0$$

The first term in the sum is the kinetic entropy, and the second is the radiation entropy. By comparison with the previous expressions, we get

$$f = \omega R = k$$

6. M. Planck, *Ann. Phys.*, (4)**1**, 73, 1900. When W. Wien in his Paris report (**II, 38,** 1900) on the theoretical laws of radiation finds my theory of irreversible radiation processes unsatisfactory because it does not contain the proof of the assertion that natural radiation is the only one that leads to irreversibility, in my opinion, he asks too much from this hypothesis. For if we could prove the hypothesis, it would no longer be a hypothesis, and one would never need to set it up at all. However, then we would not be able to deduce anything new from it. From this same standpoint the kinetic theory of gases would have to be considered unsatisfactory since there has not yet been a proof that the atomistic hypothesis is the only one which explains irreversibility, and the same objection might be raised against practically all theories arrived at inductively.

7. *Sitzungsber. k. Akad. Wiss. zu Wien (II),* **76,** 428, 1877.

or

$$\omega = \frac{k}{R} = 1.62 \times 10^{-24}$$

which means that an actual molecule is the 1.62×10^{-24} part of a gram molecule. Hence a hydrogen atom has a mass of 1.64×10^{-24} g since the gram atom of hydrogen has a mass of 1.01 g. In a gram molecule of any substance, there are $1/\omega = 6.175 \times 10^{23}$ actual molecules. O. E. Meyer[8] calculates this number to be 6.40×10^{23}, a fairly good agreement. The Loschmidt number $\mathcal{N}$, that is, the number of gas molecules in 1 cm^3 at 0°C and 1 atm pressure, is

$$\mathcal{N} = \frac{1013200}{R\omega(273)} = 2.76 \times 10^{19}$$

For this quantity Drude[9] gets = 2.1×10^{19}.

The Boltzmann-Drude constant α, that is, the mean *vis viva* of an atom at absolute temperature 1, is

$$\alpha = \tfrac{3}{2}\omega R = \tfrac{3}{2}k = 2.02 \times 10^{-16}$$

Drude[10] gets for this number $\alpha = 2.65 \times 10^{-16}$. The elementary quantum of electric charge, that is, the charge of a positive monovalent ion (or the electron) is

$$e = \epsilon\omega = 4.69 \times 10^{-10} \text{ esu}$$

F. Richarz[11] gets for this quantity $e = 1.29 \times 10^{-10}$ esu, whereas J. J. Thomson[12] gets $e = 6.5 \times 10^{-10}$ esu.

If the theory is to be fully substantiated, all these numerical relations demand not just approximate but precise agreement. The precision of the calculated numbers is definitely connected with that of the most uncertain number, namely, the radiation constant k. Greater precision is required than that of all presently known determinations of this quantity. Better determinations will be an important and necessary task for future investigation.

8. O. E. Meyer, *Die Kinetische Theorie der Gase,* 2d ed., p. 337, 1899.
9. P. Drude, *Ann. Phys.* **1**, 578, 1900.
10. e.c.
11. F. Richarz, *Wiederherstellungschir. Ann.,* **52,** 397, 1894.
12. J. J. Thomson, *Philos. Mag.,* (5)**46,** 528, 1898.

16

ON A HUERISTIC VIEWPOINT CONCERNING THE PRODUCTION AND TRANSFORMATION OF LIGHT

A. Einstein

This article was translated expressly for this Benchmark volume by R. Bruce Lindsay, Brown University, from "Über einen die Erzeugung und Verwandlung des Lichtes betreffenden heuristischen Gesichtspunkt," Ann. Phys. (Leipzig) ser. 4, **17**: 132–148 (1905), *by permission of the publisher.*

A far-reaching formal gap exists between the theoretical representation that physicists have invented for gases and other ponderable bodies on the one side and the Maxwell theory of electromagnetic processes in so-called empty space on the other. Thus while we look on the state of a body as completely determined by the positions and velocities of a very large number of atoms and electrons, we need for the determination of the electromagnetic state of a space continuous spatial functions. Hence a finite number of quantities is not sufficient for the complete determination of the electromagnetic state of the space. According to the Maxwell theory, in all electromagnetic phenomena (and accordingly also for light), the energy must be considered a continuous function of space. On the other hand, the energy of a ponderable body according to present-day conceptions must be considered as a sum over the atoms and electrons. The energy of a ponderable body cannot be resolved into arbitrarily many, arbitrarily small parts, whereas the energy of light radiation emitted from a point source of light, according to the Maxwell theory (or indeed according to any wave theory of light) is continuously distributed over an ever-increasing volume of space.

The wave theory of light operating with continuous functions of space has provided an excellent representation of pure optical phenomena and will probably never be replaced by any other theory. It must be kept in mind, however, that optical observations relate to time-average values and not to instantaneous values. In spite of the complete experimental confirmation of the wave theory of reflection, refraction, diffraction, dispersion, etc., it is conceivable that the theory of light operating with continuous functions of space may lead to disagreement with experiment if we apply it to the phenomena of the production and transformation of light.

It appears to me indeed, that observations on black-body radiation, photoluminescence, the production of cathode rays by ultraviolet light, and other types of phenomena connected with the production and transformation of light can be better understood by the assumption that light energy is discontinuously distributed in space. According to the theory examined in this paper, in a light ray emitted by a point source, the energy is not distributed continuously over a larger and larger spatial domain. It consists rather of a finite number of energy quanta localized at points in space. These quanta move without subdividing themselves and are absorbed and produced as a whole.

In what follows, I shall communicate the line of thought and present the facts that have led me to this viewpoint in the hope that it may prove useful to scientists in their investigations.

1. ON A DIFFICULTY IN THE THEORY OF BLACK-BODY RADIATION

We now consider the following case from the standpoint of the Maxwell theory of electrons. In a space surrounded by completely reflecting walls, let there be a number of gas molecules and electrons, which are able to move freely and exert conservative forces on each other if they come close enough, that is, can collide with each other like gas molecules in the kinetic theory of gases.[1]

Let a number of electrons be attached to points of space widely separated from each other. The attachment is through forces directed to the points and varying as the distance from them. These electrons are to be permitted to interact with the free gas molecules and free electrons conservatively if the latter come close to them. We give the name "resonators" to these electrons that are, so to speak, fastened to points of space. They emit electromagnetic waves of definite period and are also able to absorb them.

According to the current view of the origin of light, the radiation in the space under consideration, which for the case of dynamical equilibrium, is determined by Maxwell's theory, must be identical with black-body radiation, at least if resonators with all the frequencies under consideration are assumed to be present.

For the moment, we ignore the radiation emitted and absorbed by the resonators and inquire after the conditions that must be satisfied by the interaction of the molecules and electrons (collisions) to ensure dynamical equilibrium. The kinetic theory of gases supplies the condition for this, namely, that the mean kinetic energy of a resonator electron shall equal the mean kinetic energy of the translational motion of a gas molecule. If we resolve the motion of the resonator electron into three mutually perpendicular motions, we find for the mean energy of such an oscillatory motion

$$\bar{E} = \frac{R}{N} T$$

where R is the absolute gas constant (per mole), N the number of "real" molecules in a mole (or gram equivalent), and T the absolute temperature. In view of the equivalence of the mean kinetic and potential energies of the oscillating resonator, $\bar{E}$ is two thirds of the kinetic energy of a free monatomic molecule. If for any reason (in our case through radiation effects) the energy of a resonator were to have a

1. This assumption is equivalent to saying that the mean kinetic energies of gas molecules and electrons are equal for temperature equilibrium. With the help of this assumption, Drude calculated theoretically the ratio of the thermal and electrical conductivities of metals.

larger or smaller value than $\bar{E}$, on the average the collisions of gas molecules and electrons would lead to an energy contribution to the gas or subtraction of energy from the gas. Hence dynamical equilibrium is possible only if each resonator has the mean energy $\bar{E}$.

We reflect similarly on the interaction between the resonators and the radiation in the enclosure. Planck[2] has deduced the condition for dynamical equilibrium in this case under the assumption that the radiation can be considered to be a random process.[3] Planck found: where $\bar{E}_\nu$ is the mean energy of a resonator of natural frequency ν, L is the velocity of light, and $\rho_\nu\, d\nu$ is the energy per unit volume for that part of the radiation with frequency between ν and $\nu + d\nu$.

If the radiation energy of frequency ν is neither increased nor decreased, we must have

$$\frac{R}{N} \cdot T = \bar{E} = \bar{E}_\nu = \frac{L^3}{8\pi\nu^2}\, \rho\nu$$

$$\rho\nu = \frac{R}{N} \cdot \frac{8\pi\nu^2}{L^3} \cdot T$$

2. M. Planck, *Ann. Phys.*, **1**, 99, 1900.
3. We can formulate this assumption as follows: We develop the Z component of the electric force (Z) at any arbitrary point of the space in question between times $t = 0$ and $t = T$ (where T is greater than any time considered in the derivation) in a Fourier series

$$Z = \sum_{\nu=1}^{\nu=\infty} A_\nu \sin\left(\frac{2\pi\nu t}{T} + \alpha_\nu\right)$$

where

$$A_0 \geqslant 0 \quad \text{and} \quad 0 \leqslant \alpha_\nu \leqslant 2\pi$$

If we imagine such a development carried out at the same point of space arbitrarily often for suitably chosen initial times, we shall obtain different values for the constants A_ν and α_ν. We then get for the aggregate of all the different value combinations of A_ν and α_ν, statistical probabilities dW as follows:

$$dW = f(A_1, A_2, \ldots, \alpha_1, \alpha_2, \ldots)\, dA_1\, dA_2 \ldots d\alpha_1\, d\alpha_2 \ldots$$

The radiation will be considered random if

$$f(A_1, A_2, \ldots, \alpha_1, \alpha_2, \ldots) = F_1(A_1) \cdot F_2(A_2) \ldots f_1(\alpha_1) \cdot f_2(\alpha_2) \ldots .$$

that is, when the probability for a definite value of one of the quantities A (or α) is independent of the values possessed by the other As (or αs). The more nearly we fulfill the condition that the individual pairs of quantities A_ν and α_ν depend on emission and absorption processes of specific resonators, the greater will be the approximation to randomness in the radiation.

As a condition for dynamical equilibrium, this relation not only fails to satisfy experience but it implies that there can be no specific energy distribution between aether and matter on this picture. Thus the more we extend the frequency range of the resonators, the greater will be the radiation energy in space, and in the limit we have

$$\int_0^\infty \rho\nu d\nu = \frac{R}{N} \cdot \frac{8\pi T}{L^3} \cdot \int_0^\infty \nu^2 d\nu = \infty$$

2. PLANCK'S DETERMINATION OF THE ELEMENTARY QUANTUM

In the following we shall show that Planck's determination of the elementary quantum is to a certain extent independent of the theory he set up for black-body radiation.

The Planck formula for ρ_ν, which satisfies all experience to date, is[4]

$$\rho_\nu = \frac{\alpha\nu^3}{e^{\beta\nu/T} - 1}$$

where

$$\alpha = 6.10 \times 10^{-56}$$

$$\beta = 4.866 \times 10^{-11}$$

[*Editor's Note:* Einstein does not state the units, but it is clear that α must be (in cgs system) in erg s^4/cm^3 or g s^2/cm, and β must be in Kelvin seconds where the Kelvin is the unit of absolute temperature. This means that $\beta = h/k$, where h is Planck's constant (6.62×10^{-27} erg s), and k is the Boltzmann gas constant (1.38×10^{-16} erg/Kelvin). With these accepted values, h/k is 4.80×10^{-11} or a bit smaller than the value Einstein quoted.]

For large values of T/ν, that is, for long wavelengths and large energy density, the formula in the limit takes the form

$$\rho_\nu = \frac{\alpha}{\beta}\nu^3 T$$

It is seen that this formula agrees with that developed in paragraph 1 from the Maxwell and electron theory. By equating coefficients in both formulas, we get

$$\frac{R}{N} \cdot \frac{8\pi}{L^3} = \frac{\alpha}{\beta}$$

4. M. Planck, *Ann. Phys.*, **4**, 561, 1901.

or

$$N = \frac{\beta}{\pi} \cdot \frac{8\pi R}{L^3} = 6.17 \times 10^{23}$$

[*Editor's Note:* This is the Avogadro number, which in the most up-to-date tables is given the value of 6.025×10^{23}.]

Thus a hydrogen atom has a mass $1/N$ g = 1.62×10^{-24} g. This is exactly that found by Planck, which agrees satisfactorily with the value obtained for this quantity in other ways.

We therefore reach this conclusion: The greater the energy density and wavelength of the radiation, the more useful we find the theoretical foundations we have used. For small wavelengths and small energy densities, however, the theory fails completely.

3. ON THE ENTROPY OF RADIATION

The following discussion is contained in a famous work of W. Wien and is included here only for the sake of completeness.

Let us imagine radiation filling up volume v. We assume that the perceptible properties of this radiation are completely determined if the radiation density $\rho(\nu)$ is given for all frequencies.[5] The rays of different frequency are to be looked on as separable from each other without doing any work on each other or transferring any heat back and forth. Hence the entropy of the radiation is representable in the form

$$S = v \int_0^\infty \phi(\rho, \nu) d\nu$$

where ϕ is a function of the variables ρ and ν. This function can be reduced to one of only one variable through formulation of the assertion that through adiabatic compression of a ray between reflecting walls, its entropy is not changed. We shall not go into this matter but at once proceed to investigate how the function ϕ can be calculated from the black-body radiation law.

For a black-body radiation ρ is such a function of ν that the entropy for given energy is a maximum. Thus

$$\delta \int_0^\infty \phi(\rho, \nu) d\nu = 0$$

for

$$\delta \int_0^\infty \rho \, d\nu = 0$$

5. This assumption is arbitrary. We naturally continue to retain this very simple assumption as long as experience to the contrary does not force us to give it up.

From this it follows that for every choice of $\delta\rho$ as function of ν

$$\int_0^\infty \frac{\partial\phi}{\partial\rho} - \lambda \quad \delta\rho d\nu = 0$$

when λ is independent of ν. For black-body radiation, $\partial\phi/\partial\rho$ is also independent of ν. If the temperature increases by dT, the increase in the entropy per unit volume of the black-body radiation is

$$dS = \int_0^\infty \frac{\partial\phi}{\partial\rho} d\rho \, d\nu$$

or, since $d\phi/d\rho$ is independent of ν,

$$dS = \frac{\partial\phi}{\partial\rho} dE$$

Since dE equals the heat introduced and the process is reversible, we also have

$$dS = \frac{dE}{T}$$

Hence

$$\frac{\partial\phi}{\partial\rho} = \frac{1}{T}$$

This may be considered the law of black-body radiation. We can accordingly determine the black-body radiation law from the function ϕ, and conversely from the law, we can get ϕ by integration, taking account of the fact that ϕ vanishes for $\rho = 0$.

4. LIMITING LAW FOR THE ENTROPY OF MONOCHROMATIC RADIATION AT LOW RADIATION DENSITY

From the preceding discussion, we get the result that Wien's original radiation law

$$\rho = \alpha\nu^3 e^{-\beta\nu/T}$$

cannot be exactly correct. Experiment indeed does confirm it completely for large values of ν/T. We shall base our further calculations here on this formula, keeping in mind, however, that our results will hold only within certain limits.

From Wien's formula we get to begin with

$$\frac{1}{T} = -\frac{1}{\beta\nu} \cdot \log \frac{\rho}{\alpha\nu^3}$$

Using the relation obtained in the previous paragraph, we can write further

$$\phi\,(\rho, \nu) = -\frac{\rho}{\beta\nu}\left(\log\frac{\rho}{\alpha\nu^3} - 1\right)$$

Let us now consider radiation with energy E, whose frequency lies between ν and $\nu + d\nu$. Let the radiation occupy volume v. Then the entropy for this radiation is

$$S = v\phi(\rho, \nu)d\nu = -\frac{E}{\beta\nu}\left(\log\frac{E}{v\,\alpha\nu^3 d\nu}\right)$$

If we limit ourselves to the investigation of the dependence of the entropy on the volume occupied by the radiation and designate the entropy of the radiation by S_0 if it occupies volume v_0, we have

$$S - S_0 = \frac{E}{\rho\nu}\log\left(\frac{v}{v_0}\right)$$

This shows that the entropy of monochromatic radiation of sufficiently small density varies with the volume as the entropy of an ideal gas or a dilute solution. The equation just obtained will be interpreted in what follows in accordance with the principle introduced into physics by Ludwig Boltzmann, which says that the entropy of a system is a function of the probability of its state.

5. MOLECULAR THEORY INVESTIGATION OF THE DEPENDENCE OF THE ENTROPY OF GASES AND DILUTE SOLUTIONS ON THE VOLUME

In the calculation of entropy by means of molecular theory, the word *probability* will be applied in a sense not agreeing with its definition given in the standard calculus of probabilities. In particular, the "cases of equal probability" will frequently be hypothetically assigned in cases where the theoretical pictures being applied are sufficiently explicit so as to provide a deduction in the place of that hypothetical assignment. In another place I expect to show that in considering thermal phenomena, one can readily get by with the so-called statistical probability and hope thereby to avoid a logical difficulty, which stands in the way of the carrying through of the Boltzmann principle. Here, however, we shall give the general formulation of the principle and its application to special cases.

If there is a meaning in talking of the probability of the state of a system and if further every entropy increase can be looked on as a transition to a more probable state, the entropy S_1 of a system is a function of the probability of its instantaneous state. If then we consider two systems that are not interacting with each other, we have

$$S_1 = \phi_1(W_1)$$

$$S_2 = \phi_2(W_2)$$

If we think of these systems as forming a single system with entropy S and probability W, it then follows that

$$S = S_1 + S_2 = \phi(W)$$

and

$$W = W_1 \cdot W_2$$

The last relation expresses the fact that the states of the two systems are events that are independent of each other. From these equations we have

$$\phi(W_1 \cdot W_2) = \phi_1(W_1) + \phi_2(W_2)$$

and hence finally

$$\phi_1(W_1) = C \log(W_1) + \text{const}$$

$$\phi_2(W_2) = C \log(W_2) + \text{const}$$

$$\phi(W) = C \log(W) + \text{const}$$

The quantity C is a universal constant. From the kinetic theory of gases, it has the value R/N, wherein the constants R and N have the same meaning as in the preceding discussion in this paper. If S_0 denotes the entropy for a certain initial state of a system under consideration and W the probability of a state of entropy S, we have the general result

$$S - S_0 = \frac{R}{N} \log W$$

Let us begin by examining the following special case. Let there be a number of moving points (e.g., molecules) in a volume v_0 to be treated according to our method. In addition to these, there can be present in this space an arbitrary number of movable points of any kind. Nothing is assumed in advance about the nature of the law in accordance with which these points move except that no part of the space available and no direction have any preferred status relative to any other. The number of moving points under consideration (those first mentioned) is assumed to be so small that we can neglect their mutual interaction.

A certain entropy S_0 is associated with the system under consideration, which can be, for example, an ideal gas or a dilute solution. Let us imagine that there is a part of the whole volume v_0, which we shall designate as v, and that all n moving points are transferred to v without any change taking place in the rest of the system. Obviously, the state of the system gets another value of the entropy S, and it is our task to determine the entropy change with the help of the Boltzmann principle.

We ask how great is the probability of the last-mentioned state relative to that of the original one? In other words, how great is the probability that at a given moment, all n independently moving points in given volume v_0 will by chance find themselves in volume v?

This probability is a "statistical" probability, and for its value we clearly get

$$W = \left(\frac{v}{v_0}\right)^n$$

From this, by applying the Boltzmann principle, we get

$$S - S_0 = R\left(\frac{n}{N}\right)\log\left(\frac{v}{v_0}\right).$$

It is noteworthy of that in the derivation of this equation, from which the Boyle–Gay-Lussac law and the law of osmotic pressure can be readily derived[6] [*Editor's Note:* This derivation does not appear valid or is at any rate misleading unless we assume the first law of thermodynamics $TdS = dE + pdv$ applied to a gas and assume $dE = 0$], we need make no assumption about the law according to which the molecules move.

6. INTERPRETATION OF THE EXPRESSION FOR THE DEPENDENCE OF THE ENTROPY OF MONOCHROMATIC RADIATION AS THE VOLUME, ACCORDING TO THE BOLTZMANN PRINCIPLE

In paragraph 4 we found for the dependence of the entropy of monochromatic radiation on the volume the expression

$$S - S_0 = \frac{E}{\beta\nu}\log\left(\frac{v}{v_0}\right)$$

6. If E is the energy of the system, we get

$$-d(E - TS) = pdv = TdS$$

$$= R\frac{n}{N}\cdot T\frac{dv}{v}$$

Hence

$$pv = \frac{R}{N}nT$$

If we write this formula in the form

$$S - S_0 = \frac{R}{N} \log\left[\left(\frac{v}{v_0}\right)^{N/R \cdot E/\beta\nu}\right]$$

and compare it to the general formula expressing the Boltzmann principle

$$S - S_0 = \frac{R}{N} \log W$$

we reach the following conclusion.

If monochromatic radiation of frequency ν and energy E is enclosed in volume v_0 with reflecting walls, the probability that at an arbitrarily chosen moment of time the whole radiation energy will be found in the partial volume v of the whole volume v_0 is

$$W = \left(\frac{v}{v_0}\right)^{\frac{N}{R} \cdot \frac{E}{\beta\nu}}$$

From this we draw the further conclusion: Monochromatic radiation of small density (within the range of validity of the Wien radiation law) behaves in heat-theoretic terms as if it consisted of independent energy quanta, each with the magnitude $R\beta\nu/N$.

We now wish to compare the mean value of the energy quantum of black-body radiation with the mean kinetic energy of the motion of the center of mass of a molecule at the same temperature. The latter is $\frac{3}{2} \cdot (R/N) \cdot T$, whereas on the basis of the Wien formula we get for the mean value of the energy quantum

$$\frac{\int_0^\infty \alpha\nu^3 e^{-\beta\nu/T} d\nu}{\int_0^\infty \frac{N}{R\beta\nu} \cdot \alpha\nu^3 e^{-\beta\nu/T} d\nu} = 3\frac{R}{N} \cdot T$$

If now monochromatic radiation (of sufficiently small density) behaves with respect to the dependence of entropy on volume like a discontinuous medium, consisting of energy quanta of magnitude $R\beta\nu/N$, the suggestion is strong that we investigate whether the laws of the production and transformation of light are constituted in such a fashion that light consists of such energy quanta. We consider this question in the following paragraphs.

7. STOKES'S RULE

Let monochromatic light be transformed through photoluminescence into light of another frequency and let it be assumed that in accordance with the preceding suggestions that the exciting light as well as the light produced consists of energy quanta of magnitude $R\beta\nu/N$, where ν is the appropriate frequency. The transformation process would then be interpreted as follows: Every exciting energy quantum of frequency ν is absorbed and (at any rate, for sufficiently small distribution density of the exciting quanta) gives rise to the production of a light quantum of frequency ν_2. Actually, in the absorption of the exciting quantum, quanta of frequencies $\nu_3, \nu_4, \ldots$, may be produced at the same time, as well as energy in other forms (e.g., heat). The inbetween processes leading to such a result are not material. If the photoluminescent substance is not to be looked on as an independent producer of energy, then according to the principle of the conservation of energy, the energy of a quantum produced cannot be greater than that of the exciting light quantum. We must therefore have

$$\frac{R}{N}\beta\nu_2 \leqslant \frac{R}{N}\beta\nu_1$$

or

$$\nu_2 \leqslant \nu_1$$

This is the well-known rule of Stokes. It is to be emphasized that in weak illumination the quantity of excited light must be proportional to that of the exciting light, other conditions remaining the same, according to our conception. This is because each exciting energy quantum must excite an elementary process of the kind suggested previously, independently of the action of the other exciting energy quanta. In particular, there will be no lower limit for the intensity of the exciting light below which the light is able to act as a light exciter.

According to the conception of the phenomena set forth here, deviations from the Stokes rule are imaginable in the following situations:

1. If the number of energy quanta in unit volume that are being transformed simultaneously is so large that an energy quantum of excited light can receive its energy from several exciting energy quanta.
2. If the exciting or excited light is not of the energetic character corresponding to black-body radiation in the limit of validity of the Wien radiation law, that is, for example, if the exciting light is produced by a body so high in temperature that Wien's law no longer holds for the wavelengths in question.

The second possibility is of special interest. According to the viewpoint presented in this paper, it not out of the range of possibility that a non-Wien radiation in very small concentration should behave energetically in a fashion different from black-body radiation in the range of validity of Wien's law.

8. ON THE PRODUCTION OF CATHODE RAYS THROUGH THE IRRADIATION OF SOLID BODIES

The usual conception that light energy is distributed continuously over the space through which it passes has encountered great difficulty in explaining photoelectric phenomena. This has been mentioned in a previous paper by P. Lenard.[7]

7. P. Lenard, *Ann. Phys.*, **8**, 169, 170, 1902.

The theory that exciting light consists of energy quanta with magnitude $R\beta\nu/N$ explains the production of cathode rays by light in the following fashion. In the surface of the emitting body, the energy quanta penetrate and transform their energy in part, at any rate, into kinetic energy of electrons. On the simplest picture a light quantum gives up its entire energy to a single electron. Let us assume that this happens. We do not exclude the possibility that electrons may only partially absorb the energy of the light quanta. An electron with kinetic energy in the interior of the body when it reaches the surface will have lost a part of this kinetic energy. It is assumed in addition that each electron, in order to leave the surface, must do a certain amount of work P, which is characteristic of the body. Those electrons that are excited right at the surface of the body will leave the body with the largest normal velocity. The kinetic energy of such electrons is

$$\frac{R\beta\nu}{N} - P$$

If the body is charged to a positive potential Π and surrounded by a conductor at zero potential and if Π is just able to block the loss of electricity by the body, we must have

$$\Pi\epsilon = R\frac{\beta\nu}{N} - P$$

where ϵ is the charge on the electron. We can also write

$$\Pi E = R\,\beta\nu - P'$$

where E is the charge of a gram equivalent of a monovalent ion and P' is the potential of this quantity of negative electricity in relation to the body.[8] If we set $E = 9.6 \times 10^{-3}$, then $\Pi \times 10^{-8}$ is the potential in volts that the body receives by irradiation in a vacuum.

To see in a preliminary way whether the relation derived previously agrees in order of magnitude with experience, we set $P' = 0$ and take $\nu = 1.03 \times 10^{15}$ (corresponding to the short wavelength limit of the solar spectrum), and further $\beta = 4.806 \times 10^{-11}$. Then we get $\Pi \times 10^{7} = 4.3$ V, which agrees in order of magnitude with Lenard's results.[9]

If the formula derived previously is correct, Π, represented as a function of the frequency of the exciting light in cartesian coordinates, must plot as a straight line whose slope is independent of the substance under consideration.

So far as I can see, the point of view expressed in this paper is not in contradiction with the properties of the photoelectric effect as Lenard observed. If every energy quantum of the exciting light gives up its energy to electrons independently of all the rest, the velocity distribution of the emitted electrons, that is, the quality

8. If one assumes that the individual electron must be first pulled out of a neutral molecule by the expenditure of a certain amount of work, nothing needs to be altered in the preceding equation; P' then becomes the sum of two potentials.
9. P. Lenard, *Ann. Phys.*, **8**, 165, 184, Table 1, Fig. 2, 1902.

of the excited cathode radiation, will be independent of the intensity of the exciting light. On the other hand, the number of electrons leaving the body will be proportional to the intensity of the exciting light, other things being equal.[10]

Concerning the presumptive limits of validity of the regularities pointed out previously, comments may be made similar to those in connection with the presumptive deviations from the Stokes rule.

In the foregoing, it has been assumed that the energy of at least a part of the energy quanta of the exciting light will be completely given up to a single electron. If we do not make this plausible assumption, then in place of the preceding equation, we have

$$\Pi E + P' \leqslant R\beta\nu$$

For cathode luminescence, which forms the inverse process to that described previously, we obtain by analogous considerations

$$\Pi E + P' \geqslant R\beta\nu$$

In the substances studied by Lenard, ΠE is always considerably larger than $R\beta\nu$ since the potential through which the cathode rays have to go to be able to produce visible light in some cases amounts to several hundred and in others several thousands of volts.[11]

It must be assumed, therefore, that the kinetic energy of an electron is used in the production of many light quanta.

9. ON THE IONIZATION OF GASES BY ULTRAVIOLET LIGHT

We shall assume that in the ionization of a gas by ultraviolet light, each absorbed energy quantum is applied to the ionization of a single gas molecule. From this it follows that the ionization work (that is, the work theoretically needed for ionization) for a molecule cannot be greater than the energy of an absorbed light quantum that acts effectively. If we designate by J the theoretical ionization work per gram equivalent, we must then have

$$R\beta\nu \geqslant J$$

From Lenard's measurements, however, the longest effective wavelength for air is about 1.9×10^{-5} cm, and hence

$$R\beta\nu = 6.4 \times 10^{12} \text{ erg} \geqslant J$$

We can also get an upper limit for the ionization potential in gases at low pressure. According to J. Stark,[12] the smallest measured ionization potential (with platinum

10. P. Lenard, *Ann. Phys.*, p. 150, 166–168, 1902.
11. P. Lenard, *Ann. Phys.*, 12, 469, 1903.
12. J. Stark, *Die Elektr. in Gasen*, p. 57, Leipzig, 1902.

anodes) in air is about $10V^2$.[13] For J the upper limit would then be 9.6×10^{12} erg, which is not far from the preceding value. The theory of quanta leads indeed to another theoretical result whose experimental test appears to me to be of great importance. If every absorbed light quantum ionizes a molecule, we must have the following relation between the quantity L of light absorbed and the number j of gas molecules ionized thereby:

$$j = \frac{L}{R\beta\nu}$$

If our theory corresponds to reality, this relation must hold for every gas that shows for the appropriate frequency no noticeable absorption not accompanied by ionization.

13. In the interior of gases, the ionization potential for negative ions is, to be sure, about 5 tımes greater.

17

Reprinted from *Phys. Rev.* ser. 2, **21**:483–502 (1923)

A QUANTUM THEORY OF THE SCATTERING OF X-RAYS BY LIGHT ELEMENTS

BY ARTHUR H. COMPTON

ABSTRACT

A quantum theory of the scattering of X-rays and γ-rays by light elements. —The hypothesis is suggested that when an X-ray quantum is scattered it spends all of its energy and momentum upon some particular electron. This electron in turn scatters the ray in some definite direction. The change in momentum of the X-ray quantum due to the change in its direction of propagation results in a recoil of the scattering electron. The energy in the scattered quantum is thus less than the energy in the primary quantum by the kinetic energy of recoil of the scattering electron. The corresponding *increase in the wave-length of the scattered beam* is $\lambda_\theta - \lambda_0 = (2h/mc)\sin^2 \frac{1}{2}\theta = 0.0484 \sin^2 \frac{1}{2}\theta$, where h is the Planck constant, m is the mass of the scattering electron, c is the velocity of light, and θ is the angle between the incident and the scattered ray. Hence the increase is independent of the wave-length. *The distribution of the scattered radiation* is found, by an indirect and not quite rigid method, to be concentrated in the forward direction according to a definite law (Eq. 27). The total energy removed from the primary beam comes out less than that given by the classical Thomson theory in the ratio $1/(1 + 2\alpha)$, where $\alpha = h/mc\lambda_0 = 0.0242/\lambda_0$. Of this energy a fraction $(1 + \alpha)/(1 + 2\alpha)$ reappears as scattered radiation, while the remainder is truly absorbed and transformed into kinetic energy of recoil of the scattering electrons. Hence, if σ_0 is the *scattering absorption coefficient* according to the classical theory, the coefficient according to this theory is $\sigma = \sigma_0/(1 + 2\alpha) = \sigma_s + \sigma_a$, where σ_s is the true scattering coefficient $[(1 + \alpha)\sigma/(1 + 2\alpha)^2]$, and σ_a is the coefficient of absorption due to scattering $[\alpha\sigma/(1 + 2\alpha)^2]$. Unpublished experimental results are given which show that for graphite and the Mo–K radiation the scattered radiation is longer than the primary, the observed difference $(\lambda_{\pi/2} - \lambda_0 = .022)$ being close to the computed value .024. In the case of scattered γ-rays, the wave-length has been found to vary with θ in agreement with the theory, increasing from .022 A (primary) to .068 A ($\theta = 135°$). Also the velocity of secondary β-rays excited in light elements by γ-rays agrees with the suggestion that they are recoil electrons. As for the predicted variation of absorption with λ, Hewlett's results for carbon for wave-lengths below 0.5 A are in excellent agreement with this theory; also the predicted concentration in the forward direction is shown to be in agreement with the experimental results, both for X-rays and γ-rays. This remarkable *agreement between experiment and theory* indicates clearly that scattering is a quantum phenomenon and can be explained without introducing any new hypothesis as to the size of the electron or any new constants; also that a radiation quantum carries with it momentum as well as energy. The restriction to light elements is due to the assumption that the constraining forces acting on the scattering electrons are negligible, which is probably legitimate only for the lighter elements.

Spectrum of K-rays from Mo scattered by graphite, as compared with the spectrum of the primary rays, is given in Fig. 4, showing the change of wave-length.

Radiation from a moving isotropic radiator.—It is found that in a direction θ with the velocity, $I_\theta/I' = (1 - \beta)^2/(1 - \beta \cos \theta)^4 = (\nu_\theta/\nu')^4$. For the total radiation from a black body in motion to an observer at rest, $I/I' = (T/T')^4 = (\nu_m/\nu_m')^4$, where the primed quantities refer to the body at rest.

J. J. Thomson's classical theory of the scattering of X-rays, though supported by the early experiments of Barkla and others, has been found incapable of explaining many of the more recent experiments. This theory, based upon the usual electrodynamics, leads to the result that the energy scattered by an electron traversed by an X-ray beam of unit intensity is the same whatever may be the wave-length of the incident rays. Moreover, when the X-rays traverse a thin layer of matter, the intensity of the scattered radiation on the two sides of the layer should be the same. Experiments on the scattering of X-rays by light elements have shown that these predictions are correct when X-rays of moderate hardness are employed; but when very hard X-rays or γ-rays are employed, the scattered energy is found to be decidedly less than Thomson's theoretical value, and to be strongly concentrated on the emergent side of the scattering plate.

Several years ago the writer suggested that this reduced scattering of the very short wave-length X-rays might be the result of interference between the rays scattered by different parts of the electron, if the electron's diameter is comparable with the wave-length of the radiation. By assuming the proper radius for the electron, this hypothesis supplied a quantitative explanation of the scattering for any particular wave-length. But recent experiments have shown that the size of the electron which must thus be assumed increases with the wave-length of the X-rays employed,[1] and the conception of an electron whose size varies with the wave-length of the incident rays is difficult to defend.

Recently an even more serious difficulty with the classical theory of X-ray scattering has appeared. It has long been known that secondary γ-rays are softer than the primary rays which excite them, and recent experiments have shown that this is also true of X-rays. By a spectroscopic examination of the secondary X-rays from graphite, I have, indeed,

[1] A. H. Compton, Bull. Nat. Research Council, No. 20, p. 10 (Oct., 1922).

been able to show that only a small part, if any, of the secondary X-radiation is of the same wave-length as the primary.[1] While the energy of the secondary X-radiation is so nearly equal to that calculated from Thomson's classical theory that it is difficult to attribute it to anything other than true scattering,[2] these results show that if there is any scattering comparable in magnitude with that predicted by Thomson, it is of a greater wave-length than the primary X-rays.

Such a change in wave-length is directly counter to Thomson's theory of scattering, for this demands that the scattering electrons, radiating as they do because of their forced vibrations when traversed by a primary X-ray, shall give rise to radiation of exactly the same frequency as that of the radiation falling upon them. Nor does any modification of the theory such as the hypothesis of a large electron suggest a way out of the difficulty. This failure makes it appear improbable that a satisfactory explanation of the scattering of X-rays can be reached on the basis of the classical electrodynamics.

The Quantum Hypothesis of Scattering

According to the classical theory, each X-ray affects every electron in the matter traversed, and the scattering observed is that due to the combined effects of all the electrons. From the point of view of the quantum theory, we may suppose that any particular quantum of X-rays is not scattered by all the electrons in the radiator, but spends all of its energy upon some particular electron. This electron will in turn scatter the ray in some definite direction, at an angle with the incident beam. This bending of the path of the quantum of radiation results in a change in its momentum. As a consequence, the scattering electron will recoil with a momentum equal to the change in momentum of the X-ray. The energy in the scattered ray will be equal to that in the incident ray minus the kinetic energy of the recoil of the scattering electron; and since the scattered ray must be a complete quantum, the frequency will be reduced in the same ratio as is the energy. Thus on the quantum theory we should expect the wave-length of the scattered X-rays to be greater than that of the incident rays.

The effect of the momentum of the X-ray quantum is to set the

[1] In previous papers (Phil. Mag. **41**, 749, 1921; Phys. Rev. **18**, 96, 1921) I have defended the view that the softening of the secondary X-radiation was due to a considerable admixture of a form of fluorescent radiation. Gray (Phil. Mag. **26**, 611, 1913; Frank. Inst. Journ., Nov., 1920, p. 643) and Florance (Phil. Mag. **27**, 225, 1914) have considered that the evidence favored true scattering, and that the softening is in some way an accompaniment of the scattering process. The considerations brought forward in the present paper indicate that the latter view is the correct one.

[2] A. H. Compton, loc. cit., p. 16.

scattering electron in motion at an angle of less than 90° with the primary beam. But it is well known that the energy radiated by a moving body is greater in the direction of its motion. We should therefore expect, as is experimentally observed, that the intensity of the scattered radiation should be greater in the general direction of the primary X-rays than in the reverse direction.

The change in wave-length due to scattering.—Imagine, as in Fig. 1*A*,

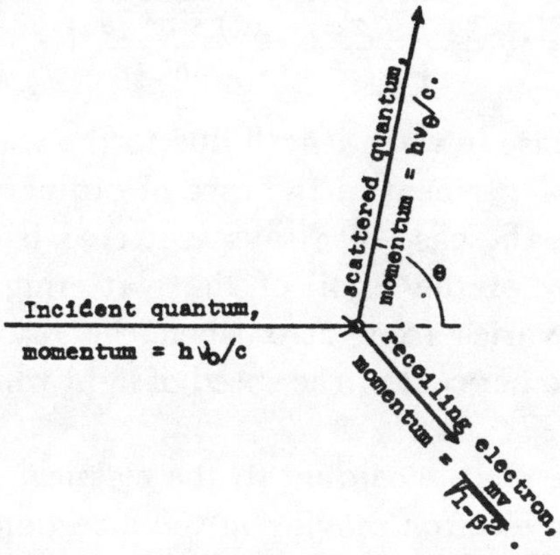

Fig. 1 A

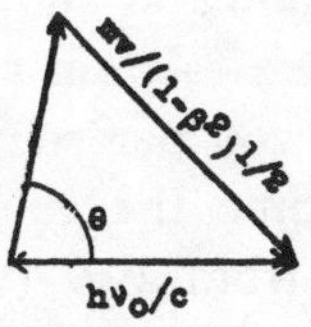

Fig. 1 B

that an X-ray quantum of frequency ν_0 is scattered by an electron of mass m. The momentum of the incident ray will be $h\nu_0/c$, where c is the velocity of light and h is Planck's constant, and that of the scattered ray is $h\nu_\theta/c$ at an angle θ with the initial momentum. The principle of the conservation of momentum accordingly demands that the momentum of recoil of the scattering electron shall equal the vector difference between the momenta of these two rays, as in Fig. 1*B*. The momentum of the electron, $m\beta c/\sqrt{1-\beta^2}$, is thus given by the relation

$$\left(\frac{m\beta c}{\sqrt{1-\beta^2}}\right)^2 = \left(\frac{h\nu_0}{c}\right)^2 + \left(\frac{h\nu_\theta}{c}\right)^2 + 2\frac{h\nu_0}{c}\cdot\frac{h\nu_\theta}{c}\cos\theta, \qquad (1)$$

where β is the ratio of the velocity of recoil of the electron to the velocity of light. But the energy $h\nu_\theta$ in the scattered quantum is equal to that of the incident quantum $h\nu_0$ less the kinetic energy of recoil of the scattering electron, *i.e.*,

$$h\nu_\theta = h\nu_0 - mc^2\left(\frac{1}{\sqrt{1-\beta^2}} - 1\right). \qquad (2)$$

We thus have two independent equations containing the two unknown quantities β and ν_θ. On solving the equations we find

$$\nu_\theta = \nu_0/(1 + 2\alpha \sin^2 \tfrac{1}{2}\theta), \qquad (3)$$

where

$$\alpha = h\nu_0/mc^2 = h/mc\lambda_0. \tag{4}$$

Or in terms of wave-length instead of frequency,

$$\lambda_\theta = \lambda_0 + (2h/mc)\sin^2 \tfrac{1}{2}\theta. \tag{5}$$

It follows from Eq. (2) that $1/(1 - \beta^2) = \{1 + \alpha[1 - (\nu_\theta/\nu_0)]\}^2$, or solving explicitly for β

$$\beta = 2\alpha \sin \tfrac{1}{2}\theta \frac{\sqrt{1 + (2\alpha + \alpha^2)\sin^2 \frac{1}{2}\theta}}{1 + 2(\alpha + \alpha^2)\sin^2 \frac{1}{2}\theta}. \tag{6}$$

Eq. (5) indicates an increase in wave-length due to the scattering process which varies from a few per cent in the case of ordinary X-rays to more than 200 per cent in the case of γ-rays scattered backward. At the same time the velocity of the recoil of the scattering electron, as calculated from Eq. (6), varies from zero when the ray is scattered directly forward to about 80 per cent of the speed of light when a γ-ray is scattered at a large angle.

It is of interest to notice that according to the classical theory, if an X-ray were scattered by an electron moving in the direction of propagation at a velocity $\beta' c$, the frequency of the ray scattered at an angle θ is given by the Doppler principle as

$$\nu_\theta = \nu_0 \Big/ \left(1 + \frac{2\beta'}{1 - \beta'}\sin^2 \tfrac{1}{2}\theta\right). \tag{7}$$

It will be seen that this is of exactly the same form as Eq. (3), derived on the hypothesis of the recoil of the scattering electron. Indeed, if $\alpha = \beta'/(1 - \beta')$ or $\beta' = \alpha/(1 + \alpha)$, the two expressions become identical. It is clear, therefore, that so far as the effect on the wave-length is concerned, we may replace the recoiling electron by a scattering electron moving in the direction of the incident beam at a velocity such that

$$\bar{\beta} = \alpha/(1 + \alpha). \tag{8}$$

We shall call $\bar{\beta}c$ the "effective velocity" of the scattering electrons.

Energy distribution from a moving, isotropic radiator.—In preparation for the investigation of the spatial distribution of the energy scattered by a recoiling electron, let us study the energy radiated from a moving, isotropic body. If an observer moving with the radiating body draws a sphere about it, the condition of isotropy means that the probability is equal for all directions of emission of each energy quantum. That is, the probability that a quantum will traverse the sphere between the angles θ' and $\theta' + d\theta'$ with the direction of motion is $\frac{1}{2}\sin\theta' d\theta'$. But the surface which the moving observer considers a sphere (Fig. $2A$) is

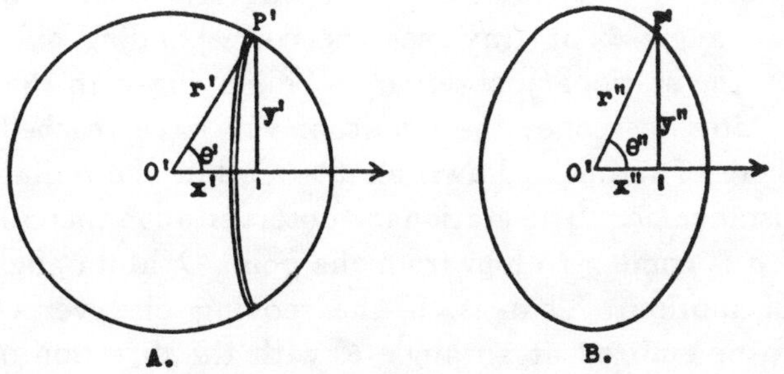

Fig. 2

considered by the stationary observer to be an oblate spheroid whose polar axis is reduced by the factor $\sqrt{1 - \beta^2}$. Consequently a quantum of radiation which traverses the sphere at the angle θ', whose tangent is y'/x' (Fig. $2A$), appears to the stationary observer to traverse the spheroid at an angle θ'' whose tangent is y''/x'' (Fig. $2B$). Since $x' = x''/\sqrt{1 - \beta^2}$ and $y' = y''$, we have

$$\tan \theta' = y'/x' = \sqrt{1 - \beta^2}\, y''/x'' = \sqrt{1 - \beta^2} \tan \theta'', \tag{9}$$

and

$$\sin \theta' = \frac{\sqrt{1 - \beta^2} \tan \theta''}{\sqrt{1 + (1 - \beta^2) \tan^2 \theta''}}. \tag{10}$$

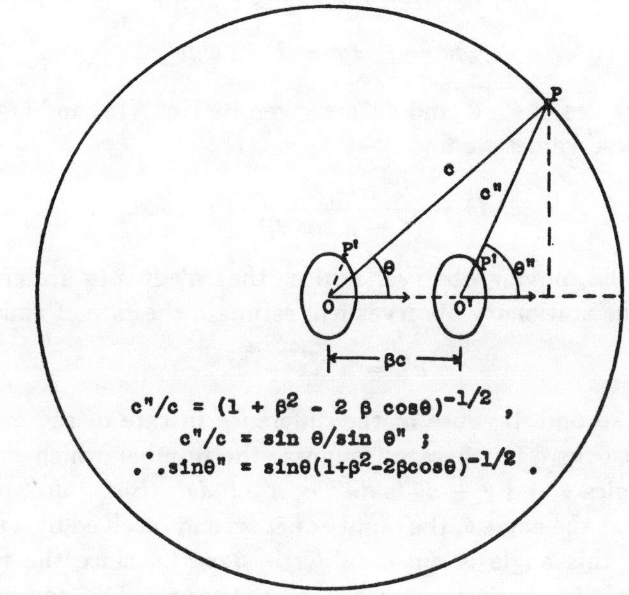

Fig. 3. The ray traversing the moving spheroid at P' at an angle θ'' reaches the stationary spherical surface drawn about O, at the point P, at an angle θ.

Imagine, as in Fig. 3, that a quantum is emitted at the instant $t = 0$, when the radiating body is at O. If it traverses the moving observer's sphere at an angle θ', it traverses the corresponding oblate spheroid, imagined by the stationary observer to be moving with the body, at an angle θ''. After 1 second, the quantum will have reached some point P on a sphere of radius c drawn about O, while the radiator will have moved a distance βc. The stationary observer at P therefore finds that the radiation is coming to him from the point O, at an angle θ with the direction of motion. That is, if the moving observer considers the quantum to be emitted at an angle θ' with the direction of motion, to the stationary observer the angle appears to be θ, where

$$\sin\theta/\sqrt{1+\beta^2-2\beta\cos\theta} = \sin\theta'', \qquad (11)$$

and θ'' is given in terms of θ' by Eq. (10). It follows that

$$\sin\theta' = \sin\theta\frac{\sqrt{1-\beta^2}}{1-\beta\cos\theta}\cdot \qquad (12)$$

On differentiating Eq. (12) we obtain

$$d\theta' = \frac{\sqrt{1-\beta^2}}{1-\beta\cos\theta}d\theta. \qquad (13)$$

The probability that a given quantum will appear to the stationary observer to be emitted between the angles θ and $\theta + d\theta$ is therefore

$$P_\theta d\theta = P_{\theta'}d\theta' = \tfrac{1}{2}\sin\theta' d\theta',$$

where the values of $\sin\theta'$ and $d\theta'$ are given by Eqs. (12) and (13). Substituting these values we find

$$P_\theta d\theta = \frac{1-\beta^2}{(1-\beta\cos\theta)^2}\cdot\tfrac{1}{2}\sin\theta d\theta. \qquad (14)$$

Suppose the moving observer notices that n' quanta are emitted per second. The stationary observer will estimate the rate of emission as

$$n'' = n'\sqrt{1-\beta^2},$$

quanta per second, because of the difference in rate of the moving and stationary clocks. Of these n'' quanta, the number which are emitted between angles θ and $\theta + d\theta$ is $dn'' = n''\cdot P_\theta d\theta$. But if dn'' per second are emitted at the angle θ, the number per second received by a stationary observer at this angle is $dn = dn''/(1 - \beta\cos\theta)$, since the radiator is approaching the observer at a velocity $\beta\cos\theta$. The energy of each quantum is, however, $h\nu_\theta$, where ν_θ is the frequency of the radiation as received by the stationary observer.[1] Thus the intensity, or the energy per unit area per unit time, of the radiation received at an angle θ and a distance R is

$$I_\theta = \frac{h\nu_\theta \cdot dn}{2\pi R^2 \sin\theta d\theta} = \frac{h\nu_\theta}{2\pi R^2 \sin\theta d\theta} \frac{n'(1-\beta^2)^{3/2}}{(1-\beta\cos\theta)^3} \tfrac{1}{2}\sin\theta d\theta$$
$$= \frac{n'h\nu_\theta}{4\pi R^2} \frac{(1-\beta^2)^{3/2}}{(1-\beta\cos\theta)^3}. \tag{15}$$

If the frequency of the oscillator emitting the radiation is measured by an observer moving with the radiator as ν', the stationary observer judges its frequency to be $\nu'' = \nu'\sqrt{1-\beta^2}$, and, in virtue of the Doppler effect, the frequency of the radiation received at an angle θ is

$$\nu_\theta = \nu''/(1-\beta\cos\theta) = \nu'[\sqrt{1-\beta^2}/(1-\beta\cos\theta)]. \tag{16}$$

Substituting this value of ν_θ in Eq. (15) we find

$$I_\theta = \frac{n'h\nu'}{4\pi R^2} \frac{(1-\beta^2)^2}{(1-\beta\cos\theta)^4}. \tag{17}$$

But the intensity of the radiation observed by the moving observer at a distance R from the source is $I' = n'h\nu'/4\pi R^2$. Thus,

$$I_\theta = I'[(1-\beta)^2/(1-\beta\cos\theta)^4] \tag{18}$$

is the intensity of the radiation received at an angle θ with the direction of motion of an isotropic radiator, which moves with a velocity βc, and which would radiate with intensity I' if it were at rest.[2]

It is interesting to note, on comparing Eqs. (16) and (18), that

$$I_\theta/I' = (\nu_\theta/\nu')^4. \tag{19}$$

[1] At first sight the assumption that the quantum which to the moving observer had energy $h\nu'$ will be $h\nu$ for the stationary observer seems inconsistent with the energy principle. When one considers, however, the work done by the moving body against the back-pressure of the radiation, it is found that the energy principle is satisfied. The conclusion reached by the present method of calculation is in exact accord with that which would be obtained according to Lorenz's equations, by considering the radiation to consist of electromagnetic waves.

[2] G. H. Livens gives for I_θ/I' the value $(1-\beta\cos\theta)^{-2}$ ("The Theory of Electricity," p. 600, 1918). At small velocities this value differs from the one here obtained by the factor $(1-\beta\cos\theta)^{-2}$. The difference is due to Livens' neglect of the concentration of the radiation in the small angles, as expressed by our Eq. (14). Cunningham ("The Principle of Relativity," p. 60, 1914) shows that if a plane wave is emitted by a radiator moving in the direction of propagation with a velocity βc, the intensity I received by a stationary observer is greater than the intensity I' estimated by the moving observer, in the ratio $(1-\beta^2)/(1-\beta)^2$, which is in accord with the value calculated according to the methods here employed.

The change in frequency given in Eq. (16) is that of the usual relativity theory. I have not noticed the publication of any result which is the equivalent of my formula (18) for the intensity of the radiation from a moving body.

This result may be obtained very simply for the total radiation from a black body, which is a special case of an isotropic radiator. For, suppose such a radiator is moving so that the frequency of maximum intensity which to a moving observer is ν_m' appears to the stationary observer to be ν_m. Then according to Wien's law, the apparent temperature T, as estimated by the stationary observer, is greater than the temperature T' for the moving observer by the ratio $T/T' = \nu_m/\nu_m'$. According to Stefan's law, however, the intensity of the total radiation from a black body is proportional to T^4; hence, if I and I' are the intensities of the radiation as measured by the stationary and the moving observers respectively,

$$I/I' = (T/T')^4 = (\nu_m/\nu_m')^4. \tag{20}$$

The agreement of this result with Eq. (19) may be taken as confirming the correctness of the latter expression.

The intensity of scattering from recoiling electrons.—We have seen that the change in frequency of the radiation scattered by the recoiling electrons is the same as if the radiation were scattered by electrons moving in the direction of propagation with an effective velocity $\bar{\beta} = \alpha/(1 + \alpha)$, where $\alpha = h/mc\lambda_0$. It seems obvious that since these two methods of calculation result in the same change in wave-length, they must also result in the same change in intensity of the scattered beam. This assumption is supported by the fact that we find, as in Eq. 19, that the change in intensity is in certain special cases a function only of the change in frequency. I have not, however, succeeded in showing rigidly that if two methods of scattering result in the same relative wave-lengths at different angles, they will also result in the same relative intensity at different angles. Nevertheless, we shall assume that this proposition is true, and shall proceed to calculate the relative intensity of the scattered beam at different angles on the hypothesis that the scattering electrons are moving in the direction of the primary beam with a velocity $\bar{\beta} = \alpha/(1 + \alpha)$. If our assumption is correct, the results of the calculation will apply also to the scattering by recoiling electrons.

To an observer moving with the scattering electron, the intensity of the scattering at an angle θ', according to the usual electrodynamics, should be proportional to $(1 + \cos^2 \theta')$, if the primary beam is unpolarized. On the quantum theory, this means that the probability that a quantum will be emitted between the angles θ' and $\theta' + d\theta'$ is proportional to $(1 + \cos^2 \theta')\cdot\sin \theta' d\theta'$, since $2\pi \sin \theta' d\theta'$ is the solid angle included between θ' and $\theta' + d\theta'$. This may be written $P_{\theta'} d\theta' = k(1 + \cos^2 \theta') \sin \theta' d\theta'$.

The factor of proportionality k may be determined by performing the integration

$$\int_0^\pi P_{\theta'} d\theta' = k\int_0^\pi (1 + \cos^2 \theta') \sin \theta' d\theta' = 1,$$

with the result that $k = 3/8$. Thus

$$P_{\theta'} d\theta' = (3/8)(1 + \cos^2 \theta') \sin \theta' d\theta' \tag{21}$$

is the probability that a quantum will be emitted at the angle θ' as measured by an observer moving with the scattering electron.

To the stationary observer, however, the quantum ejected at an angle θ' appears to move at an angle θ with the direction of the primary beam, where $\sin \theta'$ and $d\theta'$ are as given in Eqs. (12) and (13). Substituting these values in Eq. (21), we find for the probability that a given quantum will be scattered between the angles θ and $\theta + d\theta$,

$$P_\theta d\theta = \tfrac{3}{8} \sin \theta d\theta \frac{(1 - \beta^2)\{(1 + \beta^2)(1 + \cos^2 \theta) - 4\beta \cos \theta\}}{(1 - \beta \cos \theta)^4}. \tag{22}$$

Suppose the stationary observer notices that n quanta are scattered per second. In the case of the radiator emitting n'' quanta per second while approaching the observer, the n''th quantum was emitted when the radiator was nearer the observer, so that the interval between the receipt of the 1st and the n''th quantum was less than a second. That is, more quanta were received per second than were emitted in the same time. In the case of scattering, however, though we suppose that each scattering electron is moving forward, the nth quantum is scattered by an electron starting from the same position as the 1st quantum. Thus the number of quanta received per second is also n.

We have seen (Eq. 3) that the frequency of the quantum received at an angle θ is $\nu_\theta = \nu_0/(1 + 2\alpha \sin^2 \frac{1}{2}\theta) = \nu_0/\{1 + \alpha(1 - \cos \theta)\}$, where ν_0, the frequency of the incident beam, is also the frequency of the ray scattered in the direction of the incident beam. The energy scattered per second at the angle θ is thus $nh\nu_\theta P_\theta d\theta$, and the intensity, or energy per second per unit area, of the ray scattered to a distance R is

$$I_\theta = \frac{nh\nu_\theta P_\theta d\theta}{2\pi R^2 \sin \theta d\theta}$$

$$= \frac{nh}{2\pi R^2} \cdot \frac{\nu_0}{1+\alpha(1-\cos \theta)} \cdot \frac{3}{8} \cdot \frac{(1-\beta^2)\{(1+\beta^2)(1+\cos^2 \theta)-4\beta \cos \theta\}}{(1-\beta \cos \theta)^4}.$$

Substituting for β its value $\alpha/(1 + \alpha)$, and reducing, this becomes

$$I = \frac{3nh\nu_0}{16\pi R} \frac{(1 + 2\alpha)\{1 + \cos^2 \theta + 2\alpha(1 + \alpha)(1 - \cos \theta)^2\}}{(1 + \alpha - \alpha \cos \theta)^5}. \tag{23}$$

In the forward direction, where $\theta = 0$, the intensity of the scattered beam is thus

$$I_0 = \frac{3}{8\pi}\frac{nh\nu_0}{R^2}(1 + 2\alpha). \tag{24}$$

Hence

$$\frac{I_\theta}{I_0} = \frac{1}{2}\frac{1 + \cos^2\theta + 2\alpha(1 + \alpha)(1 - \cos\theta)^2}{\{1 + \alpha(1 - \cos\theta)\}^5}. \tag{25}$$

On the hypothesis of recoiling electrons, however, for a ray scattered directly forward, the velocity of recoil is zero (Eq. 6). Since in this case the scattering electron is at rest, the intensity of the scattered beam should be that calculated on the basis of the classical theory, namely,

$$I_0 = I(Ne^4/R^2m^2c^4), \tag{26}$$

where I is the intensity of the primary beam traversing the N electrons which are effective in scattering. On combining this result with Eq. (25), we find for the intensity of the X-rays scattered at an angle θ with the incident beam,

$$I_\theta = I\frac{Ne^4}{2R^2m^2c^4}\frac{1 + \cos^2\theta + 2\alpha(1 + \alpha)(1 - \cos\theta)^2}{\{1 + \alpha(1 - \cos\theta)\}^5}. \tag{27}$$

The calculation of the energy removed from the primary beam may now be made without difficulty. We have supposed that n quanta are scattered per second. But on comparing Eqs. (24) and (26), we find that

$$n = \frac{8\pi}{3}\frac{INe^4}{h\nu_0 m^2c^4(1 + 2\alpha)}.$$

The energy removed from the primary beam per second is $nh\nu_0$. If we define *the scattering absorption coefficient* as the fraction of the energy of the primary beam removed by the scattering process per unit length of path through the medium, it has the value

$$\sigma = \frac{nh\nu_0}{I} = \frac{8\pi}{3}\frac{Ne^4}{m^2c^4}\cdot\frac{1}{1 + 2\alpha} = \frac{\sigma_0}{1 + 2\alpha}, \tag{28}$$

where N is the number of scattering electrons per unit volume, and σ_0 is the scattering coefficient calculated on the basis of the classical theory.[1]

In order to determine the total energy truly scattered, we must integrate the scattered intensity over the surface of a sphere surrounding the scattering material, *i.e.*, $\epsilon_s = \int_0^\pi I_\theta \cdot 2\pi R^2 \sin\theta d\theta$. On substituting the value of I_θ from Eq. (27), and integrating, this becomes

$$\epsilon_s = \frac{8\pi}{3}\frac{INe^4}{m^2c^4}\frac{1 + \alpha}{(1 + 2\alpha)^2}.$$

[1] Cf. J. J. Thomson, "Conduction of Electricity through Gases," 2d ed., p. 325.

The *true scattering coefficient* is thus

$$\sigma_s = \frac{8\pi}{3}\frac{Ne^4}{m^2c^4}\frac{1+\alpha}{(1+2\alpha)^2} = \sigma_0 \frac{1+\alpha}{(1+2\alpha)^2}. \tag{29}$$

It is clear that the difference between the total energy removed from the primary beam and that which reappears as scattered radiation is the energy of recoil of the scattering electrons. This difference represents, therefore, a type of true absorption resulting from the scattering process. The corresponding *coefficient of true absorption due to scattering* is

$$\sigma_a = \sigma - \sigma_s = \frac{8\pi}{3}\frac{Ne^4}{m^2c^4}\frac{\alpha}{(1+2\alpha)^2} = \sigma_0 \frac{\alpha}{(1+2\alpha)^2}. \tag{30}$$

Experimental Test.

Let us now investigate the agreement of these various formulas with exper ments on the change of wave-length due to scattering, and on the magnitude of the scattering of X-rays and γ-rays by light elements.

Wave-length of the scattered rays.—If in Eq. (5) we substitute the accepted values of h, m, and c, we obtain

$$\lambda_\theta = \lambda_0 + 0.0484 \sin^2 \tfrac{1}{2}\theta, \tag{31}$$

if λ is expressed in Angström units. It is perhaps surprising that the increase should be the same for all wave-lengths. Yet, as a result of an extensive experimental study of the change in wave-length on scattering, the writer has concluded that "over the range of primary rays from 0.7 to 0.025 A, the wave-length of the secondary X-rays at 90° with the incident beam is roughly 0.03 A greater than that of the primary beam which excites it." [1] Thus the experiments support the theory in showing a wave-length increase which seems independent of the incident wave-length, and which also is of the proper order of magnitude.

A quantitative test of the accuracy of Eq. (31) is possible in the case of the characteristic K-rays from molybdenum when scattered by graphite. In Fig. 4 is shown a spectrum of the X-rays scattered by graphite at right angles with the primary beam, when the graphite is traversed by X-rays from a molybdenum target.[2] The solid line represents the spectrum of these scattered rays, and is to be compared with the broken line, which represents the spectrum of the primary rays, using the same slits and crystal, and the same potential on the tube. The primary spectrum is, of course, plotted on a much smaller scale than

[1] A. H. Compton, Bull. N. R. C., No. 20, p. 17 (1922).

[2] It is hoped to publish soon a description of the experiments on which this figure is based.

the secondary. The zero point for the spectrum of both the primary and secondary X-rays was determined by finding the position of the first order lines on both sides of the zero point.

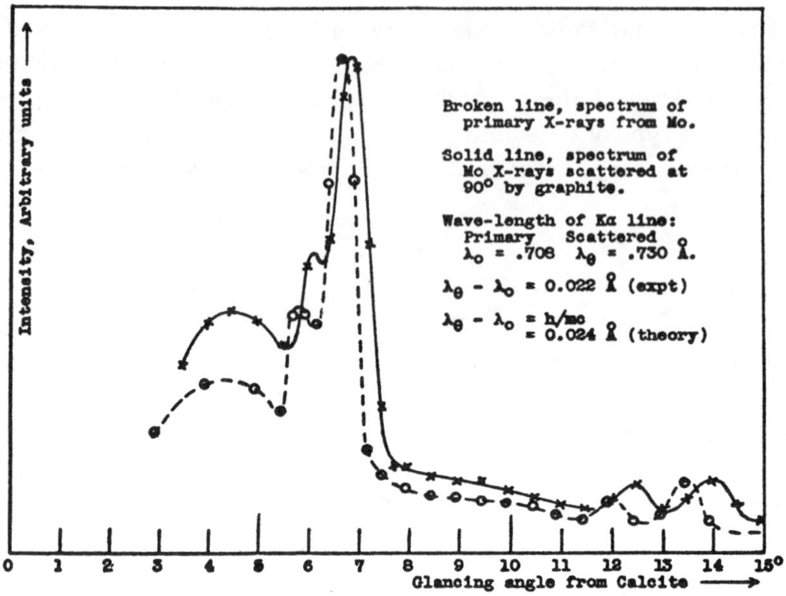

Fig. 4. Spectrum of molybdenum X-rays scattered by graphite, compared with the spectrum of the primary X-rays, showing an increase in wave-length on scattering.

It will be seen that the wave-length of the scattered rays is unquestionably greater than that of the primary rays which excite them. Thus the $K\alpha$ line from molybdenum has a wave-length 0.708 A. The wave-length of this line in the scattered beam is found in these experiments, however, to be 0.730 A. That is,

$$\lambda_\theta - \lambda_0 = 0.022 \text{ A} \quad \text{(experiment)}.$$

But according to the present theory (Eq. 5),

$$\lambda_\theta - \lambda_0 = 0.0484 \sin^2 45^\circ = 0.024 \text{ A} \quad \text{(theory)},$$

which is a very satisfactory agreement.

The variation in wave-length of the scattered beam with the angle is illustrated in the case of γ-rays. The writer has measured[1] the mass absorption coefficient in lead of the rays scattered at different angles when various substances are traversed by the hard γ-rays from RaC. The mean results for iron, aluminium and paraffin are given in column 2 of Table I. This variation in absorption coefficient corresponds to a

[1] A. H. Compton, Phil. Mag. 41, 760 (1921).

difference in wave-length at the different angles. Using the value given by Hull and Rice for the mass absorption coefficient in lead for wave-length 0.122, 3.0, remembering[1] that the characteristic fluorescent absorption τ/ρ is proportional to λ^3, and estimating the part of the absorption due to scattering by the method described below, I find for the wave-lengths corresponding to these absorption coefficients the values given in the fourth column of Table I. That this extrapolation is very

TABLE I

Wave-length of Primary and Scattered γ-rays

	Angle	μ/ρ	τ/ρ	λ obs.	λ calc.
Primary..........	0°	.076	.017	0.022 A	(0.022 A)
Scattered.........	45°	.10	.042	.030	0.029
"	90°	.21	.123	.043	0.047
"	135°	.59	.502	.068	0.063

nearly correct is indicated by the fact that it gives for the primary beam a wave-length 0.022 A. This is in good accord with the writer's value 0.025 A, calculated from the scattering of γ-rays by lead at small angles,[2] and with Ellis' measurements from his β-ray spectra, showing lines of wave-length .045, .025, .021 and .020 A, with line .020 the strongest.[3] Taking $\lambda_0 = 0.022$ A, the wave-lengths at the other angles may be calculated from Eq. (31). The results, given in the last column of Table I., and shown graphically in Fig. 5, are in satisfactory accord with the measured values. There is thus good reason for believing that Eq. (5) represents accurately the wave-length of the X-rays and γ-rays scattered by light elements.

Velocity of recoil of the scattering electrons.—The electrons which recoil in the process of the scattering of ordinary X-rays have not been observed. This is probably because their number and velocity is usually small compared with the number and velocity of the photoelectrons ejected as a result of the characteristic fluorescent absorption. I have pointed out elsewhere,[4] however, that there is good reason for believing that most of the secondary β-rays excited in light elements by the action of γ-rays are such recoil electrons. According to Eq. (6), the velocity of these electrons should vary from 0, when the γ-ray is scattered forward, to $v_{max} = \beta_{max}c = 2c\alpha[(1 + \alpha)/(1 + 2\alpha + 2\alpha^2)]$, when the γ-ray quantum

[1] Cf. L. de Broglie, Jour. de Phys. et Rad. 3, 33 (1922); A. H. Compton, Bull. N. R. C., No. 20, p. 43 (1922).
[2] A. H. Compton, Phil. Mag. 41, 777 (1921).
[3] C. D. Ellis, Proc. Roy. Soc. A, 101, 6 (1922).
[4] A. H. Compton, Bull. N. R. C., No. 20, p. 27 (1922).

is scattered backward. If for the hard γ-rays from radium C, $\alpha = 1.09$, corresponding to $\lambda = 0.022$ A, we thus obtain $\beta_{max} = 0.82$. The effective velocity of the scattering electrons is, therefore (Eq. 8), $\bar{\beta} = 0.52$. These results are in accord with the fact that the average velocity of the

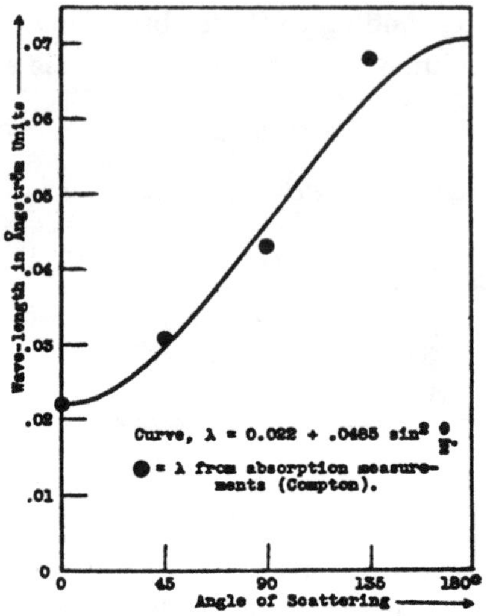

Fig. 5. The wave-length of scattered γ-rays at different angles with the primary beam, showing an increase at large angles similar to a Doppler effect.

β-rays excited by the γ-rays from radium is somewhat greater than half that of light.[1]

Absorption of X-rays due to scattering.—Valuable information concerning the magnitude of the scattering is given by the measurements of the absorption of X-rays due to scattering. Over a wide range of wave-lengths, the formula for the total mass absorption, $\mu/\rho = \kappa\lambda^3 + \sigma/\rho$, is found to hold, where μ is the linear absorption coefficient, ρ is the density, κ is a constant, and σ is the energy loss due to the scattering process. Usually the term $\kappa\lambda^3$, which represents the fluorescent absorption, is the more important; but when light elements and short wave-lengths are employed, the scattering process accounts for nearly all the energy loss. In this case, the constant κ can be determined by measurements on the longer wave-lengths, and the value of σ/ρ can then be estimated with considerable accuracy for the shorter wave-lengths from the observed values of μ/ρ.

Hewlett has measured the total absorption coefficient for carbon over a wide range of wave-lengths.[2] From his data for the longer wave-

[1] E. Rutherford, Radioactive Substances and their Radiations, p. 273.

[2] C. W. Hewlett, Phys. Rev. 17, 284 (1921).

lengths I estimate the value of κ to be 0.912, if λ is expressed in A. On subtracting the corresponding values of $\kappa\lambda^3$ from his observed values of μ/ρ, the values of σ/ρ represented by the crosses of Fig. 6 are obtained.

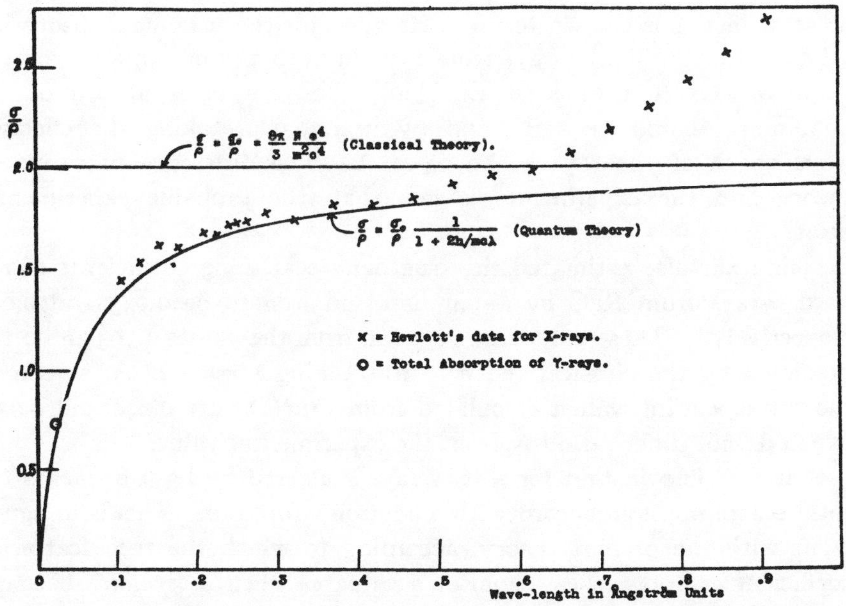

Fig. 6. The absorption in carbon due to scattering, for homogeneous X-rays.

The value of σ_0/ρ as calculated for carbon from Thomson's formula is shown by the horizontal line at $\sigma/\rho = 0.201$. The values of σ/ρ calculated from Eq. (28) are represented by the solid curve. The circle shows the experimental value of the total absorption of γ-rays by carbon, which on the present view is due wholly to the scattering process.

For wave-lengths less than 0.5 A, where the test is most significant, the agreement is perhaps within the experimental error. Experiments by Owen,[1] Crowther,[2] and Barkla and Ayers[3] show that at about 0.5 A the "excess scattering" begins to be appreciable, increasing rapidly in importance at the longer wave-lengths.[4] It is probably this effect which results in the increase of the scattering absorption above the theoretical value for the longer wave-lengths. Thus the experimental values of the absorption due to scattering seem to be in satisfactory accord with the present theory.

True absorption due to scattering has not been noticed in the case of

[1] E. A. Owen, Proc. Camb. Phil. Soc. 16, 165 (1911).
[2] J. A. Crowther, Proc. Roy. Soc. 86, 478 (1912).
[3] Barkla and Ayers, Phil. Mag. 21, 275 (1911).
[4] Cf. A. H. Compton, Washington University Studies, 8, 109 ff. (1921).

X-rays. In the case of hard γ-rays, however, Ishino has shown [1] that there is true absorption as well as scattering, and that for the lighter elements the true absorption is proportional to the atomic number. That is, this absorption is proportional to the number of electrons present, just as is the scattering. He gives for the true mass absorption coefficient of the hard γ-rays from RaC in both aluminium and iron the value 0.021. According to Eq. (30), the true mass absorption by aluminium should be 0.021 and by iron, 0.020, taking the effective wave-length of the rays to be 0.022 A. The difference between the theory and the experiments is less than the probable experimental error.

Ishino has also estimated the true mass scattering coefficients of the hard γ-rays from RaC by aluminium and iron to be 0.045 and 0.042 respectively.[2] These values are very far from the values 0.193 and 0.187 predicted by the classical theory. But taking $\lambda = 0.022$ A, as before, the corresponding values calculated from Eq. (29) are 0.040 and 0.038, which do not differ seriously from the experimental values.

It is well known that for soft X-rays scattered by light elements the total scattering is in accord with Thomson's formula. This is in agreement with the present theory, according to which the true scattering coefficient σ_s approaches Thomson's value σ_0 when $\alpha \equiv h/mc\lambda$ becomes small (Eq. 29).

The relative intensity of the X-rays scattered in different directions with the primary beam.—Our Eq. (27) predicts a concentration of the energy in the forward direction. A large number of experiments on the scattering of X-rays have shown that, except for the excess scattering at small angles, the ionization due to the scattered beam is symmetrical on the emergence and incidence sides of a scattering plate. The difference in intensity on the two sides according to Eq. (27) should, however, be noticeable. Thus if the wave-length is 0.7 A, which is probably about that used by Barkla and Ayers in their experiments on the scattering by carbon,[3] the ratio of the intensity of the rays scattered at 40° to that at 140° should be about 1.10. But their experimental ratio was 1.04, which differs from our theory by more than their probable experimental error.

It will be remembered, however, that our theory, and experiment also, indicates a difference in the wave-length of the X-rays scattered in different directions. The softer X-rays which are scattered backward are the more easily absorbed and, though of smaller intensity, may produce an

[1] M. Ishino, Phil. Mag. 33, 140 (1917).
[2] M. Ishino, loc. cit.
[3] Barkla and Ayers, loc. cit.

ionization equal to that of the beam scattered forward. Indeed, if α is small compared with unity, as is the case for ordinary X-rays, Eq. (27) may be written approximately $I_\theta/I_\theta' = (\lambda_0/\lambda_\theta)^3$, where I_θ' is the intensity of the beam scattered at the angle θ according to the classical theory. The part of the absorption which results in ionization is however proportional to λ^3. Hence if, as is usually the case, only a small part of the X-rays entering the ionization chamber is absorbed by the gas in the chamber, the ionization is also proportional to λ^3. Thus if i_θ represents the ionization due to the beam scattered at the angle θ, and if i_θ' is the corresponding ionization on the classical theory, we have $i_\theta/i_\theta' = (I_\theta/I_\theta')(\lambda_\theta/\lambda_0)^3 = 1$, or $i_\theta = i_\theta'$. That is, to a first approximation, the ionization should be the same as that on the classical theory, though the energy in the scattered beam is less. This conclusion is in good accord with the experiments which have been performed on the scattering of ordinary X-rays, if correction is made for the excess scattering which appears at small angles.

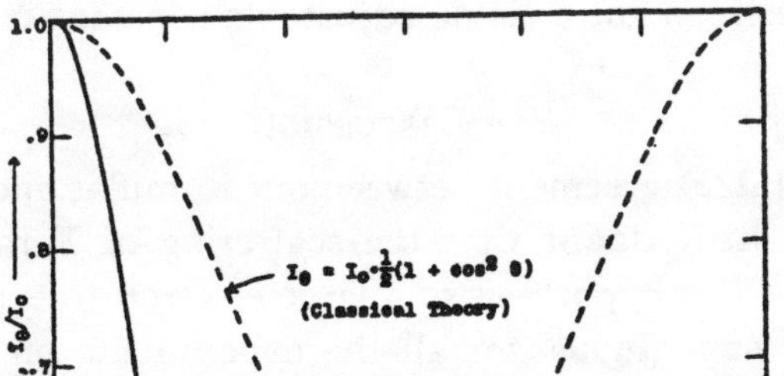

Fig. 7. Comparison of experimental and theoretical intensities of scattered γ-rays.

In the case of very short wave-lengths, however, the case is different. The writer has measured the γ-rays scattered at different angles by iron, using an ionization chamber so designed as to absorb the greater part of even the primary γ-ray beam.[1] It is not clear just how the ionization due to the γ-rays will vary with the wave-length under the conditions of the experiment, but it appears probable that the variation will not be great. If we suppose accordingly that the ionization measures the intensity of the scattered γ-ray beam, these data for the intensity are represented by the circles in Fig. 7. The experiments showed that the intensity at 90° was 0.074 times that predicted by the classical theory, or 0.037 I_0, where I_0 is the intensity of the scattering at the angle $\theta = 0$ as calculated on either the classical or the quantum theory. The absolute intensities of the scattered beam are accordingly plotted using I_0 as the unit. The solid curve shows the intensity in the same units, calculated according to Eq. (27). As before, the wave-length of the γ-rays is taken as 0.022 A. The beautiful agreement between the theoretical and the experimental values of the scattering is the more striking when one notices that there is not a single adjustable constant connecting the two sets of values.

Discussion

This remarkable agreement between our formulas and the experiments can leave but little doubt that the scattering of X-rays is a quantum phenomenon. The hypothesis of a large electron to explain these effects is accordingly superfluous, for all the experiments on X-ray scattering to which this hypothesis has been applied are now seen to be explicable from the point of view of the quantum theory without introducing any new hypotheses or constants. In addition, the present theory accounts satisfactorily for the change in wave-length due to scattering, which was left unaccounted for on the hypothesis of the large electron. From the standpoint of the scattering of X-rays and γ-rays, therefore, there is no longer any support for the hypothesis of an electron whose diameter is comparable with the wave-length of hard X-rays.

The present theory depends essentially upon the assumption that each electron which is effective in the scattering scatters a complete quantum. It involves also the hypothesis that the quanta of radiation are received from definite directions and are scattered in definite directions. The experimental support of the theory indicates very convincingly that a radiation quantum carries with it directed momentum as well as energy.

Emphasis has been laid upon the fact that in its present form the

[1] A. H. Compton, Phil. Mag. **41**, 758 (1921).

quantum theory of scattering applies only to light elements. The reason for this restriction is that we have tacitly assumed that there are no forces of constraint acting upon the scattering electrons. This assumption is probably legitimate in the case of the very light elements, but cannot be true for the heavy elements. For if the kinetic energy of recoil of an electron is less than the energy required to remove the electron from the atom, there is no chance for the electron to recoil in the manner we have supposed. The conditions of scattering in such a case remain to be investigated.

The manner in which interference occurs, as for example in the cases of excess scattering and X-ray reflection, is not yet clear. Perhaps if an electron is bound in the atom too firmly to recoil, the incident quantum of radiation may spread itself over a large number of electrons, distributing its energy and momentum among them, thus making interference possible. In any case, the problem of scattering is so closely allied with those of reflection and interference that a study of the problem may very possibly shed some light upon the difficult question of the relation between interference and the quantum theory.

Many of the ideas involved in this paper have been developed in discussion with Professor G. E. M. Jauncey of this department.

18

EXPERIMENTAL TEST OF THE THEORY OF BOHR, KRAMERS, AND SLATER

W. Bothe and H. Geiger

This article was translated expressly for this Benchmark volume by R. Bruce Lindsay, Brown University, from "Experimentelles zur Theorie von Bohr, Kramers und Slater," Naturwissenschaften **13**: 440–441 (1925), *by permission of the publisher.*

N. Bohr, H. A. Kramers, and J. C. Slater[1] have recently given an explanation of the Compton effect, which is of fundamental significance for our understanding of the nature of radiation. The most significant difference from the older representation is the assumption that in an elementary scattering process, the principles of conservation of energy and momentum do not hold in general, but only on the average for a large number of scattering processes. Bohr thought of this point of view as the logical consequence of more general hypotheses on the interaction between radiation and matter through which the junction of quantum theory with the classical wave theory of radiation could be realized.

A decision between the two interpretations of the Compton effect can be obtained experimentally by recording with point counters (an $h\nu$ counter and a charge e counter) the scattered radiation and associated recoil electrons in hydrogen.

The accompanying diagram indicates schematically the equipment used in our experiment.

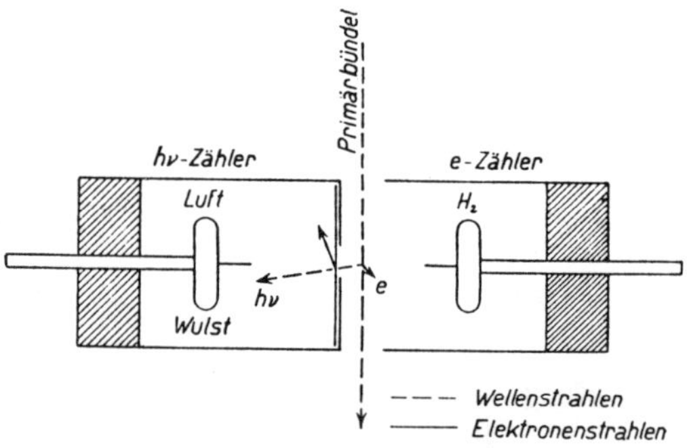

(Reprinted from *Naturwissenschaften* **13**: 441 (1975); copyright ©1925 by Springer-Verlag, Berlin, Heidelberg, New York)

In the hydrogen chamber placed very close to the $h\nu$ counter, a very small fraction of the incident primary radiation is scattered and is recorded in the $h\nu$ counter, insofar as this scattered radiation is transformed in the platinum foil surrounding the counter into electron-radiation-producing ions. The associated recoil electrons

1. N. Bohr, H. A. Kramers, and J. C. Slater, *Philos. Mag.*, **47**, 785–802, 1924.

were then indicated by the *e* counter. In an ideal experimental arrangement, the viewpoint of Compton and Debye would lead to the appearance of an *e* count simultaneously with each $h\nu$ count. In reality, because of the unavoidable incompleteness of the experimental arrangement, we could expect a coincidence in only about 10 $h\nu$ counts, as several trials confirmed. On the other hand, on the Bohr viewpoint, coincidences are practically excluded entirely. Our observations gave a coincidence for about every 11 $h\nu$ counts if we deduced for the coincidences to be expected on pure chance whose frequency can be readily computed from the magnitude of the critical time intervals within which two counts can be considered as coinciding. The probability that all observed coincidences could be simulated by a chance coincidence of counts turns out to be scarcely 10^{-5} in the most unfavorable case. Hence we conclude that our results definitely favor the earlier hypothesis (of Compton and Debye).

The precise observation of the coincidences was essential for the success of the investigation since obviously we must always expect a certain number of more or less sharp chance coincidences. A considerable difficulty arose from the fact that in its usual form, the point counter operated with variable delays up to 1/100 s. The reason for this was that the ions leading to the current impulse were produced at a variable distance from the point, depending on the direction of the entering radiation particles (quanta), and hence a perceptible time could elapse before they reached the effective vicinity of the point after passing through the relatively weak field. Such delays could be avoided in practice by superposing on the point field a homogeneous supplementary field produced by a metal pad placed behind the point. By this field the ions from every part of the counter space could be brought to the point in the shortest possible time. When the delays were cut down in this way, the critical time interval could be reduced to 1/1000 s. The measurement of time intervals of this minuteness could be carried out with an accuracy of 1/10,000 s.

The complete publication will appear shortly in the *Zeitschrift für Physik.*

Part V
THE ELECTRON

Editor's Comments on Papers 19 and 20

19 WIEDEMANN and EBERT
Excerpt from *On Electrical Discharges*

20 THOMSON
Cathode Rays

An awareness of the association of electrical phenomena with the atomic point of view dates back to Michael Faraday's (1791–1867) experiments on electrolysis in 1832. As a consequence of these experiments, Faraday proposed that a specific amount of electricity is attached to each ion transferred in a solution. In 1874 the Irish physicist George Johnston Stoney (1826–1911) suggested that a specific name be assigned to represent the postulated fundamental unit of electricity associated with the positive hydrogen ion. Work by Faraday, Geissler, Hittorf, and Goldstein on discharges in gases at low pressure led to the discovery during the 1870s of cathode rays, which appear to move in straight lines from cathode to anode in the discharge tube. By 1880 William Crookes (1832–1919) had proposed on the basis of the deflection of these rays by a magnetic field that they are actually streams of fast-moving particles that have mass and carry negative electric charges. The electrified-particle theory was not accepted by all scientists of that time, particularly those in the German universities. Some believed the cathode rays were a form of short-wavelength light. Among those holding this view were two German physicists, Eilhard Wiedemann (1852–1928) and his assistant Hermann Ebert (1861–1913). Although their assumption as to the character of cathode rays was subsequently proven incorrect by the work of Kaufmann, J. J. Thomson, and others, they did make an important contribution to the dynamical description of cathode rays by making calorimetric measurements of their energies for the first time. The portion of their 1892 paper containing these results is reproduced here as Paper 19, "The Energy of Cathode Rays," with explanatory editorial notes.

Eilhard Wiedemann was born in Berlin and spent most of his professional career at the University of Erlangen. He was the son of Gustav Wiedemann (1826–1899), who is remembered for his contribution to the formulation of the Wiedemann-Franz law relating electrical to thermal conductivity of metals (see Paper 31). Hermann Ebert was born in Leipzig and took his doctorate at Erlangen where he was assistant to Wiedemann for several years. During that period he and Wiedemann carried out the studies that are reported in Paper 19. Ebert was a skillful theoretical and experimental physicist and subsequently went on to hold professorships at Kiel and the Technical University of Munich. The significance of the Wiedemann-Ebert experiment stems from its revelation of the very large amounts of concentrated kinetic energy that the cathode rays possess. At applied voltages of up to 10,000 V in the discharge tube, the rays were shown to have energies per unit cross-sectional area of the order 10^5 times greater than direct sunlight. Hence a glimpse was obtained for the first time of the high degree of energy intensity that could be achieved by the transfer of electrical energy to the kinetic energy of charged particles on the atomic scale.

The year 1895 was important in physics since that was the year Röntgen discovered x-rays. This discovery coupled with the still-unsettled questions concerning the ultimate nature of cathode rays led the eminent British physicist J. J. Thomson to undertake a careful study of electrical discharges in gases. Sir John Joseph Thomson (1856–1940) was born near Manchester, England, to a family of modest means He first obtained an engineering degree at Owens College but was advised to go on to Cambridge and study mathematics. He graduated from Cambridge in 1880 as Second Wrangler in the mathematics tripos and was almost immediately offered a fellowship. This marked the beginning of a lifelong association with that university. In 1884 at the age of twenty-eight Thomson was appointed to the Cavendish professorship, which had recently been vacated by Lord Rayleigh. Although a mathematician by training, Thomson quickly took to experimental physics. For the first time research students in large numbers began to work at the Cavendish Laboratory and under Thomson's efficient direction to achieve increasingly significant results. Thomson was awarded the Nobel Prize in physics in 1906 for his work on the discharge of electricity through gases.

With the aid of improved vacuum techniques, Thomson was able to obtain more sharply defined cathode-ray beams than had ever been studied before. He was able to measure with considerable accuracy the ratio of the mass to the charge of the cathode rays on the assumption that they are really streams of particles, and he found that this ratio is always independent of the nature of the gas through which the discharge passes. He finally concluded that the cathode rays are indeed electrified particles representing constituents of the hitherto presumed indivisible atoms. At first he referred to the electrified particle as a "corpuscle" and considered that its charge might be larger than the fundamental unit of electricity postulated earlier as that carried by the hydrogen ion and to which Stoney had suggested in 1891 that the name "electron" be attached. However, as a consequence of additional later experiments involving the photoelectric effect and the emission of particles from surfaces of hot metals, showing that the ratio of charge to mass is here the same as for cathode rays, it was concluded that all the electrified particles in question are really one and the same and furthermore are constituents of all normal atoms. The historical record generally accepts the discovery of the electron as dating from the publication of Thomson's paper in 1897 "Cathode Rays" (Paper 20), even though it was not until several years later that scientific convention began using the term Stoney had suggested earlier. An accurate measurement of the charge and thereby of the mass of the electron was not to be completed until R. A. Millikan (1868–1953), an American, performed the oil-drop experiment in 1912.

At the same time that electrons were being explored experimentally, theorists were attempting to modify the electromagnetic field equations of Maxwell to account for the effects of the field on electrified particles. In this connection we call attention to the work of the eminent Dutch physicist Hendrik Anton Lorentz (1853–1928) who was stimulated to enter the field of electrodynamics through his interest in both kinetic theory and electromagnetism. His familiarity with the atomic point of view prompted him to examine more carefully the relationship between the particle sources of radiation, e.g., electrons, and the surrounding medium, the so-called ether of space. In what was a major step for the 1890s Lorentz proposed that electric currents be treated as the flow of electrons or other charged particles. He modified Maxwell's equations by introducing the force on these charges due to electric and magnetic fields, the so-called Lorentz force. He generalized his theory to include effects on and by bound charges in dielectrics. All this, of course, involved consideration of the relation between the energy of the particles and the energy in the field.

An important part of Lorentz's treatment was his accounting for the modification produced in the state of the aether of space by the velocity and acceleration of the charges. The derivation of the forces with which the aether acts on the charged particles is contained in Chapter Four of Lorentz's extensive memoir *Theory of a System of Charged Particles Moving Through the Aether without Entraining it* (Lorentz 1892). Lorentz's electron theory proved most fruitful in dealing with a wide range of problems involving the interaction of matter and radiation. One of the more important of these was the introduction of the equations transforming the electromagnetic equation as seen from the moving electron to those applying to the laboratory frame. These transformation equations were somewhat later to prove of great importance in the formulation of the restricted principle of relativity by Einstein; they are still referred to as the "Lorentz-Einstein transformation equations." Often they also carry the name of George Francis Fitzgerald (1851–1901), the Irish physicist who made extensive researches into optics and electromagnetic theory and who was the author of the so-called Fitzgerald contraction hypothesis in relativity.

The classical description of the electron that treated the particle as a localizable rigid body obeying the classical laws of mechanics and electromagnetism was to prove short lived as the emergence of the quantum theory soon produced radical changes. The key steps that forced new views on physicists resulted from efforts to construct a geometrical model of the atom incorporating the electron and to explain the dynamical behavior of electrons in this new model. The quantization of energy and momentum became decisive considerations as will be pointed out in the next group of papers.

REFERENCE

Lorentz, H. A., 1892. "Theory of a System of Charged Particles Moving Through the Aether Without Entraining It," *Collected Papers of H. A. Lorentz,* Vol. II. The Hague: Martinus Nijhoff, 1936, pp. 228–255.

19

ON ELECTRICAL DISCHARGES

E. Wiedemann and E. Ebert

This excerpt was translated expressly for this Benchmark volume by R. Bruce Lindsay, Brown University, from "Über electrische Entladungen," Sitzungsber. Phys.-Med. Soc. Erlangen **24**: 84–127 (1892).

[*Editorial Introduction:* The article opens with a review of the whole problem of electrical discharges through gases at low pressure. The authors aim is to lay the foundation for the investigation of both the qualitative and quantitative explanation of the phenomena observed in such discharges. There is much preliminary discussion of electrical oscillations, mainly qualitative in character. It is of interest to note the authors' assumption that cathode rays are analogous to very short wavelength light rays, thus adopting a viewpoint challenged by J. J. Thomson in his epoch-making work of a few years later. We reprint here only the section on the energy of cathode rays; the authors' measurement of this was presumably the first such to be performed.]

31. DETERMINATION OF THE ENERGY OF CATHODE RAYS

The lower part *B* of the discharge tube (see attached figure) was placed in a calorimeter filled with turpentine along with a thermometer. The calorimeter was shielded from radiation by a double-walled metal mantle. The tube remained in connection with the pump. The pressure was determined from the number of evacu-

Fig. 2.

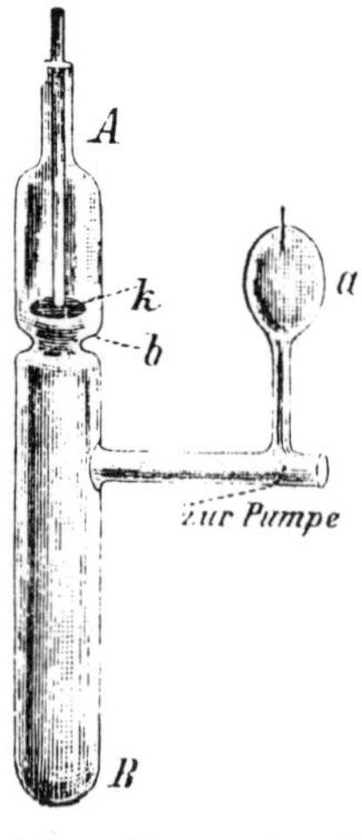

[Reprinted from *Sitzungsber. Phys.-Med. Soc. Erlangen* **24**:103 (1892).]

ations up to the time of use of the discharge, each evacuation having been (at higher pressure) calibrated by a manometer.

[*Editor's Note:* Evidently their manometer did not go down to the very low pressures involved in the discharge.]

The equivalent water values of the calorimeter, the thermometer, the immersed part of the glass tube as well as that of the turpentine were determined in the usual way. The negative electricity was conducted to plate *h* (the cathode). The anode was connected to earth.

a. For pressure between 0.0038 and 0.0014 cm of mercury (i.e., between 50 and 18 dyn/cm^2), the whole lower part of the tube immersed in the turpentine glowed with an intense green light. The observed heat flow to the calorimeter amounted to 3g cal/min. The thermal effect of the cathode rays was accordingly 5×10^{-2} g cal/s or 2×10^5 erg/s.
b. The discharge pictures (343/s) were completely sharp in the rotating mirror. We can therefore assume that the duration of the discharge is not quite 1/1000 of the time between two discharges. This is indeed the upper limit.
c. If we wish to compare the energy of the cathode rays with that of light rays emitted from a continuous, luminous body, we must take into account that the former are emitted during only about 1/1000 part of the total time of observation. If, therefore, the cathode rays with the same intensity as those we observed were to be emitted continuously, the heat produced by them would necessarily be 1000 times greater than that observed, i.e., 3000 g cal/min.
d. In comparing the energy of cathode rays with that of light rays, we must consider the bundles of rays in the two cases to have the same cross-section, e.g., 1 cm^2. At the cathode the cathode ray radiation had a diameter of about 1 mm. Suppose we take the cross-section as 1 mm^2. Then in 1 cm^2 we should have $3 \times 10^3 \times 10^2 = 3 \times 10^5$ g cal/min or 5×10^3 g cal/s cm^2.

32. COMPARISON WITH THE TOTAL DISCHARGE ENERGY

We now investigate how large a part of the energy used in the discharge as a whole turns up in the cathode rays. The potential difference between the ends of the discharge tube was found to be 41 cm$^{1/2}$ g$^{1/2}$ s^{-1} = 12,300 V, and the total charge flowing per second was 3.4×10^{-4} C. Hence the work done by the current in one second was $1.23 \times 10^4 \times 3.4 \times 10^{-4}$ W = 4 J/s. The total energy associated with the flow of current in the discharge tube is therefore about 1 g cal/s or 20 times as large as that of the cathode rays themselves.

33. COMPARISON OF THE ENERGY OF THE CATHODE RAYS WITH THE ENERGY OF SUNLIGHT

To form an appreciation of the energy of cathode rays, we shall compare this to the energy of sunlight.

According to the latest estimate, sunshine outside the earth's atmosphere produces 3 g cal/cm^2/min at normal incidence or 0.05 g cal/cm^{2}s.

[*Editor's Note:* The latest (1970) estimate replaces this with 0.033 g cal/cm^2/s.]

By passage through the atmosphere in the latitude of Germany, about 30% of this is lost. Hence at the earth's surface there falls about 0.035 g cal/s/cm^2, or 1.5×10^6 erg/s cm^2. Hence the energy effect of the cathode rays in our experiment was

$$\frac{5 \times 10^3}{3.5 \times 10^2} = 1.4 \times 10^5$$

times greater than that of the brightest sunlight. This result makes quite understandable the fact that a beam of cathode rays falling on a platinum foil can heat it to white heat quite quickly.

[*Editor's Note:* The authors then go on to estimate from the foregoing the amplitude of cathode rays considered as of the same wave nature as ordinary light. They arrive at the result 10^{-7} cm. But this is not meaningful on the interpretation of cathode rays as a beam of particles. The remainder of the paper concerns other properties of cathode rays such as the pressure they exert on objects on which they fall, the luminescence they produce, and the effect of a magnetic field on them.]

20

Reprinted from *Philos. Mag.* ser. 5, **44**:293–316 (1897)

CATHODE RAYS

J. J. Thomson*

THE experiments † discussed in this paper were undertaken in the hope of gaining some information as to the nature of the Cathode Rays. The most diverse opinions are held as to these rays; according to the almost unanimous opinion of German physicists they are due to some process in the æther to which—inasmuch as in a uniform magnetic field their course is circular and not rectilinear—no phenomenon hitherto observed is analogous: another view of these rays is that, so far from being wholly ætherial, they are in fact wholly material, and that they mark the paths of particles of matter charged with negative electricity. It would seem at first sight that it ought not to be difficult to discriminate between views so different, yet experience shows that this is not the case, as amongst the physicists who have most deeply studied the subject can be found supporters of either theory.

The electrified-particle theory has for purposes of research a great advantage over the ætherial theory, since it is definite and its consequences can be predicted; with the ætherial theory it is impossible to predict what will happen under any given circumstances, as on this theory we are dealing with hitherto

* Communicated by the Author.

† Some of these experiments have already been described in a paper read before the Cambridge Philosophical Society (Proceedings, vol. ix. 1897), and in a Friday Evening Discourse at the Royal Institution ('Electrician,' May 21, 1897).

unobserved phenomena in the æther, of whose laws we are ignorant.

The following experiments were made to test some of the consequences of the electrified-particle theory.

Charge carried by the Cathode Rays.

If these rays are negatively electrified particles, then when they enter an enclosure they ought to carry into it a charge of negative electricity. This has been proved to be the case by Perrin, who placed in front of a plane cathode two coaxial metallic cylinders which were insulated from each other: the outer of these cylinders was connected with the earth, the inner with a gold-leaf electroscope. These cylinders were closed except for two small holes, one in each cylinder, placed so that the cathode rays could pass through them into the inside of the inner cylinder. Perrin found that when the rays passed into the inner cylinder the electroscope received a charge of negative electricity, while no charge went to the electroscope when the rays were deflected by a magnet so as no longer to pass through the hole.

This experiment proves that something charged with negative electricity is shot off from the cathode, travelling at right angles to it, and that this something is deflected by a magnet; it is open, however, to the objection that it does not prove that the cause of the electrification in the electroscope has anything to do with the cathode rays. Now the supporters of the ætherial theory do not deny that electrified particles are shot off from the cathode; they deny, however, that these charged particles have any more to do with the cathode rays than a rifle-ball has with the flash when a rifle is fired. I have therefore repeated Perrin's experiment in a form which is not open to this objection. The arrangement used was as follows:— Two coaxial cylinders (fig. 1) with slits in them are placed in a bulb connected with the discharge-tube; the cathode rays from the cathode A pass into the bulb through a slit in a metal plug fitted into the neck of the tube; this plug is connected with the anode and is put to earth. The cathode rays thus do not fall upon the cylinders unless they are deflected by a magnet. The outer cylinder is connected with the earth, the inner with the electrometer. When the cathode rays (whose path was traced by the phosphorescence on the glass) did not fall on the slit, the electrical charge sent to the electrometer when the induction-coil producing the rays was set in action was small and irregular; when, however, the rays were bent by a magnet so as to fall on the slit there was a large charge of negative electricity sent to the electrometer. I was surprised at the magnitude of the charge; on some occasions enough negative electricity went through the narrow slit into the inner cylinder in one second to alter the potential of a capacity of 1·5 microfarads by 20 volts. If the rays were so

Fig. 1.

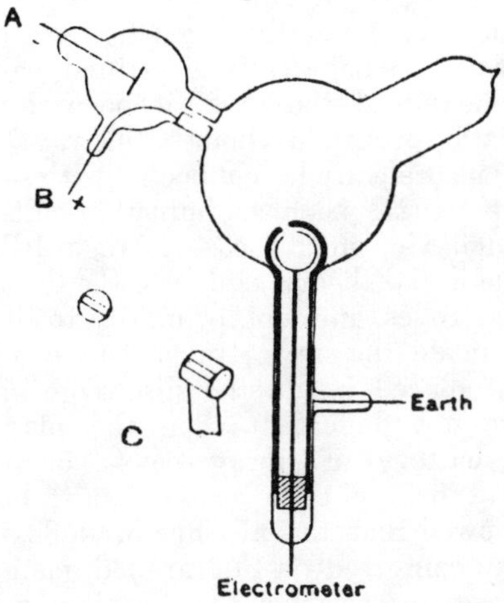

much bent by the magnet that they overshot the slits in the cylinder, the charge passing into the cylinder fell again to a very small fraction of its value when the aim was true. Thus this experiment shows that however we twist and deflect the cathode rays by magnetic forces, the negative electrification follows the same path as the rays, and that this negative electrification is indissolubly connected with the cathode rays.

When the rays are turned by the magnet so as to pass through the slit into the inner cylinder, the deflexion of the electrometer connected with this cylinder increases up to a certain value, and then remains stationary although the rays continue to pour into the cylinder. This is due to the fact that the gas in the bulb becomes a conductor of electricity when the cathode rays pass through it, and thus, though the inner cylinder is perfectly insulated when the rays are not passing, yet as soon as the rays pass through the bulb the air between the inner cylinder and the outer one becomes a conductor, and the electricity escapes from the inner cylinder to the earth. Thus the charge within the inner cylinder does not go on continually increasing; the cylinder settles down into a state of equilibrium in which the rate at which it gains negative electricity from the rays is equal to the rate at which it loses it by conduction through the air. If the inner cylinder has initially a positive charge it rapidly loses that charge and acquires a negative one; while if the initial charge is a negative one, the cylinder will leak if the initial negative potential is numerically greater than the equilibrium value.

Deflexion of the Cathode Rays by an Electrostatic Field.

An objection very generally urged against the view that the cathode rays are negatively electrified particles, is that hitherto no deflexion of the rays has been observed under a small electrostatic force, and though the rays are deflected when they pass near electrodes connected with sources of large differences of potential, such as induction-coils or electrical machines, the deflexion in this case is regarded by the supporters of the ætherial theory as due to the discharge passing between the electrodes, and not primarily to the electrostatic field. Hertz made the rays travel between two parallel plates of metal placed inside the discharge-tube, but found that they were not deflected when the plates were connected with a battery of storage-cells; on repeating this experiment I at first got the same result, but subsequent experiments showed that the absence of deflexion is due to the conductivity conferred on the rarefied gas by the cathode rays. On measuring this conductivity it was found that it diminished very rapidly as the exhaustion increased; it seemed then that on trying Hertz's experiment at very high exhaustions there might be a chance of detecting the deflexion of the cathode rays by an electrostatic force.

The apparatus used is represented in fig. 2.

Fig. 2.

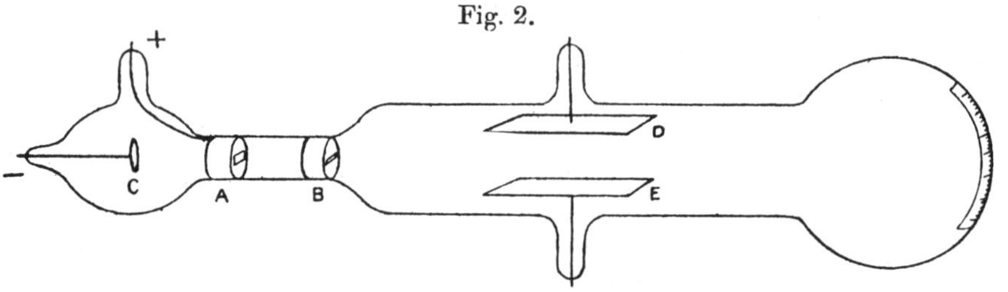

The rays from the cathode C pass through a slit in the anode A, which is a metal plug fitting tightly into the tube and connected with the earth; after passing through a second slit in another earth-connected metal plug B, they travel between two parallel aluminium plates about 5 cm. long by 2 broad and at a distance of 1·5 cm. apart; they then fall on the end of the tube and produce a narrow well-defined phosphorescent patch. A scale pasted on the outside of the tube serves to measure the deflexion of this patch.

At high exhaustions the rays were deflected when the two aluminium plates were connected with the terminals of a battery of small storage-cells; the rays were depressed when the upper plate was connected with the negative pole of the battery, the lower with the positive, and raised when the upper plate was connected with the positive, the lower with the negative pole. The deflexion was proportional to the difference of potential between the plates, and I could detect the deflexion when the potential-difference was as small as two volts. It was only when the vacuum was a good one that the deflexion took place, but that the absence of deflexion is due to the conductivity of the medium is shown by what takes place when the vacuum has just arrived at the stage at which the deflexion begins. At this stage there is a deflexion of the rays when the plates are first connected with the terminals of the battery, but if this connexion is maintained the patch of phosphorescence gradually creeps back to its undeflected position. This is just what would happen if the space between the plates were a conductor, though a very bad one, for then the positive and negative ions between the plates would slowly diffuse, until the positive plate became coated with negative ions, the negative plate with positive ones; thus the electric intensity between the plates would vanish and the cathode rays be free from electrostatic force. Another illustration of this is afforded by what happens when the pressure is low enough to show the deflexion and a large difference of potential, say 200 volts, is established between the plates; under these circumstances there is a large deflexion of the cathode rays, but the medium under the large electromotive force breaks down every now and then and a bright discharge passes between the plates; when this occurs the phosphorescent patch produced by the cathode rays jumps back to its undeflected position. When the cathode rays are deflected by the electrostatic field, the phosphorescent band breaks up into several bright bands separated by comparatively dark spaces; the phenomena are exactly analogous to those observed by Birkeland when the cathode rays are deflected by a magnet, and called by him the magnetic spectrum.

A series of measurements of the deflexion of the rays by the electrostatic force under various circumstances will be found later on in the part of the paper which deals with the velocity of the rays and the ratio of the mass of the electrified particles to the charge carried by them. It may, however, be mentioned here that the deflexion gets smaller as the pressure diminishes, and when in consequence the potential-difference in the tube in the neighbourhood of the cathode increases.

Conductivity of a Gas through which Cathode Rays are passing.

The conductivity of the gas was investigated by means of the apparatus shown in fig. 2. The upper plate D was connected with one terminal of a battery of small storage-cells, the other terminal of which was connected with the earth; the other plate E was connected with one of the coatings of a condenser of one microfarad capacity, the other coating of which was to earth; one pair of quadrants of an electrometer was also connected with E, the other pair of quadrants being to earth. When the cathode rays are passing between the plates the two pairs of quadrants of the electrometer are first connected with each other, and then the connexion between them was broken. If the space between the plates were a non-conductor, the potential of the pair of quadrants not connected with the earth would remain zero and the needle of the electrometer would not move; if, however, the space between the plates were a conductor, then the potential of the lower plate would approach that of the upper, and the needle of the electrometer would be deflected. There is always a deflexion of the electrometer, showing that a current passes between the plates. The magnitude of the current depends very greatly upon the pressure of the gas; so much so, indeed, that it is difficult to obtain consistent readings in consequence of the changes which always occur in the pressure when the discharge passes through the tube.

We shall first take the case when the pressure is only just low enough to allow the phosphorescent patch to appear at the end of the tube; in this case the relation between the current between the plates and the initial difference of potential is represented by the curve shown in fig. 3. In this

Fig. 3.

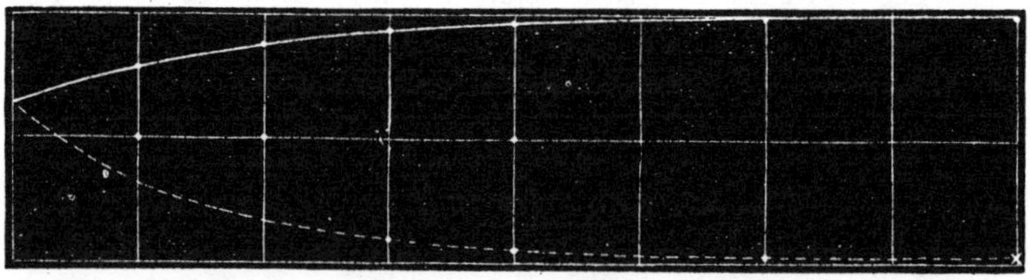

figure the abscissæ represent the initial difference of potential between the plates, each division representing two volts, and the ordinates the rise in potential of the lower plate in one minute each division again representing two volts. The quantity of electricity which has passed between the plates in one minute is the quantity required to raise 1 microfarad to the potential-difference shown by the curve. The upper and lower curve relates to the case when the upper plate is connected with the negative and positive pole respectively of the battery.

Even when there is no initial difference of potential between the plates the lower plate acquires a negative charge from the impact on it of some of the cathode rays.

We see from the curve that the current between the plates soon reaches a value where it is only slightly affected by an increase in the potential-difference between the plates; this is a feature common to conduction through gases traversed by Röntgen rays, by uranium rays, by ultra-violet light, and, as we now see, by cathode rays. The rate of leak is not greatly different whether the upper plate be initially positively or negatively electrified.

The current between the plates only lasts for a short time; it ceases long before the potential of the lower plate approaches that of the upper. Thus, for example, when the potential of the upper plate was about 400 volts above that of the earth, the potential of the lower plate never rose above 6 volts: similarly, if the upper plate were connected with the negative pole of the battery, the fall in potential of the lower plate was very small in comparison with the potential-difference between the upper plate and the earth.

These results are what we should expect if the gas between the plates and the plug B (fig. 2) were a very much better conductor than the gas between the plates, for the lower plate will be in a steady state when the current coming to it from the upper plate is equal to the current going from it to the plug: now if the conductivity of the gas between the plate and the plug is much greater than that between the plates, a small difference of potential between the lower plate and the plug will be consistent with a large potential-difference between the plates.

So far we have been considering the case when the pressure is as high as is consistent with the cathode rays reaching the end of the tube; we shall now go to the other extreme and consider the case when the pressure is as low as is consistent with the passage of a discharge through the bulb. In this case, when the plates are not connected with the battery we get a negative charge communicated to the lower plate, but only very slowly in comparison with the effect in the previous case. When the upper plate is connected with the negative pole of a battery, this current to the lower plate is only slightly increased even when the difference of potential is as much as 400 volts: a small potential-difference of about

20 volts seems slightly to decrease the rate of leak. Potential-differences much exceeding 400 volts cannot be used, as though the dielectric between the plates is able to sustain them for some little time, yet after a time an intensely bright arc flashes across between the plates and liberates so much gas as to spoil the vacuum. The lines in the spectrum of this glare are chiefly mercury lines; its passage leaves very peculiar markings on the aluminium plates.

If the upper plate was charged positively, then the negative charge communicated to the lower plate was diminished, and stopped when the potential-difference between the plates was about 20 volts; but at the lowest pressure, however great (up to 400 volts) the potential-difference, there was no leak of positive electricity to the lower plate at all comparable with the leak of negative electricity to this plate when the two plates were disconnected from the battery. In fact at this very low pressure all the facts are consistent with the view that the effects are due to the negatively electrified particles travelling along the cathode rays, the rest of the gas possessing little conductivity. Some experiments were made with a tube similar to that shown in fig. 2, with the exception that the second plug B was absent, so that a much greater number of cathode rays passed between the plates. When the upper plate was connected with the positive pole of the battery a luminous discharge with well-marked striations passed between the upper plate and the earth-connected plug through which the cathode rays were streaming; this occurred even though the potential-difference between the plate and the plug did not exceed 20 volts. Thus it seems that if we supply cathode rays from an external source to the cathode a small potential-difference is sufficient to produce the characteristic discharge through a gas.

Magnetic Deflexion of the Cathode Rays in Different Gases.

The deflexion of the cathode rays by the magnetic field was studied with the aid of the apparatus shown in fig. 4. The cathode was placed in a side-tube fastened on to a bell-jar; the opening between this tube and the bell-jar was closed by a metallic plug with a slit in it; this plug was connected with the earth and was used as the anode. The cathode rays passed through the slit in this plug into the bell-jar, passing in front of a vertical plate of glass ruled into small squares. The bell-jar was placed between two large parallel coils arranged as a Helmholtz galvanometer. The course of the rays was determined by taking photographs of the bell-jar when the cathode rays were passing through it; the divisions on the plate enabled the path of the rays to be determined. Under the action of the magnetic field the narrow beam of cathode rays spreads out into a broad fan-shaped luminosity in the gas. The luminosity in this fan is not uniformly

Fig. 4.

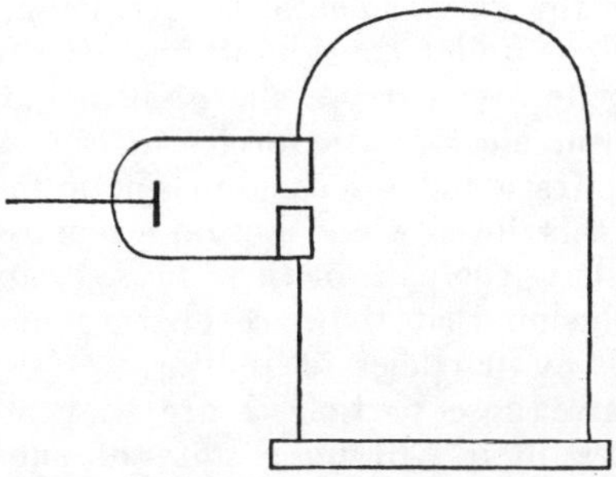

distributed, but is condensed along certain lines. The phosphorescence on the glass is also not uniformly distributed; it is much spread out, showing that the beam consists of rays which are not all deflected to the same extent by the magnet. The luminosity on the glass is crossed by bands along which the luminosity is very much greater than in the adjacent parts. These bright and dark bands are called by Birkeland, who first observed them, the magnetic spectrum. The brightest spots on the glass are by no means always the terminations of the brightest streaks of luminosity in the gas; in fact, in some cases a very bright spot on the glass is not connected with the cathode by any appreciable luminosity, though there may be plenty of luminosity in other parts of the gas. One very interesting point brought out by the photographs is that in a given magnetic field, and with a given mean potential-difference between the terminals, the path of the rays is independent of the nature of the gas. Photographs were taken of the discharge in hydrogen, air, carbonic acid, methyl iodide, *i. e.*, in gases whose densities range from 1 to 70, and yet, not only were the paths of the most deflected rays the same in all cases, but even the details, such as the distribution of the bright and dark spaces, were the same; in fact, the photographs could hardly be distinguished from each other. It is to be ncted that the pressures were not the same; the pressures in the different gases were adjusted so that the mean potential-differences between the cathode and the anode were the same in all the gases. When the pressure of a gas is lowered, the potential-difference between the terminals increases, and the deflexion of the rays produced by a magnet diminishes, or at any rate the deflexion of the rays when the phosphorescence is a maximum diminishes. If an air-break is inserted an effect of the same kind is produced.

In the experiments with different gases, the pressures were as high as was consistent with the appearance of the phosphorescence on the glass, so as to ensure having as much as possible of the gas under consideration in the tube.

As the cathode rays carry a charge of negative electricity, are deflected by an electrostatic force as if they were negatively electrified, and are acted on by a magnetic force in just the way in which this force would act on a negatively electrified body moving along the path of these rays, I can see no escape from the conclusion that they are charges of negative electricity carried by particles of matter. The question next arises, What are these particles? are they atoms, or molecules, or matter in a still finer state of subdivision? To throw some light on this point, I have made a series of measurements of the ratio of the mass of these particles to the charge carried by it. To determine this quantity, I have used two independent methods. The first of these is as follows:—Suppose we consider a bundle of homogeneous cathode rays. Let m be the mass of each of the particles, e the charge carried by it. Let N be the number of particles passing across any section of the beam in a given time; then Q the quantity of electricity carried by these particles is given by the equation

$$Ne=Q.$$

We can measure Q if we receive the cathode rays in the inside of a vessel connected with an electrometer. When these rays strike against a solid body, the temperature of the body is raised; the kinetic energy of the moving particles being converted into heat; if we suppose that all this energy is converted into heat, then if we measure the increase in the temperature of a body of known thermal capacity caused by the impact of these rays, we can determine W, the kinetic energy of the particles, and if v is the velocity of the particles,

$$\tfrac{1}{2}Nmv^2=W.$$

If ρ is the radius of curvature of the path of these rays in a uniform magnetic field H, then

$$\frac{mv}{e}=H\rho=I,$$

where I is written for $H\rho$ for the sake of brevity. From these equations we get

$$\frac{1}{2}\frac{m}{e}v^2 = \frac{W}{Q}.$$

$$v = \frac{2W}{QI},$$

$$\frac{m}{e} = \frac{I^2 Q}{2W}.$$

Thus, if we know the values of Q, W, and I, we can deduce the values of v and m/e.

To measure these quantities, I have used tubes of three different types. The first I tried is like that represented in fig. 2, except that the plates E and D are absent, and two coaxial cylinders are fastened to the end of the tube. The rays from the cathode C fall on the metal plug B, which is connected with the earth, and serves for the anode; a horizontal slit is cut in this plug. The cathode rays pass through this slit, and then strike against the two coaxial cylinders at the end of the tube; slits are cut in these cylinders, so that the cathode rays pass into the inside of the inner cylinder. The outer cylinder is connected with the earth, the inner cylinder, which is insulated from the outer one, is connected with an electrometer, the deflexion of which measures Q, the quantity of electricity brought into the inner cylinder by the rays. A thermo-electric couple is placed behind the slit in the inner cylinder; this couple is made of very thin strips of iron and copper fastened to very fine iron and copper wires. These wires passed through the cylinders, being insulated from them, and through the glass to the outside of the tube, where they were connected with a low-resistance galvanometer, the deflexion of which gave data for calculating the rise of temperature of the junction produced by the impact against it of the cathode rays. The strips of iron and copper were large enough to ensure that every cathode ray which entered the inner cylinder struck against the junction. In some of the tubes the strips of iron and copper were placed end to end, so that some of the rays struck against the iron, and others against the copper; in others, the strip of one metal was placed in front of the other; no difference, however, could be detected between the results got with these two arrangements. The strips of iron and copper were weighed, and the thermal capacity of the junction calculated. In one set of junctions this capacity was 5×10^{-3}, in another 3×10^{-3}. If we assume that the cathode rays which strike against the junction give their energy up to it, the deflexion of the galvanometer gives us W or $\frac{1}{2}Nmv^2$.

The value of I, *i. e.*, Hρ, where ρ is the curvature of the path of the rays in a magnetic field of strength H was found as follows :—The tube was fixed between two large circular coils placed parallel to each other, and separated by a distance equal to the radius of either ; these coils produce a uniform magnetic field, the strength of which is got by measuring with an ammeter the strength of the current passing through them. The cathode rays are thus in a uniform field, so that their path is circular. Suppose that the rays, when deflected by a magnet, strike against the glass of the tube at E

Fig. 5.

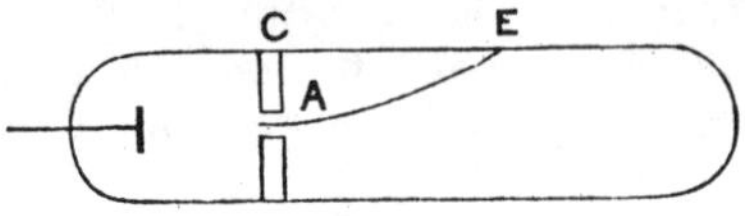

(fig. 5), then, if ρ is the radius of the circular path of the rays,

$$2\rho = \frac{CE^2}{AC} + AC ;$$

thus, if we measure CE and AC we have the means of determining the radius of curvature of the path of the rays.

The determination of ρ is rendered to some extent uncertain, in consequence of the pencil of rays spreading out under the action of the magnetic field, so that the phosphorescent patch at E is several millimetres long ; thus values of ρ differing appreciably from each other will be got by taking E at different points of this phosphorescent patch. Part of this patch was, however, generally considerably brighter than the rest ; when this was the case, E was taken as the brightest point ; when such a point of maximum brightness did not exist, the middle of the patch was taken for E. The uncertainty in the value of ρ thus introduced amounted sometimes to about 20 per cent.; by this I mean that if we took E first at one extremity of the patch and then at the other, we should get values of ρ differing by this amount.

The measurement of Q, the quantity of electricity which enters the inner cylinder, is complicated by the cathode rays making the gas through which they pass a conductor, so that though the insulation of the inner cylinder was perfect when the rays were off, it was not so when they were passing through the space between the cylinders ; this caused some of the charge communicated to the inner cylinder to leak away so that the actual charge given to the cylinder by the cathode rays was larger than that indicated by the electrometer.

To make the error from this cause as small as possible, the inner cylinder was connected to the largest capacity available, 1·5 microfarad, and the rays were only kept on for a short time, about 1 or 2 seconds, so that the alteration in potential of the inner cylinder was not large, ranging in the various experiments from about ·5 to 5 volts. Another reason why it is necessary to limit the duration of the rays to as short a time as possible, is to avoid the correction for the loss of heat from the thermo-electric junction by conduction along the wires; the rise in temperature of the junction was of the order 2° C.; a series of experiments showed that with the same tube and the same gaseous pressure Q and W were proportional to each other when the rays were not kept on too long.

Tubes of this kind gave satisfactory results, the chief drawback being that sometimes in consequence of the charging up of the glass of the tube, a secondary discharge started from the cylinder to the walls of the tube, and the cylinders were surrounded by glow; when this glow appeared, the readings were very irregular; the glow could, however, be got rid of by pumping and letting the tube rest for some time. The results got with this tube are given in the Table under the heading Tube 1.

The second type of tube was like that used for photographing the path of the rays (fig. 4); double cylinders with a thermo-electric junction like those used in the previous tube were placed in the line of fire of the rays, the inside of the bell-jar was lined with copper gauze connected with the earth. This tube gave very satisfactory results; we were never troubled with any glow round the cylinders, and the readings were most concordant; the only drawback was that as some of the connexions had to be made with sealing-wax, it was not possible to get the highest exhaustions with this tube, so that the range of pressure for this tube is less than that for tube 1. The results got with this tube are given in the Table under the heading Tube 2.

The third type of tube was similar to the first, except that the openings in the two cylinders were made very much smaller; in this tube the slits in the cylinders were replaced by small holes, about 1·5 millim. in diameter. In consequence of the smallness of the openings, the magnitude of the effects was very much reduced; in order to get measurable results it was necessary to reduce the capacity of the condenser in connexion with the inner cylinder to ·15 microfarad, and to make the galvanometer exceedingly sensitive, as the rise in temperature of the thermo-electric junction was in these experiments only about ·5° C. on the average. The results obtained in this tube are given in the Table under the heading Tube 3.

The results of a series of measurements with these tubes are given in the following Table :—

Gas.	Value of W/Q.	I.	m/e	v.
		Tube 1.		
Air	$4\cdot6 \times 10^{11}$	230	$\cdot57 \times 10^{-7}$	4×10^{9}
Air	$1\cdot8 \times 10^{12}$	350	$\cdot34 \times 10^{-7}$	1×10^{10}
Air	$6\cdot1 \times 10^{11}$	230	$\cdot43 \times 10^{-7}$	$5\cdot4 \times 10^{9}$
Air	$2\cdot5 \times 10^{12}$	400	$\cdot32 \times 10^{-7}$	$1\cdot2 \times 10^{10}$
Air	$5\cdot5 \times 10^{11}$	230	$\cdot48 \times 10^{-7}$	$4\cdot8 \times 10^{9}$
Air	1×10^{12}	285	$\cdot4 \times 10^{-7}$	7×10^{9}
Air	1×10^{12}	285	$\cdot4 \times 10^{-7}$	7×10^{9}
Hydrogen	6×10^{12}	205	$\cdot35 \times 10^{-7}$	6×10^{9}
Hydrogen	$2\cdot1 \times 10^{12}$	460	$\cdot5 \times 10^{-7}$	$9\cdot2 \times 10^{9}$
Carbonic acid	$8\cdot4 \times 10^{11}$	260	$\cdot4 \times 10^{-7}$	$7\cdot5 \times 10^{9}$
Carbonic acid	$1\cdot47 \times 10^{12}$	340	$\cdot4 \times 10^{-7}$	$8\cdot5 \times 10^{9}$
Carbonic acid	$3\cdot0 \times 10^{12}$	480	$\cdot39 \times 10^{-7}$	$1\cdot3 \times 10^{10}$
		Tube 2.		
Air	$2\cdot8 \times 10^{11}$	175	$\cdot53 \times 10^{-7}$	$3\cdot3 \times 10^{9}$
Air	$4\cdot4 \times 10^{11}$	195	$\cdot47 \times 10^{-7}$	$4\cdot1 \times 10^{9}$
Air	$3\cdot5 \times 10^{11}$	181	$\cdot47 \times 10^{-7}$	$3\cdot8 \times 10^{9}$
Hydrogen	$2\cdot8 \times 10^{11}$	175	$\cdot53 \times 10^{-7}$	$3\cdot3 \times 10^{9}$
Air	$2\cdot5 \times 10^{11}$	160	$\cdot51 \times 10^{-7}$	$3\cdot1 \times 10^{9}$
Carbonic acid	2×10^{11}	148	$\cdot54 \times 10^{-7}$	$2\cdot5 \times 10^{9}$
Air	$1\cdot8 \times 10^{11}$	151	$\cdot63 \times 10^{-7}$	$2\cdot3 \times 10^{9}$
Hydrogen	$2\cdot8 \times 10^{11}$	175	$\cdot53 \times 10^{-7}$	$3\cdot3 \times 10^{9}$
Hydrogen	$4\cdot4 \times 10^{11}$	201	$\cdot46 \times 10^{-7}$	$4\cdot4 \times 10^{9}$
Air	$2\cdot5 \times 10^{11}$	176	$\cdot61 \times 10^{-7}$	$2\cdot8 \times 10^{9}$
Air	$4\cdot2 \times 10^{11}$	200	$\cdot48 \times 10^{-7}$	$4\cdot1 \times 10^{9}$
		Tube 3.		
Air	$2\cdot5 \times 10^{11}$	220	$\cdot9 \times 10^{-7}$	$2\cdot4 \times 10^{9}$
Air	$3\cdot5 \times 10^{11}$	225	$\cdot7 \times 10^{-7}$	$3\cdot2 \times 10^{9}$
Hydrogen	3×10^{11}	250	$1\cdot0 \times 10^{-7}$	$2\cdot5 \times 10^{9}$

It will be noticed that the value of m/e is considerably greater for Tube 3, where the opening is a small hole, than for Tubes 1 and 2, where the opening is a slit of much greater area. I am of opinion that the values of m/e got from Tubes 1 and 2 are too small, in consequence of the leakage from the inner cylinder to the outer by the gas being rendered a conductor by the passage of the cathode rays.

It will be seen from these tables that the value of m/e is independent of the nature of the gas. Thus, for the first tube the mean for air is $\cdot 40 \times 10^{-7}$, for hydrogen $\cdot 42 \times 10^{-7}$, and for carbonic acid gas $\cdot 4 \times 10^{-7}$; for the second tube the mean for air is $\cdot 52 \times 10^{-7}$, for hydrogen $\cdot 50 \times 10^{-7}$, and for carbonic acid gas $\cdot 54 \times 10^{-7}$.

Experiments were tried with electrodes made of iron instead of aluminium; this altered the appearance of the discharge and the value of v at the same pressure, the values of m/e were, however, the same in the two tubes; the effect produced by different metals on the appearance of the discharge will be described later on.

In all the preceding experiments, the cathode rays were first deflected from the cylinder by a magnet, and it was then found that there was no deflexion either of the electrometer or the galvanometer, so that the deflexions observed were entirely due to the cathode rays; when the glow mentioned previously surrounded the cylinders there was a deflexion of the electrometer even when the cathode rays were deflected from the cylinder.

Before proceeding to discuss the results of these measurements I shall describe another method of measuring the quantities m/e and v of an entirely different kind from the preceding; this method is based upon the deflexion of the cathode rays in an electrostatic field. If we measure the deflexion experienced by the rays when traversing a given length under a uniform electric intensity, and the deflexion of the rays when they traverse a given distance under a uniform magnetic field, we can find the values of m/e and v in the following way:—

Let the space passed over by the rays under a uniform electric intensity F be l, the time taken for the rays to traverse this space is l/v, the velocity in the direction of F is therefore

$$\frac{Fe}{m}\frac{l}{v},$$

so that θ, the angle through which the rays are deflected when they leave the electric field and enter a region free from electric force, is given by the equation

$$\theta = \frac{Fe}{m}\frac{l}{v^2}.$$

If, instead of the electric intensity, the rays are acted on by a magnetic force H at right angles to the rays, and extending across the distance l, the velocity at right angles to the original path of the rays is

$$\frac{Hev}{m}\frac{l}{v},$$

so that ϕ, the angle through which the rays are deflected when they leave the magnetic field, is given by the equation

$$\phi = \frac{He}{m}\frac{l}{v}.$$

From these equations we get

$$v = \frac{\phi}{\theta}\frac{F}{H}$$

and

$$\frac{m}{e} = \frac{H^2\theta \, . \, l}{F\phi^2}.$$

In the actual experiments H was adjusted so that $\phi = \theta$; in this case the equations become

$$v = \frac{F}{H},$$

$$\frac{m}{e} = \frac{H^2 l}{F\theta}.$$

The apparatus used to measure v and m/e by this means is that represented in fig. 2. The electric field was produced by connecting the two aluminium plates to the terminals of a battery of storage-cells. The phosphorescent patch at the end of the tube was deflected, and the deflexion measured by a scale pasted to the end of the tube. As it was necessary to darken the room to see the phosphorescent patch, a needle coated with luminous paint was placed so that by a screw it could be moved up and down the scale; this needle could be seen when the room was darkened, and it was moved until it coincided with the phosphorescent patch. Thus, when light was admitted, the deflexion of the phosphorescent patch could be measured.

The magnetic field was produced by placing outside the tube two coils whose diameter was equal to the length of the plates; the coils were placed so that they covered the space occupied by the plates, the distance between the coils was equal to the radius of either. The mean value of the magnetic force over the length l was determined in the following way: a narrow coil C whose length was l, connected with a ballistic galvanometer, was placed between the coils; the plane of the windings of C was parallel to the planes of the coils; the cross section of the coil was a rectangle 5 cm. by 1 cm. A given current was sent through the outer coils and the kick α of the galvanometer observed when this current was reversed. The coil C was then placed at the centre of two very large coils, so as to be in a field of uniform magnetic force: the current through the large coils was reversed and the kick β of the galvanometer again observed; by comparing α and β we can get the mean value of the magnetic force over a length l; this was found to be

$$60 \times \iota,$$

where ι is the current flowing through the coils.

A series of experiments was made to see if the electrostatic deflexion was proportional to the electric intensity between the plates; this was found to be the case. In the following experiments the current through the coils was adjusted so that the electrostatic deflexion was the same as the magnetic:—

Gas.	θ.	H.	F.	l.	m/e.	v.
Air	8/110	5·5	$1{\cdot}5 \times 10^{10}$	5	$1{\cdot}3 \times 10^{-7}$	$2{\cdot}8 \times 10^{9}$
Air	9·5/110	5·4	$1{\cdot}5 \times 10^{10}$	5	$1{\cdot}1 \times 10^{-7}$	$2{\cdot}8 \times 10^{9}$
Air	13/110	6·6	$1{\cdot}5 \times 10^{10}$	5	$1{\cdot}2 \times 10^{-7}$	$2{\cdot}3 \times 10^{9}$
Hydrogen	9/110	6·3	$1{\cdot}5 \times 10^{10}$	5	$1{\cdot}5 \times 10^{-7}$	$2{\cdot}5 \times 10^{9}$
Carbonic acid...	11/110	6·9	$1{\cdot}5 \times 10^{10}$	5	$1{\cdot}5 \times 10^{-7}$	$2{\cdot}2 \times 10^{9}$
Air	6/110	5	$1{\cdot}8 \times 10^{10}$	5	$1{\cdot}3 \times 10^{-7}$	$3{\cdot}6 \times 10^{9}$
Air	7/110	3·6	1×10^{10}	5	$1{\cdot}1 \times 10^{-7}$	$2{\cdot}8 \times 10^{9}$

The cathode in the first five experiments was aluminium, in the last two experiments it was made of platinum; in the last experiment Sir William Crookes's method of getting rid of the mercury vapour by inserting tubes of pounded sulphur, sulphur iodide, and copper filings between the bulb and the pump was adopted. In the calculation of m/e and v no allowance has been made for the magnetic force due to the coil in the region outside the plates; in this region the magnetic force will be in the opposite direction to that between the plates, and will tend to bend the cathode rays in the opposite direction: thus the effective value of H will be smaller than the value used in the equations, so that the values of m/e are larger, and those of v less than they would be if this correction were applied. This method of determining the values of m/e and v is much less laborious and probably more accurate than the former method; it cannot, however, be used over so wide a range of pressures.

From these determinations we see that the value of m/e is independent of the nature of the gas, and that its value 10^{-7} is very small compared with the value 10^{-4}, which is the smallest value of this quantity previously known, and which is the value for the hydrogen ion in electrolysis.

Thus for the carriers of the electricity in the cathode rays m/e is very small compared with its value in electrolysis. The smallness of m/e may be due to the smallness of m or the largeness of e, or to a combination of these two. That the carriers of the charges in the cathode rays are small compared with ordinary molecules is shown, I think, by Lenard's results as to the rate at which the brightness of the phosphorescence produced by these rays diminishes with the length of path travelled by the ray. If we regard this phosphorescence as due to the impact of the charged particles, the distance through which the rays must travel before the phosphorescence fades to a given fraction (say $1/e$, where $e=2{\cdot}71$) of its original intensity, will be some moderate multiple of the mean free path. Now Lenard found that this distance depends solely upon the density of the medium, and not upon its chemical nature or physical state. In air at atmospheric pressure the distance was about half a centimetre, and this must be comparable with the mean free path of the carriers through air at atmospheric pressure. But the mean free path of the molecules of air is a quantity of quite a different order. The carrier, then, must be small compared with ordinary molecules.

The two fundamental points about these carriers seem to me to be (1) that these carriers are the same whatever the gas through which the discharge passes, (2) that the mean free paths depend upon nothing but the density of the medium traversed by these rays.

It might be supposed that the independence of the mass of the carriers of the gas through which the discharge passes was due to the mass concerned being the quasi mass which a charged body possesses in virtue of the electric field set up in its neighbourhood; moving the body involves the production of a varying electric field, and, therefore, of a certain amount of energy which is proportional to the square of the velocity. This causes the charged body to behave as if its mass were increased by a quantity, which for a charged sphere is $\frac{1}{5} e^2/\mu a$ ('Recent Researches in Electricity and Magnetism'), where e is the charge and a the radius of the sphere. If we assume that it is this mass which we are concerned with in the cathode rays, since m/e would vary as e/a, it affords no clue to the explanation of either of the properties (1 and 2) of these rays. This is not by any means the only objection to this hypothesis, which I only mention to show that it has not been overlooked.

The explanation which seems to me to account in the most simple and straightforward manner for the facts is founded on a view of the constitution of the chemical elements which has been favourably entertained by many chemists: this view is that the atoms of the different chemical elements are different aggregations of atoms of the same kind. In the form in which this hypothesis was enunciated by Prout, the atoms of the different elements were hydrogen atoms; in this precise form the hypothesis is not tenable, but if we substitute for hydrogen some unknown primordial substance X, there is nothing known which is inconsistent with this hypothesis, which is one that has been recently supported by Sir Norman Lockyer for reasons derived from the study of the stellar spectra.

If, in the very intense electric field in the neighbourhood of the cathode, the molecules of the gas are dissociated and are split up, not into the ordinary chemical atoms, but into these primordial atoms, which we shall for brevity call corpuscles; and if these corpuscles are charged with electricity and projected from the cathode by the electric field, they would behave exactly like the cathode rays. They would evidently give a value of m/e which is independent of the nature of the gas and its pressure, for the carriers are the same whatever the gas may be; again, the mean free paths of these corpuscles would depend solely upon the density of the medium through which they pass. For the molecules of the medium are composed of a number of such corpuscles separated by considerable spaces; now the collision between a single corpuscle and the molecule will not be between the corpuscles and the molecule as a whole, but between this corpuscle and the individual corpuscles which form the molecule; thus the number of collisions the particle makes as it moves through a crowd of these molecules will be proportional, not to the number of the molecules in the crowd, but to the number of the individual corpuscles. The mean free path is inversely proportional to the number of collisions in unit time, and so is inversely proportional to the number of corpuscles in unit volume; now as these corpuscles are all of the same mass, the number of corpuscles in unit volume will be proportional to the mass of unit volume, that is the mean free path will be inversely proportional to the density of the gas. We see, too, that so long as the distance between neighbouring corpuscles is large compared with the linear dimensions of a corpuscle the mean free path will be independent of the way they are arranged, provided the number in unit volume remains constant, that is the mean free path will depend only on the density of the medium traversed by the corpuscles, and will be independent of its chemical nature and physical state: this from Lenard's very remarkable measurements of the absorption of the cathode rays by various media, must be a property possessed by the carriers of the charges in the cathode rays.

Thus on this view we have in the cathode rays matter in a new state, a state in which the subdivision of matter is carried very much further than in the ordinary gaseous state: a state in which all matter—that is, matter derived from different sources such as hydrogen, oxygen, &c.—is of one and the same kind; this matter being the substance from which all the chemical elements are built up.

With appliances of ordinary magnitude, the quantity of matter produced by means of the dissociation at the cathode is so small as to almost to preclude the possibility of any direct chemical investigation of its properties. Thus the coil I used would, I calculate, if kept going uninterruptedly night and day for a year, produce only about one three-millionth part of a gramme of this substance.

The smallness of the value of m/e is, I think, due to the largeness of e as well as the smallness of m. There seems to me to be some evidence that the charges carried by the corpuscles in the atom are large compared with those carried by the ions of an electrolyte. In the molecule of HCl, for example, I picture the components of the hydrogen atoms as held together by a great number of tubes of electrostatic force; the components of the chlorine atom are similarly held together, while only one stray tube binds the hydrogen atom to the chlorine atom. The reason for attributing this high charge to the constituents of the atom is derived from the values of the specific inductive capacity of gases: we may imagine that the specific inductive capacity of a gas is due to the setting in the electric field of the electric doublet formed by the two oppositely electrified atoms which form the molecule of the gas. The measurements of the specific inductive capacity show, however, that this is very approximately an additive quantity: that is, that we can assign a certain value to each element, and find the specific inductive capacity of HCl by adding the value for hydrogen to the value for chlorine; the value of H_2O by adding twice the value for hydrogen to the value for oxygen, and so on. Now the electrical moment of the doublet formed by a positive charge on one atom of the molecule and a negative charge on the other atom would not be an additive property; if, however, each atom had a definite electrical moment, and this were large compared with the electrical moment of the two atoms in the molecule, then the electrical moment of any compound, and hence its specific inductive capacity, would be an additive property. For the electrical moment of the atom, however, to be large compared with that of the molecule, the charge on the corpuscles would have to be very large compared with those on the ion.

If we regard the chemical atom as an aggregation of a number of primordial atoms, the problem of finding the configurations of stable equilibrium for a number of equal particles acting on each other according to some law of force —whether that of Boscovich, where the force between them is a repulsion when they are separated by less than a certain critical distance, and an attraction when they are separated by a greater distance, or even the simpler case of a number of mutually repellent particles held together by a central force —is of great interest in connexion with the relation between the properties of an element and its atomic weight. Unfortunately the equations which determine the stability of such a collection of particles increase so rapidly in complexity with the number of particles that a general mathematical investigation is scarcely possible. We can, however, obtain a good deal of insight into the general laws which govern such configurations by the use of models, the simplest of which is the floating magnets of Professor Mayer. In this model the magnets arrange themselves in equilibrium under their mutual repulsions and a central attraction caused by the pole of a large magnet placed above the floating magnets.

A study of the forms taken by these magnets seems to me to be suggestive in relation to the periodic law. Mayer showed that when the number of floating magnets did not exceed 5 they arranged themselves at the corners of a regular polygon— 5 at the corners of a pentagon, 4 at the corners of a square, and so on. When the number exceeds 5, however, this law no longer holds: thus 6 magnets do not arrange themselves at the corners of a hexagon, but divide into two systems, consisting of 1 in the middle surrounded by 5 at the corners of a pentagon. For 8 we have two in the inside and 6 outside; this arrangement in two systems, an inner and an outer, lasts up to 18 magnets. After this we have three systems: an inner, a middle, and an outer; for a still larger number of magnets we have four systems, and so on.

Mayer found the arrangement of magnets was as follows:—

1.	2.	3.	4.	5.
1.5 1.6 1.7	2.6 2.7	3.7 3.8	4.8 4.9	5.9
1.5.9 1.6.9 1.6.10 1.6.11	2.7.10 2.8.10 2.7.11	3.7.10 3.7.11 3.8.10 3.8.11 3.8.12 3.8.13	4.8.12 4.8.13 4.9.12 4.9.13	5.9.12 5.9.13
1.5.9.12 1.5.9.13 1.6.9.12 1.6.10.12 1.6.10.13 1.6.11.12 1.6.11.13 1.6.11.14 1.6.11.15 1.7.12.14	2.7.10.15 2.7.12.14	3.7.12.13 3.7.12.14 3.7.13.14 3.7.13.15	4.9.13.14 4.9.13.15 4.9.14.15	

where, for example, 1.6.10.12 means an arrangement with one magnet in the middle, then a ring of six, then a ring of ten, and a ring of twelve outside.

Now suppose that a certain property is associated with two magnets forming a group by themselves; we should have this property with 2 magnets, again with 8 and 9, again with 19 and 20, and again with 34, 35, and so on. If we regard the system of magnets as a model of an atom, the number of magnets being proportional to the atomic weight, we should have this property occurring in elements of atomic weight 2, (8, 9), 19, 20, (34, 35). Again, any property conferred by three magnets forming a system by themselves would occur with atomic weights 3, 10, and 11; 20, 21, 22, 23, and 24; 35, 36, 37 and 39; in fact, we should have something quite analogous to the periodic law, the first series corresponding to the arrangement of the magnets in a single group, the second series to the arrangement in two groups, the third series in three groups, and so on.

Velocity of the Cathode Rays.

The velocity of the cathode rays is variable, depending upon the potential-difference between the cathode and anode, which is a function of the pressure of the gas—the velocity increases as the exhaustion improves; the measurements given above show, however, that at all the pressures at which experiments were made the velocity exceeded 10^9 cm./sec. This velocity is much greater than the value 2×10^7 which I previously obtained (Phil. Mag. Oct. 1894) by measuring directly the interval which separated the appearance of luminosity at two places on the walls of the tube situated at different distances from the cathode.

In my earlier experiments the pressure was higher than in the experiments described in this paper, so that the velocity of the cathode rays would on this account be less. The difference between the two results is, however, too great to be wholly explained in this way, and I attribute the difference to the glass requiring to be bombarded by the rays for a finite time before becoming phosphorescent, this time depending upon the intensity of the bombardment. As this time diminishes with the intensity of bombardment, the appearance of phosphorescence at the piece of glass most removed from the cathode would be delayed beyond the time taken for the rays to pass from one place to the other by the difference in time taken by the glass to become luminous; the apparent velocity measured in this way would thus be less than the true velocity. In the former experiments endeavours were made to diminish this effect by making the rays strike the glass at the greater distance from the cathode less obliquely than they struck the glass nearer to the cathode; the obliquity was adjusted until the brightness of the phosphorescence was approximately equal in the two cases. In view, however, of the discrepancy between the results obtained in this way and those obtained by the later method, I think that it was not successful in eliminating the lag caused by the finite time required by the gas to light up.

Experiments with Electrodes of Different Materials.

In the experiments described in this paper the electrodes were generally made of aluminium. Some experiments, however, were made with iron and platinum electrodes.

Though the value of m/e came out the same whatever the material of the electrode, the appearance of the discharge varied greatly; and as the measurements showed, the potential-difference between the cathode and anode depended greatly upon the metal used for the electrode; the pressure being the same in all cases.

To test this point further I used a tube like that shown in fig. 6, where *a, b, c* are cathodes made of different metals, the anodes being in all cases platinum wires. The cathodes were disks of aluminium, iron, lead, tin, copper, mercury, sodium amalgam, and silver chloride; the potential-difference

Fig. 6.

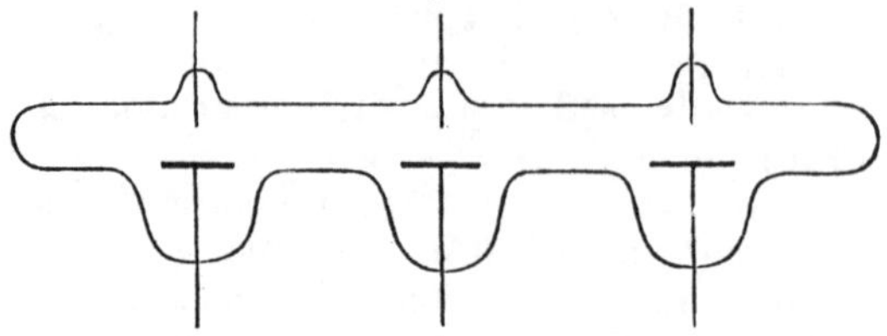

between the cathode and anode was measured by Lord Kelvin's vertical voltmeter, and also by measuring the length of spark in air which, when placed in parallel with the anode and cathode, seemed to allow the discharge to go as often through the spark-gap as through the tube. With this arrangement the pressures were the same for all the cathodes. The potential-difference between the anode and cathode and the equivalent spark-length depended greatly upon the nature of the cathode. The extent of the variation in potential may be estimated from the following table:—

Cathode.	Mean Potential-Difference between Cathode and Anode.
Aluminium	1800 volts.
Lead	2100 "
Tin	2400 "
Copper	2600 "
Iron....................	2900 "

The potential-difference when the cathode was made of sodium amalgam or silver chloride was less even than that of aluminium.

The order of many of the metals changed about very capriciously, experiments made at intervals of a few minutes frequently giving quite different results. From the abrupt way in which these changes take place I am inclined to think that gas absorbed by the electrode has considerable influence on the passage of the discharge.

I have much pleasure in thanking Mr. Everitt for the assistance he has given me in the preceding investigation.

Cambridge, Aug. 7, 1897.

Part VI
ATOMIC STRUCTURE

Editor's Comments on Papers 21, 22, and 23

Following the discovery of the electron, Thomson turned his attention to the structure of the atom. He realized that the atom must contain electrons whose negative charges must be balanced by equal amounts of positive charge. At first he did not have any solid experimental evidence concerning the actual numbers of positive and negative charges in a given atom, but after many x-ray and electron-scattering experiments, he concluded that n, the number of electrons, lies somewhere between $A/10$ and $2A$, where A is the atomic weight. His model of the atom portrayed these electrons as being distributed throughout the entire volume of the atom and circulating in orbital rings through a medium characterized by a uniform distribution of positive charge. Even though this so-called plum-pudding model did not endure long, the Thomson technique of probing the interior of the atom by scattering methods did point the way to eventual success. In an experiment suggested by Rutherford (Ernest Lord Rutherford, 1871–1937), Marsden and Geiger obtained scattering patterns with α-particles as projectiles. From the observation that several particles are deflected through angles between 90 and 180°, Rutherford concluded that a much better fit to the data could be obtained by assuming that the mass of the atom is concentrated in a very small, positively charged nucleus with radius of about 10^{-13} cm. Surrounding the nucleus is a cloud of negatively charged electrons whose outermost orbits specify the effective boundary of the atom at a radial distance of about 10^{-8} cm from the nucleus. (A brief biographical sketch of Rutherford is given in Part VIII, Nuclear Energy).

The dynamics of the newly discovered nuclear or planetary atom quickly challenged physicists. The principal experimental data that any satisfactory theory had to explain were the emission of discrete line spectra from gaseous elements in electrical discharge tubes. The lines of hydrogen that had been studied earlier by Balmer, Rydberg, and Ritz appear to show a regularity in the relationships among observed frequencies that is expressed by an empirical formula containing integer variables. The stage was set for a very major contribution in the development of physics. It came about through the bold and creative insight of Niels Bohr (1885–1962).

Bohr, the son of a professor of physiology at the University of Copenhagen, received his education at that institution and completed the doctorate in physics in 1906. He spent some time in England at the Universities of Cambridge and Manchester and there made the acquaintance of J. J. Thomson and Rutherford. During his stay at Rutherford's laboratory in Manchester in 1912, Bohr became intrigued by the problem of atomic stability and initiated the theoretical researches on the Balmer series and atomic structure that led to the Nobel Prize in 1922. Bohr's lifetime career centered on his directorship of the world-renowned Institute for Theoretical Physics in Copenhagen, where he exerted tremendous influence on twentieth-century science. This influence encompassed not only the acquisition of knowledge through the research of the many students and colleagues under his inspired leadership but also the search for a satisfactory philosophical interpretation of the new physics. Following World War II Bohr recognized both the potential and the peril of the nuclear age, one which he himself and John Wheeler of the United States had helped to initiate through their formulation of the model explaining the fission process in U^{235}. While continuing the directorship of his institute he also exerted widespread international influence to promote peaceful uses of nuclear energy.

The previous attempts to theoretically explain spectra radiation had been based on classical electromagnetic theory applied to the periodic motions of the electron in orbit around the nucleus. The classical theory predicted a continuous increase in the frequency of the radiation emitted by the atom as the accelerating electron loses energy and spirals inward with increasing frequency of rotation toward the nucleus. There was no way to account for the apparently stable configuration of Rutherford's nuclear atom and the experimentally observed discrete spectral lines. Bohr's new theory (1913), which is reproduced as Paper 21, was keyed to two fundamentally novel postulates concerning the behavior of atomic systems. These hypotheses, which drew on Planck's quantum idea, had the effect of ruling that the stable energy states of an atom are characterized by a set of discrete or quantized values and that when an atom passes from one state to another, the frequency f of the emitted radiation is related to the difference between the energies

E_2 and E_1 of the two states, respectively, according to the formula: $hf = E_2 - E_1$, where h is Planck's constant (see Paper 15). Bohr further showed that his results imply the quantization of angular momentum in integral multiples of Planck's constant. Bohr's theory was to have considerable success in explaining the major features of spectroscopy of atoms whose electronic structure can be reduced to an equivalent single electron outside an inner core of electrons, and the spectroscopy of singly ionized helium, doubly ionized lithium, etc. It was not able to account precisely for multielectron spectra even in the simple case of helium. Nevertheless, it is important to reflect on the significance of the Rutherford-Bohr era in relation to the theme of the present collection. The traditionally held view of atomists, dating back to the time of Democritus, that the atom was a homogeneous, irreducible unit of matter was now being upset. A new system of organized matter consisting of particles of inconceivably small dimensions had to be reckoned with. Atoms, which now appeared to be the gross structural form taken by this organized level of matter, are governed by dynamical laws quite different from those applying to large levels of organized matter such as planets orbiting around the sun. The conservation-of-energy principle still seemed inviolable, at least when averaged over large numbers of atoms (and later proved applicable to individual events); but the discreteness of radiant energy that Planck and Einstein had discovered now appeared to be a part of a much more general graininess applying to the energy states of all matter at the atomic level. The concepts of atomism and energy were undergoing dramatic revisions and at the same time were becoming more and more closely intertwined. This ferment continues to the present day.

One year after the publication of Bohr's theory, James Franck (1882–1964) and Gustav Hertz (1887–1975) performed an important experiment that further confirmed the existence of discrete atomic energy levels. They subjected mercury vapor to bombardment by a stream of electrons of known and controllable kinetic energy. They discovered that when the accelerating potential reaches 4.9 V the electron current abruptly drops. As the voltage is subsequently built up to multiples of 4.9, similar drops in current occur. Initially Franck and Hertz interpreted the results as indicating that the 4.9 V represents the ionization potential of the mercury atom and that an inelastic collision between an electron possessing exactly this energy and the mercury atom would cause the complete removal from the atom of an electron from the stable quantized energy level by a resonance-absorption process. Using the Bohr energy-frequency formula, they calculated the energy of 4.9 electron volts to correspond very closely with an observed resonance line in the mercury spectrum wavelength of 2536 Å. However, about five years later they came to realize that the 4.9 electron volts corresponds only to the transition to the next energy level rather than complete ionization. In Paper 22, they summarize their own work and that of others (performed during the war years in the United States) and reinterpreted their original 1914 results. The modification means that when an electron collides with at atom and possesses a kinetic energy exactly equal to the energy difference between any two quantized orbits, the electron's energy is transferred to the atom with a resulting excitation to the higher state. The idea was therefore established that a change of energy state by quantized amounts is not restricted to radiative emission or absorption processes but can occur as a consequence of inelastic mechanical collisions. The Franck-Hertz experiment was to become a prototype for the study of levels of organized matter within the nucleus itself. Inelastic collisions between particles possessing energies ranging from thousands to millions of electron volts and nuclei have been used to induce resonance transitions between nuclear-energy states.

James Franck was born in Hamburg, Germany, and received his Ph.D. degree from the University of Berlin. It was while at Berlin that he collaborated with Gustav Hertz, also a graduate of that university, on the inelastic electron collision experiments. Franck came to the United States in 1935 and held professorships at the John Hopkins University and the University of Chicago. Hertz subsequently served as a director of the Siemens Co. research laboratories until the end of World War II. He spent some time in the USSR conducting research after World War II and returned to Leipzig, East Germany, as director of the Physics Institute in 1955. In 1925 Franck and Hertz were honored for their electron-collision experiments by the award of the Nobel Prize in physics.

In 1916 Einstein undertook a different approach to the spectral-line-intensity problem. He recognized that any complete description of the transition process must take into account the energy density of the radiation field possessing a frequency equal to that of the transition. Paper 23 gives the details of his analysis. He assumed that there is a probability of emission from the higher state dependent on both an intrinsic factor analogous to the radioactive-decay constant and on the external radiation field density appropriate to the transition frequency. In addition, there is a probability of absorption with transition from the lower to higher state dependent also on the radiation field density. By invoking the Boltzmann distribution law (see pages **116** and **394**) to describe the ratio of atoms in the two states when at equilibrium, Einstein was able to obtain Planck's radiation law (see Paper 15) with the quantity hf representing the difference between the two energy states. This result had the effect of replacing physical causality by indeterminancy in the description of the behavior of individual atoms and anticipated the subsequent probabalistic interpretations of quantum mechanics. A second important point was the indicated dependence of emission on the intensity of the radiation field environment at that frequency. The contribution to emission by stimulation from the ambient radiation field has found great practical application in the operation of the laser.

21

Reprinted from pages 1–19 of *Philos. Mag.* ser. 6, **26**:1–25 (1913)

ON THE CONSTITUTION OF ATOMS AND MOLECULES

N. Bohr*

Introduction.

IN order to explain the results of experiments on scattering of α rays by matter Prof. Rutherford† has given a theory of the structure of atoms. According to this theory, the atoms consist of a positively charged nucleus surrounded by a system of electrons kept together by attractive forces from the nucleus; the total negative charge of the electrons is equal to the positive charge of the nucleus. Further, the nucleus is assumed to be the seat of the essential part of the mass of the atom, and to have linear dimensions exceedingly small compared with the linear dimensions of the whole atom. The number of electrons in an atom is deduced to be approximately equal to half the atomic weight. Great interest is to be attributed to this atom-model; for, as Rutherford has shown, the assumption of the existence of nuclei, as those in question, seems to be necessary in order to account for the results of the experiments on large angle scattering of the α rays‡.

In an attempt to explain some of the properties of matter on the basis of this atom-model we meet, however, with difficulties of a serious nature arising from the apparent

* Communicated by Prof. E. Rutherford, F.R.S.
† E. Rutherford, Phil. Mag. xxi. p. 669 (1911).
‡ See also Geiger and Marsden, Phil. Mag. April 1913.

instability of the system of electrons: difficulties purposely avoided in atom-models previously considered, for instance, in the one proposed by Sir J. J. Thomson *. According to the theory of the latter the atom consists of a sphere of uniform positive electrification, inside which the electrons move in circular orbits.

The principal difference between the atom-models proposed by Thomson and Rutherford consists in the circumstance that the forces acting on the electrons in the atom-model of Thomson allow of certain configurations and motions of the electrons for which the system is in a stable equilibrium; such configurations, however, apparently do not exist for the second atom-model. The nature of the difference in question will perhaps be most clearly seen by noticing that among the quantities characterizing the first atom a quantity appears—the radius of the positive sphere—of dimensions of a length and of the same order of magnitude as the linear extension of the atom, while such a length does not appear among the quantities characterizing the second atom, viz. the charges and masses of the electrons and the positive nucleus; nor can it be determined solely by help of the latter quantities.

The way of considering a problem of this kind has, however, undergone essential alterations in recent years owing to the development of the theory of the energy radiation, and the direct affirmation of the new assumptions introduced in this theory, found by experiments on very different phenomena such as specific heats, photoelectric effect, Röntgen-rays, &c. The result of the discussion of these questions seems to be a general acknowledgment of the inadequacy of the classical electrodynamics in describing the behaviour of systems of atomic size †. Whatever the alteration in the laws of motion of the electrons may be, it seems necessary to introduce in the laws in question a quantity foreign to the classical electrodynamics, *i. e.* Planck's constant, or as it often is called the elementary quantum of action. By the introduction of this quantity the question of the stable configuration of the electrons in the atoms is essentially changed, as this constant is of such dimensions and magnitude that it, together with the mass and charge of the particles, can determine a length of the order of magnitude required.

This paper is an attempt to show that the application of the above ideas to Rutherford's atom-model affords a basis

* J. J. Thomson, Phil. Mag. vii. p. 237 (1904).

† See f. inst., 'Théorie du rayonnement et les quanta.' Rapports de la réunion à Bruxelles, Nov. 1911. Paris, 1912.

for a theory of the constitution of atoms. It will further be shown that from this theory we are led to a theory of the constitution of molecules.

In the present first part of the paper the mechanism of the binding of electrons by a positive nucleus is discussed in relation to Planck's theory. It will be shown that it is possible from the point of view taken to account in a simple way for the law of the line spectrum of hydrogen. Further, reasons are given for a principal hypothesis on which the considerations contained in the following parts are based.

I wish here to express my thanks to Prof. Rutherford for his kind and encouraging interest in this work.

Part I.—Binding of Electrons by Positive Nuclei.

§ 1. *General Considerations.*

The inadequacy of the classical electrodynamics in accounting for the properties of atoms from an atom-model as Rutherford's, will appear very clearly if we consider a simple system consisting of a positively charged nucleus of very small dimensions and an electron describing closed orbits around it. For simplicity, let us assume that the mass of the electron is negligibly small in comparison with that of the nucleus, and further, that the velocity of the electron is small compared with that of light.

Let us at first assume that there is no energy radiation. In this case the electron will describe stationary elliptical orbits. The frequency of revolution ω and the major-axis of the orbit $2a$ will depend on the amount of energy W which must be transferred to the system in order to remove the electron to an infinitely great distance apart from the nucleus. Denoting the charge of the electron and of the nucleus by $-e$ and E respectively and the mass of the electron by m, we thus get

$$\omega = \frac{\sqrt{2}}{\pi}\,\frac{W^{\frac{3}{2}}}{eE\sqrt{m}}, \quad 2a = \frac{eE}{W}. \quad . \quad . \quad . \quad (1)$$

Further, it can easily be shown that the mean value of the kinetic energy of the electron taken for a whole revolution is equal to W. We see that if the value of W is not given, there will be no values of ω and a characteristic for the system in question.

Let us now, however, take the effect of the energy radiation into account, calculated in the ordinary way from the acceleration of the electron. In this case the electron will no longer describe stationary orbits. W will continuously increase, and the electron will approach the nucleus describing orbits of smaller and smaller dimensions, and with greater and greater frequency ; the electron on the average gaining in kinetic energy at the same time as the whole system loses energy. This process will go on until the dimensions of the orbit are of the same order of magnitude as the dimensions of the electron or those of the nucleus. A simple calculation shows that the energy radiated out during the process considered will be enormously great compared with that radiated out by ordinary molecular processes.

It is obvious that the behaviour of such a system will be very different from that of an atomic system occurring in nature. In the first place, the actual atoms in their permanent state seem to have absolutely fixed dimensions and frequencies. Further, if we consider any molecular process, the result seems always to be that after a certain amount of energy characteristic for the systems in question is radiated out, the systems will again settle down in a stable state of equilibrium, in which the distances apart of the particles are of the same order of magnitude as before the process.

Now the essential point in Planck's theory of radiation is that the energy radiation from an atomic system does not take place in the continuous way assumed in the ordinary electrodynamics, but that it, on the contrary, takes place in distinctly separated emissions, the amount of energy radiated out from an atomic vibrator of frequency ν in a single emission being equal to $\tau h\nu$, where τ is an entire number, and h is a universal constant*.

Returning to the simple case of an electron and a positive nucleus considered above, let us assume that the electron at the beginning of the interaction with the nucleus was at a great distance apart from the nucleus, and had no sensible velocity relative to the latter. Let us further assume that the electron after the interaction has taken place has settled down in a stationary orbit around the nucleus. We shall, for reasons referred to later, assume that the orbit in question is circular ; this assumption will, however, make no alteration in the calculations for systems containing only a single electron.

Let us now assume that, during the binding of the electron, a homogeneous radiation is emitted of a frequency ν, equal to half the frequency of revolution of the electron in its final

* See f. inst., M. Planck, *Ann. d. Phys.* xxxi. p. 758 (1910) ; xxxvii. p. 642 (1912) ; *Verh. deutsch. Phys. Ges.* 1911, p. 138.

orbit; then, from Planck's theory, we might expect that the amount of energy emitted by the process considered is equal to $\tau h\nu$, where h is Planck's constant and τ an entire number. If we assume that the radiation emitted is homogeneous, the second assumption concerning the frequency of the radiation suggests itself, since the frequency of revolution of the electron at the beginning of the emission is 0. The question, however, of the rigorous validity of both assumptions, and also of the application made of Planck's theory, will be more closely discussed in § 3.

Putting

$$W = \tau h \frac{\omega}{2}, \quad \ldots \ldots \quad (2)$$

we get by help of the formula (1)

$$W = \frac{2\pi^2 m e^2 E^2}{\tau^2 h^2}, \quad \omega = \frac{4\pi^2 m e^2 E^2}{\tau^3 h^3}, \quad 2a = \frac{\tau^2 h^2}{2\pi^2 m e E}. \quad (3)$$

If in these expressions we give τ different values, we get a series of values for W, ω, and a corresponding to a series of configurations of the system. According to the above considerations, we are led to assume that these configurations will correspond to states of the system in which there is no radiation of energy; states which consequently will be stationary as long as the system is not disturbed from outside. We see that the value of W is greatest if τ has its smallest value 1. This case will therefore correspond to the most stable state of the system, *i. e.* will correspond to the binding of the electron for the breaking up of which the greatest amount of energy is required.

Putting in the above expressions $\tau = 1$ and $E = e$, and introducing the experimental values

$$e = 4{\cdot}7 \,.\, 10^{-10}, \quad \frac{e}{m} = 5{\cdot}31 \,.\, 10^{17}, \quad h = 6{\cdot}5 \,.\, 10^{-27},$$

we get

$$2a = 1{\cdot}1 \,.\, 10^{-8} \text{ cm.}, \quad \omega = 6{\cdot}2 \,.\, 10^{15} \frac{1}{\text{sec.}}, \quad \frac{W}{e} = 13 \text{ volt.}$$

We see that these values are of the same order of magnitude as the linear dimensions of the atoms, the optical frequencies, and the ionization-potentials.

The general importance of Planck's theory for the discussion of the behaviour of atomic systems was originally pointed out by Einstein*. The considerations of Einstein

* A. Einstein, *Ann. d. Phys.* xvii. p. 132 (1905); xx. p. 199 (1906); xxii. p. 180 (1907).

have been developed and applied on a number of different phenomena, especially by Stark, Nernst, and Sommerfield. The agreement as to the order of magnitude between values observed for the frequencies and dimensions of the atoms, and values for these quantities calculated by considerations similar to those given above, has been the subject of much discussion. It was first pointed out by Haas*, in an attempt to explain the meaning and the value of Planck's constant on the basis of J. J. Thomson's atom-model, by help of the linear dimensions and frequency of an hydrogen atom.

Systems of the kind considered in this paper, in which the forces between the particles vary inversely as the square of the distance, are discussed in relation to Planck's theory by J. W. Nicholson †. In a series of papers this author has shown that it seems to be possible to account for lines of hitherto unknown origin in the spectra of the stellar nebulæ and that of the solar corona, by assuming the presence in these bodies of certain hypothetical elements of exactly indicated constitution. The atoms of these elements are supposed to consist simply of a ring of a few electrons surrounding a positive nucleus of negligibly small dimensions. The ratios between the frequencies corresponding to the lines in question are compared with the ratios between the frequencies corresponding to different modes of vibration of the ring of electrons. Nicholson has obtained a relation to Planck's theory showing that the ratios between the wave-length of different sets of lines of the coronal spectrum can be accounted for with great accuracy by assuming that the ratio between the energy of the system and the frequency of rotation of the ring is equal to an entire multiple of Planck's constant. The quantity Nicholson refers to as the energy is equal to twice the quantity which we have denoted above by W. In the latest paper cited Nicholson has found it necessary to give the theory a more complicated form, still, however, representing the ratio of energy to frequency by a simple function of whole numbers.

The excellent agreement between the calculated and observed values of the ratios between the wave-lengths in question seems a strong argument in favour of the validity of the foundation of Nicholson's calculations. Serious

* A. E. Haas, *Jahrb. d. Rad. u. El.* vii. p. 261 (1910). See further, A. Schidlof, *Ann. d. Phys.* xxxv. p. 90 (1911); E. Wertheimer, *Phys. Zeitschr.* xii. p. 409 (1911), *Verh. deutsch. Phys. Ges.* 1912, p. 431; F. A. Lindemann, *Verh. deutsch. Phys. Ges.* 1911, pp. 482, 1107; F. Haber, *Verh. deutsch. Phys. Ges.* 1911, p. 1117.

† J. W. Nicholson, Month. Not. Roy. Astr. Soc. lxxii. pp. 49, 139, 677, 693, 729 (1912).

objections, however, may be raised against the theory. These objections are intimately connected with the problem of the homogeneity of the radiation emitted. In Nicholson's calculations the frequency of lines in a line-spectrum is identified with the frequency of vibration of a mechanical system in a distinctly indicated state of equilibrium. As a relation from Planck's theory is used, we might expect that the radiation is sent out in quanta; but systems like those considered, in which the frequency is a function of the energy, cannot emit a finite amount of a homogeneous radiation; for, as soon as the emission of radiation is started, the energy and also the frequency of the system are altered. Further, according to the calculation of Nicholson, the systems are unstable for some modes of vibration. Apart from such objections—which may be only formal (see p. 23)—it must be remarked, that the theory in the form given does not seem to be able to account for the well-known laws of Balmer and Rydberg connecting the frequencies of the lines in the line-spectra of the ordinary elements.

It will now be attempted to show that the difficulties in question disappear if we consider the problems from the point of view taken in this paper. Before proceeding it may be useful to restate briefly the ideas characterizing the calculations on p. 5. The principal assumptions used are:

(1) That the dynamical equilibrium of the systems in the stationary states can be discussed by help of the ordinary mechanics, while the passing of the systems between different stationary states cannot be treated on that basis.

(2) That the latter process is followed by the emission of a *homogeneous* radiation, for which the relation between the frequency and the amount of energy emitted is the one given by Planck's theory.

The first assumption seems to present itself; for it is known that the ordinary mechanics cannot have an absolute validity, but will only hold in calculations of certain mean values of the motion of the electrons. On the other hand, in the calculations of the dynamical equilibrium in a stationary state in which there is no relative displacement of the particles, we need not distinguish between the actual motions and their mean values. The second assumption is in obvious contrast to the ordinary ideas of electrodynamics, but appears to be necessary in order to account for experimental facts.

In the calculations on page 5 we have further made use

of the more special assumptions, viz. that the different stationary states correspond to the emission of a different number of Planck's energy-quanta, and that the frequency of the radiation emitted during the passing of the system from a state in which no energy is yet radiated out to one of the stationary states, is equal to half the frequency of revolution of the electron in the latter state. We can, however (see § 3), also arrive at the expressions (3) for the stationary states by using assumptions of somewhat different form. We shall, therefore, postpone the discussion of the special assumptions, and first show how by the help of the above principal assumptions, and of the expressions (3) for the stationary states, we can account for the line-spectrum of hydrogen.

§ 2. *Emission of Line-spectra.*

Spectrum of Hydrogen.—General evidence indicates that an atom of hydrogen consists simply of a single electron rotating round a positive nucleus of charge e*. The re-formation of a hydrogen atom, when the electron has been removed to great distances away from the nucleus—*e. g.* by the effect of electrical discharge in a vacuum tube—will accordingly correspond to the binding of an electron by a positive nucleus considered on p. 5. If in (3) we put $E=e$, we get for the total amount of energy radiated out by the formation of one of the stationary states,

$$W_\tau=\frac{2\pi^2me^4}{h^2\tau^2}.$$

The amount of energy emitted by the passing of the system from a state corresponding to $\tau=\tau_1$ to one corresponding to $\tau=\tau_2$, is consequently

$$W_{\tau_2}-W_{\tau_1}=\frac{2\pi^2me^4}{h^2}\left(\frac{1}{\tau_2^2}-\frac{1}{\tau_1^2}\right).$$

If now we suppose that the radiation in question is homogeneous, and that the amount of energy emitted is equal to $h\nu$, where ν is the frequency of the radiation, we get

$$W_{\tau_2}-W_{\tau_1}=h\nu,$$

* See f. inst. N. Bohr, Phil. Mag. xxv. p. 24 (1913). The conclusion drawn in the paper cited is strongly supported by the fact that hydrogen, in the experiments on positive rays of Sir J. J. Thomson, is the only element which never occurs with a positive charge corresponding to the loss of more than one electron (comp. Phil. Mag. xxiv. p. 672 (1912)).

and from this

$$\nu = \frac{2\pi^2 me^4}{h^3}\left(\frac{1}{\tau_2^2} - \frac{1}{\tau_1^2}\right). \quad . \quad . \quad . \quad . \quad . \quad . \quad (4)$$

We see that this expression accounts for the law connecting the lines in the spectrum of hydrogen. If we put $\tau_2 = 2$ and let τ_1 vary, we get the ordinary Balmer series. If we put $\tau_2 = 3$, we get the series in the ultra-red observed by Paschen* and previously suspected by Ritz. If we put $\tau_2 = 1$ and $\tau_2 = 4, 5, \ldots$, we get series respectively in the extreme ultra-violet and the extreme ultra-red, which are not observed, but the existence of which may be expected.

The agreement in question is quantitative as well as qualitative. Putting

$$e = 4{\cdot}7 \,.\, 10^{-10}, \quad \frac{e}{m} = 5{\cdot}31 \,.\, 10^{17}, \quad \text{and} \quad h = 6{\cdot}5 \,.\, 10^{-27},$$

we get

$$\frac{2\pi^2 me^4}{h^3} = 3{\cdot}1 \,.\, 10^{15}.$$

The observed value for the factor outside the bracket in the formula (4) is

$$3{\cdot}290 \,.\, 10^{15}.$$

The agreement between the theoretical and observed values is inside the uncertainty due to experimental errors in the constants entering in the expression for the theoretical value. We shall in § 3 return to consider the possible importance of the agreement in question.

It may be remarked that the fact, that it has not been possible to observe more than 12 lines of the Balmer series in experiments with vacuum tubes, while 33 lines are observed in the spectra of some celestial bodies, is just what we should expect from the above theory. According to the equation (3) the diameter of the orbit of the electron in the different stationary states is proportional to τ^2. For $\tau = 12$ the diameter is equal to $1{\cdot}6 \,.\, 10^{-6}$ cm., or equal to the mean distance between the molecules in a gas at a pressure of about 7 mm. mercury; for $\tau = 33$ the diameter is equal to $1{\cdot}2 \,.\, 10^{-5}$ cm., corresponding to the mean distance of the molecules at a pressure of about 0·02 mm. mercury. According to the theory the necessary condition for the appearance of a great number of lines is therefore a very small density of the gas; for simultaneously to obtain an

* F. Paschen, *Ann. d. Phys.* xxvii. p. 565 (1908).

intensity sufficient for observation the space filled with the gas must be very great. If the theory is right, we may therefore never expect to be able in experiments with vacuum tubes to observe the lines corresponding to high numbers of the Balmer series of the emission spectrum of hydrogen; it might, however, be possible to observe the lines by investigation of the absorption spectrum of this gas (see § 4).

It will be observed that we in the above way do not obtain other series of lines, generally ascribed to hydrogen; for instance, the series first observed by Pickering* in the spectrum of the star ζ Puppis, and the set of series recently found by Fowler† by experiments with vacuum tubes containing a mixture of hydrogen and helium. We shall, however, see that, by help of the above theory, we can account naturally for these series of lines if we ascribe them to helium.

A neutral atom of the latter element consists, according to Rutherford's theory, of a positive nucleus of charge $2e$ and two electrons. Now considering the binding of a single electron by a helium nucleus, we get, putting $E=2e$ in the expressions (3) on page 5, and proceeding in exactly the same way as above,

$$\nu = \frac{8\pi^2 me^4}{h^3}\left(\frac{1}{\tau_2^2} - \frac{1}{\tau_1^2}\right) = \frac{2\pi^2 me^4}{h^3}\left(\frac{1}{\left(\frac{\tau_2}{2}\right)^2} - \frac{1}{\left(\frac{\tau_1}{2}\right)^2}\right).$$

If we in this formula put $\tau_2=1$ or $\tau_2=2$, we get series of lines in the extreme ultra-violet. If we put $\tau_2=3$, and let τ_1 vary, we get a series which includes 2 of the series observed by Fowler, and denoted by him as the first and second principal series of the hydrogen spectrum. If we put $\tau_2=4$, we get the series observed by Pickering in the spectrum of ζ Puppis. Every second of the lines in this series is identical with a line in the Balmer series of the hydrogen spectrum; the presence of hydrogen in the star in question may therefore account for the fact that these lines are of a greater intensity than the rest of the lines in the series. The series is also observed in the experiments of Fowler, and denoted in his paper as the Sharp series of the hydrogen spectrum. If we finally in the above formula put $\tau_2=5, 6, ..$, we get series, the strong lines of which are to be expected in the ultra-red.

The reason why the spectrum considered is not observed in

* E. C. Pickering, Astrophys. J. iv. p. 369 (1896); v. p. 92 (1897).
† A. Fowler, Month. Not. Roy. Astr. Soc. lxxiii. Dec. 1912.

ordinary helium tubes may be that in such tubes the ionization of helium is not so complete as in the star considered or in the experiments of Fowler, where a strong discharge was sent through a mixture of hydrogen and helium. The condition for the appearance of the spectrum is, according to the above theory, that helium atoms are present in a state in which they have lost both their electrons. Now we must assume that the amount of energy to be used in removing the second electron from a helium atom is much greater than that to be used in removing the first. Further, it is known from experiments on positive rays, that hydrogen atoms can acquire a negative charge; therefore the presence of hydrogen in the experiments of Fowler may effect that more electrons are removed from some of the helium atoms than would be the case if only helium were present.

Spectra of other substances.—In case of systems containing more electrons we must—in conformity with the result of experiments—expect more complicated laws for the line-spectra than those considered. I shall try to show that the point of view taken above allows, at any rate, a certain understanding of the laws observed.

According to Rydberg's theory—with the generalization given by Ritz *—the frequency corresponding to the lines of the spectrum of an element can be expressed by

$$\nu = F_r(\tau_1) - F_s(\tau_2),$$

where τ_1 and τ_2 are entire numbers, and F_1, F_2, F_3, ... are functions of τ which approximately are equal to $\frac{K}{(\tau+a_1)^2}$, $\frac{K}{(\tau+a_2)^2}$, ... K is a universal constant, equal to the factor outside the bracket in the formula (4) for the spectrum of hydrogen. The different series appear if we put τ_1 or τ_2 equal to a fixed number and let the other vary.

The circumstance that the frequency can be written as a difference between two functions of entire numbers suggests an origin of the lines in the spectra in question similar to the one we have assumed for hydrogen; *i. e.* that the lines correspond to a radiation emitted during the passing of the system between two different stationary states. For systems containing more than one electron the detailed discussion may be very complicated, as there will be many different configurations of the electrons which can be taken into consideration as stationary states. This may account for the different sets of series in the line spectra emitted from the

* W. Ritz, *Phys. Zeitschr.* ix. p. 521 (1908).

substances in question. Here I shall only try to show how, by help of the theory, it can be simply explained that the constant K entering in Rydberg's formula is the same for all substances.

Let us assume that the spectrum in question corresponds to the radiation emitted during the binding of an electron; and let us further assume that the system including the electron considered is neutral. The force on the electron, when at a great distance apart from the nucleus and the electrons previously bound, will be very nearly the same as in the above case of the binding of an electron by a hydrogen nucleus. The energy corresponding to one of the stationary states will therefore for τ great be very nearly equal to that given by the expression (3) on p. 5, if we put $E=e$. For τ great we consequently get

$$\lim (\tau^2 . F_1(\tau)) = \lim (\tau^2 . F_2(\tau)) = \ldots . = \frac{2\pi^2 me^4}{h^3},$$

in conformity with Rydberg's theory.

§ 3. *General Considerations continued.*

We shall now return to the discussion (see p. 7) of the special assumptions used in deducing the expressions (3) on p. 5 for the stationary states of a system consisting of an electron rotating round a nucleus.

For one, we have assumed that the different stationary states correspond to an emission of a different number of energy-quanta. Considering systems in which the frequency is a function of the energy, this assumption, however, may be regarded as improbable; for as soon as one quantum is sent out the frequency is altered. We shall now see that we can leave the assumption used and still retain the equation (2) on p. 5, and thereby the formal analogy with Planck's theory.

Firstly, it will be observed that it has not been necessary, in order to account for the law of the spectra by help of the expressions (3) for the stationary states, to assume that in any case a radiation is sent out corresponding to more than a single energy-quantum, $h\nu$. Further information on the frequency of the radiation may be obtained by comparing calculations of the energy radiation in the region of slow vibrations based on the above assumptions with calculations based on the ordinary mechanics. As is known, calculations on the latter basis are in agreement with experiments on the energy radiation in the named region.

Let us assume that the ratio between the total amount of

energy emitted and the frequency of revolution of the electron for the different stationary states is given by the equation $W=f(\tau) . h\omega$, instead of by the equation (2). Proceeding in the same way as above, we get in this case instead of (3)

$$W=\frac{\pi^2 me^2 E^2}{2h^2 f^2(\tau)}, \quad \omega=\frac{\pi^2 me^2 E^2}{2h^3 f^3(\tau)}.$$

Assuming as above that the amount of energy emitted during the passing of the system from a state corresponding to $\tau=\tau_1$ to one for which $\tau=\tau_2$ is equal to $h\nu$, we get instead of (4)

$$\nu=\frac{\pi^2 me^2 E^2}{2h^3}\left(\frac{1}{f^2(\tau_2)}-\frac{1}{f^2(\tau_1)}\right).$$

We see that in order to get an expression of the same form as the Balmer series we must put $f(\tau)=c\tau$.

In order to determine c let us now consider the passing of the system between two successive stationary states corresponding to $\tau=N$ and $\tau=N-1$; introducing $f(\tau)=c\tau$, we get for the frequency of the radiation emitted

$$\nu=\frac{\pi^2 me^2 E^2}{2c^2 h^3} \cdot \frac{2N-1}{N^2(N-1)^2}.$$

For the frequency of revolution of the electron before and after the emission we have

$$\omega_N=\frac{\pi^2 me^2 E^2}{2c^3 h^3 N^3} \text{ and } \omega_{N-1}=\frac{\pi^2 me^2 E^2}{2c^3 h^3 (N-1)^3}.$$

If N is great the ratio between the frequency before and after the emission will be very near equal to 1; and according to the ordinary electrodynamics we should therefore expect that the ratio between the frequency of radiation and the frequency of revolution also is very nearly equal to 1. This condition will only be satisfied if $c=\frac{1}{2}$. Putting $f(\tau)=\frac{\tau}{2}$, we, however, again arrive at the equation (2) and consequently at the expression (3) for the stationary states.

If we consider the passing of the system between two states corresponding to $\tau=N$ and $\tau=N-n$, where n is small compared with N, we get with the same approximation as above, putting $f(\tau)=\frac{\tau}{2}$,

$$\nu=n\omega.$$

The possibility of an emission of a radiation of such a frequency may also be interpreted from analogy with the ordinary electrodynamics, as an electron rotating round a nucleus in an elliptical orbit will emit a radiation which according to Fourier's theorem can be resolved into homogeneous components, the frequencies of which are $n\omega$, if ω is the frequency of revolution of the electron.

We are thus led to assume that the interpretation of the equation (2) is not that the different stationary states correspond to an emission of different numbers of energy-quanta, but that the frequency of the energy emitted during the passing of the system from a state in which no energy is yet radiated out to one of the different stationary states, is equal to different multiples of $\frac{\omega}{2}$, where ω is the frequency of revolution of the electron in the state considered. From this assumption we get exactly the same expressions as before for the stationary states, and from these by help of the principal assumptions on p. 7 the same expression for the law of the hydrogen spectrum. Consequently we may regard our preliminary considerations on p. 5 only as a simple form of representing the results of the theory.

Before we leave the discussion of this question, we shall for a moment return to the question of the significance of the agreement between the observed and calculated values of the constant entering in the expressions (4) for the Balmer series of the hydrogen spectrum. From the above consideration it will follow that, taking the starting-point in the form of the law of the hydrogen spectrum and assuming that the different lines correspond to a homogeneous radiation emitted during the passing between different stationary states, we shall arrive at exactly the same expression for the constant in question as that given by (4), if we only assume (1) that the radiation is sent out in quanta $h\nu$, and (2) that the frequency of the radiation emitted during the passing of the system between successive stationary states will coincide with the frequency of revolution of the electron in the region of slow vibrations.

As all the assumptions used in this latter way of representing the theory are of what we may call a qualitative character, we are justified in expecting—if the whole way of considering is a sound one—an absolute agreement between the values calculated and observed for the constant in question, and not only an approximate agreement. The formula (4) may therefore be of value in the discussion of the results of experimental determinations of the constants e, m, and h.

While there obviously can be no question of a mechanical foundation of the calculations given in this paper, it is, however, possible to give a very simple interpretation of the result of the calculation on p. 5 by help of symbols taken from the ordinary mechanics. Denoting the angular momentum of the electron round the nucleus by M, we have immediately for a circular orbit $\pi M = \frac{T}{\omega}$, where ω is the frequency of revolution and T the kinetic energy of the electron; for a circular orbit we further have T = W (see p. 3) and from (2), p. 5, we consequently get

$$M = \tau M_0,$$

where

$$M_0 = \frac{h}{2\pi} = 1 \cdot 04 \times 10^{-27}.$$

If we therefore assume that the orbit of the electron in the stationary states is circular, the result of the calculation on p. 5 can be expressed by the simple condition: that the angular momentum of the electron round the nucleus in a stationary state of the system is equal to an entire multiple of a universal value, independent of the charge on the nucleus. The possible importance of the angular momentum in the discussion of atomic systems in relation to Planck's theory is emphasized by Nicholson *.

The great number of different stationary states we do not observe except by investigation of the emission and absorption of radiation. In most of the other physical phenomena, however, we only observe the atoms of the matter in a single distinct state, *i. e.* the state of the atoms at low temperature. From the preceding considerations we are immediately led to the assumption that the "permanent" state is the one among the stationary states during the formation of which the greatest amount of energy is emitted. According to the equation (3) on p. 5, this state is the one which corresponds to $\tau = 1$.

§ 4. *Absorption of Radiation.*

In order to account for Kirchhoff's law it is necessary to introduce assumptions on the mechanism of absorption of radiation which correspond to those we have used considering the emission. Thus we must assume that a system consisting of a nucleus and an electron rotating round it under certain circumstances can absorb a radiation of a frequency equal to the frequency of the homogeneous radiation emitted during

* J. W. Nicholson, *loc. cit.* p. 679.

the passing of the system between different stationary states. Let us consider the radiation emitted during the passing of the system between two stationary states A_1 and A_2 corresponding to values for τ equal to τ_1 and τ_2, $\tau_1 > \tau_2$. As the necessary condition for an emission of the radiation in question was the presence of systems in the state A_1, we must assume that the necessary condition for an absorption of the radiation is the presence of systems in the state A_2.

These considerations seem to be in conformity with experiments on absorption in gases. In hydrogen gas at ordinary conditions for instance there is no absorption of a radiation of a frequency corresponding to the line-spectrum of this gas; such an absorption is only observed in hydrogen gas in a luminous state. This is what we should expect according to the above. We have on p. 9 assumed that the radiation in question was emitted during the passing of the systems between stationary states corresponding to $\tau \geqq 2$. The state of the atoms in hydrogen gas at ordinary conditions should, however, correspond to $\tau=1$; furthermore, hydrogen atoms at ordinary conditions combine into molecules, *i. e.* into systems in which the electrons have frequencies different from those in the atoms (see Part III.). From the circumstance that certain substances in a non-luminous state, as, for instance, sodium vapour, absorb radiation corresponding to lines in the line-spectra of the substances, we may, on the other hand, conclude that the lines in question are emitted during the passing of the system between two states, one of which is the permanent state.

How much the above considerations differ from an interpretation based on the ordinary electrodynamics is perhaps most clearly shown by the fact that we have been forced to assume that a system of electrons will absorb a radiation of a frequency different from the frequency of vibration of the electrons calculated in the ordinary way. It may in this connexion be of interest to mention a generalization of the considerations to which we are led by experiments on the photo-electric effect, and which may be able to throw some light on the problem in question. Let us consider a state of the system in which the electron is free, *i. e.* in which the electron possesses kinetic energy sufficient to remove to infinite distances from the nucleus. If we assume that the motion of the electron is governed by the ordinary mechanics and that there is no (sensible) energy radiation, the total energy of the system—as in the above considered stationary states—will be constant. Further, there will be perfect continuity between the two kinds of states, as the difference between frequency and dimensions of the systems in successive stationary states will diminish without limit if τ increases. In the following considerations we shall for the sake of brevity refer to the two kinds of states in question as "mechanical" states; by this notation only emphasizing the assumption that the motion of the electron in both cases can be accounted for by the ordinary mechanics.

Tracing the analogy between the two kinds of mechanical states, we might now expect the possibility of an absorption of radiation, not only corresponding to the passing of the system between two different stationary states, but also corresponding to the passing between one of the stationary states and a state in which the electron is free; and as above, we might expect that the frequency of this radiation was determined by the equation $E = h\nu$, where E is the difference between the total energy of the system in the two states. As it will be seen, such an absorption of radiation is just what is observed in experiments on ionization by ultra-violet light and by Röntgen rays. Obviously, we get in this way the same expression for the kinetic energy of an electron ejected from an atom by photo-electric effect as that deduced by Einstein *, *i. e.* $T = h\nu - W$, where T is the kinetic energy of the electron ejected, and W the total amount of energy emitted during the original binding of the electron.

The above considerations may further account for the result of some experiments of R. W. Wood † on absorption of light by sodium vapour. In these experiments, an absorption corresponding to a very great number of lines in the principal series of the sodium spectrum is observed, and in addition a continuous absorption which begins at the head of the series and extends to the extreme ultra-violet. This is exactly what we should expect according to the analogy in question, and, as we shall see, a closer consideration of the above experiments allows us to trace the analogy still further. As mentioned on p. 9 the radii of the orbits of the electrons will for stationary states corresponding to high values for τ be very great compared with ordinary atomic dimensions. This circumstance was used as an explanation of the non-appearance in experiments with vacuum-tubes of lines corresponding to the higher numbers in the Balmer series of the hydrogen spectrum. This is also in conformity with experiments on the emission spectrum of sodium; in the principal series of the emission spectrum of this substance

* A. Einstein, *Ann. d. Phys.* xvii. p. 146 (1905).
† R. W. Wood, Physical Optics, p. 513 (1911).

rather few lines are observed. Now in Wood's experiments the pressure was not very low, and the states corresponding to high values for τ could therefore not appear ; yet in the absorption spectrum about 50 lines were detected. In the experiments in question we consequently observe an absorption of radiation which is not accompanied by a complete transition between two different stationary states. According to the present theory we must assume that this absorption is followed by an emission of energy during which the systems pass back to the original stationary state. If there are no collisions between the different systems this energy will be emitted as a radiation of the same frequency as that absorbed, and there will be no true absorption but only a scattering of the original radiation ; a true absorption will not occur unless the energy in question is transformed by collisions into kinetic energy of free particles. In analogy we may now from the above experiments conclude that a bound electron—also in cases in which there is no ionization—will have an absorbing (scattering) influence on a homogeneous radiation, as soon as the frequency of the radiation is greater than W/h, where W is the total amount of energy emitted during the binding of the electron. This would be highly in favour of a theory of absorption as the one sketched above, as there can in such a case be no question of a coincidence of the frequency of the radiation and a characteristic frequency of vibration of the electron. It will further be seen that the assumption, that there will be an absorption (scattering) of any radiation corresponding to a transition between two different mechanical states, is in perfect analogy with the assumption generally used that a free electron will have an absorbing (scattering) influence on light of any frequency. Corresponding considerations will hold for the emission of radiation.

In analogy to the assumption used in this paper that the emission of line-spectra is due to the re-formation of atoms after one or more of the lightly bound electrons are removed, we may assume that the homogeneous Röntgen radiation is emitted during the settling down of the systems after one of the firmly bound electrons escapes, *e.g.* by impact of cathode particles *. In the next part of this paper, dealing with the constitution of atoms, we shall consider the question more closely and try to show that a calculation based on this assumption is in quantitative agreement with the results of experiments : here we shall only mention briefly a problem with which we meet in such a calculation.

* Compare J. J. Thomson, Phil. Mag. xxiii. p. 456 (1912).

Experiments on the phenomena of X-rays suggest that not only the emission and absorption of radiation cannot be treated by the help of the ordinary electrodynamics, but not even the result of a collision between two electrons of which the one is bound in an atom. This is perhaps most clearly shown by some very instructive calculations on the energy of β-particles emitted from radioactive substances recently published by Rutherford *. These calculations strongly suggest that an electron of great velocity in passing through an atom and colliding with the electrons bound will loose energy in distinct finite quanta. As is immediately seen, this is very different from what we might expect if the result of the collisions was governed by the usual mechanical laws. The failure of the classical mechanics in such a problem might also be expected beforehand from the absence of anything like equipartition of kinetic energy between free electrons and electrons bound in atoms. From the point of view of the "mechanical" states we see, however, that the following assumption—which is in accord with the above analogy—might be able to account for the result of Rutherford's calculation and for the absence of equipartition of kinetic energy: two colliding electrons, bound or free, will, after the collision as well as before, be in mechanical states. Obviously, the introduction of such an assumption would not make any alteration necessary in the classical treatment of a collision between two free particles. But, considering a collision between a free and a bound electron, it would follow that the bound electron by the collision could not acquire a less amount of energy than the difference in energy corresponding to successive stationary states, and consequently that the free electron which collides with it could not lose a less amount.

The preliminary and hypothetical character of the above considerations needs not to be emphasized. The intention, however, has been to show that the sketched generalization of the theory of the stationary states possibly may afford a simple basis of representing a number of experimental facts which cannot be explained by help of the ordinary electrodynamics, and that the assumptions used do not seem to be inconsistent with experiments on phenomena for which a satisfactory explanation has been given by the classical dynamics and the wave theory of light.

* E. Rutherford, Phil. Mag. xxiv. pp. 453 & 893 (1912).

22

THE CONFIRMATION OF BOHR'S ATOMIC THEORY OF OPTICAL SPECTRA BY THE INVESTIGATION OF INELASTIC COLLISIONS OF SLOW ELECTRONS WITH GAS MOLECULES

J. Franck and G. Hertz

This excerpt was translated expressly for this Benchmark volume by R. Bruce Lindsay, Brown University, from "Die Bestätigung der Borschen Atom-Theorie in optischen Spectren durch Untersuchung der unelastischen Zusammenstösse langsamer Elektronen mit Gasmolecülen," Phys. Z. **20**:132–143 (1919), *by permission of the publisher.*

In a series of researches carried out since the middle of 1914 and reported in this journal under the heading of kinetics of electrons, we have been able to show that in monatomic gases without electron affinity like the rare gases and metallic vapors, electrons in collision with these atoms are elastically reflected until their velocity has attained a certain minimum value. After they have attained this critical velocity, they can lose their whole velocity on collision. In helium, neon, and mercury vapor, we were able to show further that the energy threshold corresponding to this critical velocity agrees with that for which we could observe ionization as a result of the electron collisions in the gas. We therefore concluded that the energy contributed to the atom quantumwise by the colliding electron is used for the ionization of the atom. Stark[1] as well as ourselves for long suspected a connection between the ionization potential and the energy quantum. The possibility presented itself of observing this experimentally in the case of mercury vapor. If we set the minimum energy transferred equal to $h\nu$, it turns out that of the gases previously investigated,

1. J. Stark, *Phys. Zeitschift* **9**, 85, 1908. We also call attention here to the report given by Stark in *Jahrbuch der Radioaktivität und Elektronik* **13**, 395, 1916. Stark was concerned in part with the same problem. We shall not give here a closer inspection of Stark's completely different consideration since our investigation has been concerned with only the connection between the unexcited atom and the Bohr theory. We take the same point of view with respect to the important work of Stark on the atlas of the spectra of the chemical elements, summarized in a report in *Jahrbuch der Radioaktivität und Elektronik,* **14**, 139, 1917. The discussion there was about the arrangement of the series of ionized molecules. It is only where nonionized molecules are considered that points of contract with our work occur. Finally, we must call attention also to Lenard's well-known work on the emission of light by flames containing metal vapors and concerning the arc spectra that Becher reported on in *Jahrbuch der Radioaktivität und Elektronik,* **13**, 139, 1916. The problems considered there were handled, however, from quite a different point of view from ours.

it was only in the case of mercury vapor for which the energy corresponding to 4.9 ± 0.1 electron volt was transferred that the investigation could be carried out in the part of the ultraviolet attainable to measurement (with the frequency of the emitted light corresponding to this energy). In fact, the calculated frequency agreed with that observed, namely, the line with wavelength 2536 Å. This work demonstrated that the kinetic energy of the colliding electrons is related to the excitation of transitions of outer electrons in the atom with the utilization of the energy $h\nu$. The next question to arise was this: How does the energy transfer take place if the colliding electrons have kinetic energy in excess of $h\nu$? This question was answered by an experimental investigation carried out by Åkeson in Lenard's laboratory.[2] The net result of this was that the colliding electron can retain the balance of energy over $h\nu$ as kinetic energy. For example, in the case of mercury, an energy loss of 5 electron volts was found. This suggested strongly that the Bohr theory might be tested in this way. However, application of the experimental results to the Bohr theory showed the existence of the following discrepancies. According to Bohr[3] ionization of the normal, unexcited atom takes place when an electron from the outermost quantized orbit is ejected to infinity. Expressed optically, this means that the ionization takes place through the contribution to the atom of energy $h\nu_\infty$, where ν_∞ is the ultraviolet-limiting frequency of the absorption series of the unexcited gas. The emission of light of this frequency occurs when the electron springs back from infinity to its original quantized orbit. In mercury, however, the line 2536 Å is not the ultraviolet-limiting frequency but rather the line of longest wavelength for which absorption in mercury vapor takes place at room temperature. According to Paschen's work[4], this is the first red member of the series 1.5 **S**-**mp**$_2$, with the ultraviolet limit corresponding to 1187.98 Å. The true ionization potential must therefore be at $h\nu_{1189}$ or 10.4 e V. On the other hand, at 4.9 eV, only the transition to the next quantum orbit can take place, with emission of the line with wavelength 2536 Å. The situation is the same with helium. Here we have measured 20.5 ± 0.5 eV for the first inelastic collision and also for the ionization potential. Here also from the Bohr theory an energy transfer of 21 eV is necessary to excite the first red line of the series extending into the far ultraviolet (corresponding to the Lyman series in hydrogen), whereas ionization should not take place until the energy 28 eV is reached, corresponding to the ultraviolet limit of the series. There is a similar discrepancy in hydrogen. Bohr and later van der Bijl[5] have pointed out that perhaps the observed ionization is of secondary origin since a photoelectric effect could be produced at the electrodes by the ultraviolet radiation produced in the experimental tube. This objection has proved to be valid in the case of mercury vapor. In helium and neon, on the other hand, we have shown from our current-voltage curves that the sources of the newly formed electrons lie in the immediate neighborhood of those places at which the primary electrons suffer their inelastic collisions. Hence in this case the photoelectric effect does not

2. N. Åkeson, *Sitzungsber. Heidelberg Akad. Math.—Nat. Klasse, Abt. Al,* 1914 *21.Abh.*
3. N. Bohr, *Philos. Mag.,* **26,** 1, 476, 851, 1913; **27,** 506, 914; **29,** 332, 1915; **30,** 394, 1915.
4. F. Paschen, *Ann. Phys.,* **42,** 840, 1913.
5. A. J. van der Bijl, *Phys. Rev.,* **9,** 173, 1917.

provide an explanation of this phenomenon. Besides, from the standpoint of collision ionization theory, it was completely inexplicable how in the rare gases and in metal vapors at several atmospheres pressure and small potential drops, the electron could gain the energy necessary for ionization if in the large number of elastic collisions an electron suffers until it has by repeated reflections fallen through the ionization, it still has not attained the ionization potential due to inelastic collisions. This last reason moved us to retain for the time being the idea of the identity of the work of ionization with the smallest energy quantum transferred by collision in the neutral atom. We postponed the solution of the difficulty until we had had a chance to perform further experiments on other rare gases and other metal vapors. During the intervening period in which we were unable to perform our own experiments, the field has been taken over primarily by American physicists, who much extended early work with the result that the Bohr theory has been confirmed in this domain of phenomena as conclusively as in experiments with x-ray spectra and on the fine structure of the helium lines.

[*Editor's Note:* The rest of the paper is given over to detailed results of collision experiments in metal vapors like mercury, zinc, cadmium, magnesium, barium, and the alkalies as well as polyatomic gases. There is a detailed bibliography We omit this part but include the final section of summarizing observations.]

To generalize the results obtained to date from the hypothesis of inelastic collisions of slow electrons with gas molecules, we can say that the laws attesting the validity of the Bohr theory have been confirmed in large measure. The following phenomena are susceptible to experimental investigation by the method of collisions:

1. Every transition corresponding to an absorption or emission line, accordingly every transition of an electron in the outer quantized orbits of an atom, from one quantized orbit to another, can be brought about by the collision with a free electron in which the kinetic energy of the electron suffers a change of $h\nu$.
2. The transitions that occur depend on the state of excitation of the atom.
3. For a normal unexcited atom, the possible transitions are identical with those given by the absorption series of the unexcited atom.
4. The ultraviolet-limiting frequencies of the series multiplied by the action constant h give the ionization energies of the atom. The ionization energy of the unexcited atom is given by the limiting frequency of the absorption series of the atom.
5. The investigation of inelastic collisions of the light emission associated with them provides a new method for the ordering of the lines in a series and for the determination of their coupling with each other. The series limits are available to direct observation.
6. The thoroughgoing analogy of optical absorption with the quantized energy loss of the colliding electrons provides evidence of the validity of the first Planck hypothesis of quantized absorption.

23

QUANTUM THEORY OF THE EMISSION AND ABSORPTION OF RADIATION

A. Einstein

This article was translated expressly for this Benchmark volume by R. Bruce Lindsay, Brown University, from "Strahlungs–Emission und Absorption nach der Quantentheorie," Verh. Dtsch. Phys. Ges. **18**:318–323 (1916), *by permission of the publisher.*

When Planck created the quantum theory and established his radiation law sixteen years ago, he proceeded as follows. He calculated the mean energy $\bar{E}$ of a resonator as a function of the temperature in accordance with the fundamental principle of quantum theory that he had introduced. He then used this to calculate the radiation density ρ as a function of the frequency ν and the temperature. For the latter he employed the relation between radiation density and resonator energy $\bar{E}$, set up in accordance with electromagnetic theory, namely,

$$\bar{E} = \frac{c^3 \rho}{8\pi\nu^2} \tag{1}$$

[*Editor's Note:* The term c is the velocity of light in free space.]

His deduction was of unsurpassed boldness but splendidly confirmed. The confirmation included not only the radiation formula itself but also the value of the elementary quantum derived from it. The value of $\bar{E}$ calculated from the quantum theory was confirmed by later investigations on specific heat. At the same time there was confirmation of equation 1, deduced from purely electromagnetic considerations. However, an unsatisfactory feature remained, namely, that the electromagnetic-mechanical considerations leading to equation 1 are not compatible with the fundamental idea of the quantum theory. It is not surprising that Planck himself and all other theoreticians who have concerned themselves with the matter have tried without ceasing to reform the theory so as to have it based on consistent grounds.

Since the great success of the Bohr theory of spectra, there seems to be no longer any doubt that the fundamental idea of the quantum theory must be retained. Accordingly it appears that the unity of the theory must be established by replacing the electromagnetic-mechanical considerations, which led Planck to equation 1, by purely quantum theory assumptions about the interaction of matter and radiation. In the endeavor to carry out such a program, the following reflection forced itself on me. It is attractive in its simplicity and generality.

1. THE PLANCK RESONATOR IN A RADIATION FIELD

According to the classical theory, the behavior of a monochromatic resonator in a radiation field can be easily understood by the use of the following method first applied to Brownian motion. Let E be the instantaneous energy present in the resonator. We ask what will be the energy after the expiration of time τ, which is long compared with the period of oscillation of the resonator but still sufficiently short so that the percentage change in E during τ can be treated as infinitesimal. We distinguish between two types of changes. The first is the change

$$\Delta_1 E = -AE\tau$$

which is brought about by radiation. The second is a change $\Delta_2 E$, which is brought about by the contribution of energy to the resonator by the electric field. This second type of change increases as the radiation density ρ increases and has a sign and value dependent on chance. An electromagnetic-statistical argument provides the mean-value relation

$$\overline{\Delta_2 E} = B\rho\tau$$

The constants A and B may be calculated in well-known fashion. We call $\Delta_1 E$ the energy change by emission of radiation and $\Delta_2 E$ the energy change by absorption of radiation. Since the mean value of E taken over many resonators must be independent of the time, we have

$$\overline{E + \Delta_1 E + \Delta_2 E} = E$$

from which we deduce

$$\bar{E} = \frac{B}{A}\rho$$

If now we calculate A and B in the usual way for a monochromatic resonator, using electromagnetic theory and mechanics, we come out with equation 1.

We shall now introduce a corresponding consideration based on the quantum theory without making any specific assumptions about the entities assumed to interact with the radiation. In what follows, these entities will be called "molecules."

2. QUANTUM THEORY AND RADIATION

Consider a gas all of whose molecules are identical and imagine that it is in static equilibrium with temperature radiation. Let each molecule be able to exist only in a discrete series of states $Z_1, Z_2, \ldots$, having the energy values $\epsilon_1, \epsilon_2, \ldots$. In analogy with statistical mechanics or directly from the Boltzmann principle or thermodynamics, it follows that the probability of the state Z_n, which here means the relative number of molecules in state Z_n, is given by

$$W_n = p_n e^{-\epsilon_n/kT} \tag{2}$$

where k is Boltzmann's gas constant, and p_n is the statistical weight of the state Z_n, that is, a constant characteristic of the quantum state of the molecule in question, independent of the temperature.

We shall now assume that a molecule can make a transition from state Z_n to state Z_m by absorption of radiation of frequency $\nu = \nu_{nm}$, and can also go from state Z_m to Z_n by the emission of radiation of the same frequency. In such cases the energy transferred is $\epsilon_m - \epsilon_n$. In general, this will be possible for every combination of indices n and m. For each of these elementary processes taking place at temperature equilibrium, statistical equilibrium will have to hold also. We shall therefore be able to restrict our attention to a single elementary process associated with a definite index pair (n, m).

At temperature equilibrium, just as many molecules will go from state Z_n to state Z_m by absorption of radiation as go from state Z_m to Z_n by emission. For these transitions we set up simple hypotheses to which we are led by the limiting case of the classical theory, which we have already briefly sketched.

We shall distinguish between two kinds of transitions:

a. *Emission*

Emission will be a transition from state Z_m to state Z_n with the emission of radiation energy $\epsilon_m - \epsilon_n$. This transition takes place without external influence. We can hardly think of it otherwise than as similar to radioactive decay. The number of transitions per unit time will be set equal to

$$A_m{}^n \, N_m$$

where $A_m{}^n$ is a constant characteristic of the states Z_m and Z_n, and N_m is the number of molecules in the state Z_m.

b. *Irradiation*

Irradiation is conditioned by the radiative field in which the molecule is situated. We shall assume that the irradiation is proportional to the radiation density ρ of the effective frequency. In the case of the resonators, they can just as well suffer a loss of energy as a gain of energy. This means that they can just as well suffer a transition $Z_m \to Z_n$ as a transition $Z_n \to Z_m$. The number of transitions $Z_n \to Z_m$ in unit time is therefore expressible in the form

$$B_n{}^m \, N_n \rho$$

and the number of transitions $Z_m \to Z_n$ by

$$B_m{}^n \, N_m \rho$$

where $B_n{}^m$ and $B_m{}^n$ are constants that are related to the pair of states Z_n and Z_m.

The conditions for statistical equilibrium with respect to the reactions $Z_n \to Z_m$ and $Z_m \to Z_n$ may then be written in the form

$$A_m{}^n \, N_m + B_m{}^n \, N_m \rho = B_n{}^m \, N_n \rho \tag{3}$$

On the other hand, equation 2 yields

$$\frac{N_n}{N_m} = \frac{p_n}{p_m} e^{\frac{\epsilon_m - \epsilon_n}{kT}} \tag{4}$$

From (3) and (4), we get

$$A_m^n\, p_m = \rho \left(B_n^m p_n e^{\frac{\epsilon_m - \epsilon_n}{kT}} - B_m^n\, p_m\right) \tag{5}$$

ρ is the radiation density for that frequency ν that is emitted in the transition $Z_n \rightarrow Z_m$ and absorbed in the transition $Z_m \rightarrow Z_n$. Our equation 5 gives the relation between T and ρ for this frequency. If we assume that ρ goes to infinity with T, we must necessarily have

$$B_n^m\, p_n = B_m^n\, p_m \tag{6}$$

If for convenience we further set

$$\frac{A_m^n}{B_m^n} = \alpha_{mn} \tag{7}$$

we obtain

$$\rho = \frac{\alpha_{mn}}{e^{\frac{\epsilon_m - \epsilon_n}{kT}} - 1} \tag{5a}$$

This is Planck's relation between ρ and T with the constants left undetermined. The constants A_m^n and B_m^n could be directly calculated if we were in possession of an electrodynamics and mechanics modified in the sense of the quantum theory. From the fact that ρ must be a universal function of T and ν, it follows that α_{mn} and $\epsilon_m - \epsilon_n$ cannot depend on the particular properties of the molecule but must depend only on the effective frequency ν. It follows further from the Wien displacement law that α_{mn} must be proportional to the third power of ν and $\epsilon_m - \epsilon_n$ proportional to the first power. Accordingly we must have

$$\epsilon_m - \epsilon_n = h\nu \tag{8}$$

where h is a constant.

I am naturally perfectly willing to admit that the three hypotheses concerning emission and absorption that lead to Planck's radiation formula cannot be considered certain results. But the simplicity of the hypotheses and the generality with which the method is developed as well as its natural connection with the limiting case of the Planck linear oscillator (in the sense of classical electrodynamics and mechanics) lead me to believe that it is very probable that this will form the fundamental basis of the future theory. The fact that the statistical law assumed for the emission of radiation is none other than the Rutherford law of radioactive decay and that the result expressed in (8) combined with (5a) is identical with the second fundamental hypothesis of the Bohr theory of spectra, both speak in favor of the viewpoint developed here.

3. REMARK ON THE LAW OF PHOTOCHEMICAL EQUIVALENCE

The law of photochemical equivalence affects our train of thought in the following way. Let us imagine a gas whose temperature is so low that in the thermal radiation the frequency ν corresponding to the transition from state Z_m to state Z_n does not arise to any appreciable extent.

According to equations 2 and 5a, the state Z_m will occur with extraordinary rarity as compared to the state Z_n, and we shall therefore assume that almost all the molecules of the gas are in state Z_n. Now let us imagine that it is possible for a molecule to get into state Z_m by a process other than the transition $Z_n \rightarrow Z_m$ previously discussed, namely, by an elementary chemical process, for example, due to a monomolecular dissociation. It will be further assumed that the reaction velocity of this dissociation process is large compared with the velocity (or frequency of occurrence) of the reaction transition $Z_m \rightarrow Z_n$.

What happens now, if we irradiate the gas with an effective frequency? Molecules will go continuously from the state Z_n to the state Z_m through absorption of radiation energy $\epsilon_m - \epsilon_n = h\nu$. Only rarely will these molecules go back to the state Z_n with the emission of radiation or stimulated to do so by radiation. For the most part they will suffer chemical dissociations corresponding to the assumed greater reaction velocity of this process. Hence for each dissociating molecule, practically the radiation energy $h\nu$ will be absorbed, as the equivalence law requires.

It is essential to this conception that the light absorption does not lead directly to molecular dissociations, but acts rather through the agency of the quantum state Z_m. We are therefore not permitted to distinguish between radiation absorption that is chemically effective and that which is not chemically effective. Light absorption and chemical reaction appear as independent processes.

Part VII

ENERGY AND RELATIVITY

Editor's Comments on Papers 24 and 25

24 EINSTEIN
Elementary Derivation of the Equivalence of Mass and Energy

25 PLANCK
The Principle of Relativity and the Fundamental Equations of Mechanics

In Einstein's first fundamental paper on the special or restricted theory of relativity entitled "'Zur Elektrodynamik bewegter Körper" (On the Electrodynamics of Moving Bodies), though the relativistic modification of mass in mechanics was introduced, the important relation between what Einstein later called "intrinsic mass" and energy was not stressed. However, this connection *was* definitely pointed out in a subsequent article by Einstein with the title "Ist die Trägheit eines Körpers von seinem Energieinhalt abhängig" (Does the Inertia of a Body Depend on Its Energy Content). This paper essentially presented the famous relation between intrinsic mass and energy in the formula $E = mc^2$, where m is the intrinsic mass (the mass independent of velocity), E the corresponding energy, and c the velocity of light in free space (approximately 3×10^{10} cm/s). The significance of this result was not fully grasped at the time. Einstein returned to it, however, in numerous later papers and demonstrated it from several points of view. One of the clearest and most detailed presentations is that in his Josiah Willard Gibbs Lecture before the American Mathematical Society in 1934. We reprint here the resulting paper as Paper 24. In this article the author shows that the mass-energy relation follows directly from the application of the principles of the conservation of momentum and energy to the relativistic dynamics worked out in his 1905 paper. The reader of the paper should be careful to note that the velocities of all particles are expressed as ratios of the actual velocity to that of light in free space. Thus in the equations of this paper u^2 always appears in place of the more conventional u^2/c^2.

It will be recalled that prior to Einstein's 1905 paper, H. A. Lorentz, in his article entitled "Electromagnetic Phenomena in a System Moving with any Velocity Smaller Than That of Light" (1904), had discovered the transformation equations between two coordinate systems moving with respect to each other at constant velocity, which are necessary to keep the equations of the electromagnetic field invariant with respect to the transformation. Einstein used these equations in his 1905 paper to satisfy his relativity principle not only for electromagnetic phenomena but for the whole of mechanics as well. Einstein's bold extrapolation raised many questions at the time of its first appearance. It was possibly for this reason that Max Planck turned his attention to the problem of seeing precisely how the Einstein relativity principle would affect the fundamental equations of mechanics, particularly with respect to the role of energy. The result was Planck's 1906 paper, which we reprint here in full in English translation as Paper 25. The most interesting point in Planck's paper is that he finally shows how the Lagrangian function must appear in relativity theory in order to lead to the appropriate Lagrangian equations. By setting up the relativistic Hamiltonian function, he also writes the canonical equations of Hamilton in relativistic form. The reader should note the difference in his terminology from more modern usage. It is probable that this paper of Planck made it possible for the Einstein theory of relativity to receive a somewhat more favorable welcome at the hands of theoretical physicists during the early part of the twentieth century.

REFERENCES

Einstein, A., 1905*a*. "Zur Elektrodynamik bewegter Körper," (On the Electrodynamics of Moving Bodies), *Ann. der Phys.* (4)**17,** 891.

Einstein, A., 1905*b*. "Ist die Trägheit eines Körpers von seinem Energieinhalt abhängig" (Does the Inertia of a Body Depend on Its Energy Content), *Ann. der Phys.* (4)**18,** 639.

Lorentz, H. A., 1904. "Electromagnetic Phenomena in a System Moving with any Velocity Smaller Than That of Light," *Proc. Acad. of Sci., Amsterdam,* **6,** 809.

24

Reprinted from *Am. Math. Soc. Bull.* 41:223–230 (1935)

ELEMENTARY DERIVATION OF THE EQUIVALENCE OF MASS AND ENERGY*

BY ALBERT EINSTEIN

The special theory of relativity grew out of the Maxwell electromagnetic equations. So it came about that even in the derivation of the mechanical concepts and their relations the consideration of those of the electromagnetic field has played an essential role. The question as to the independence of those relations is a natural one because the Lorentz transformation, the real basis of the special relativity theory, in itself has nothing to do with the Maxwell theory and because we do not know the extent to which the energy concepts of the Maxwell theory can be maintained in the face of the data of molecular physics. In the following considerations, except for the Lorentz transformation, we will depend only on the assumption of the conservation principles for impulse and energy.

We begin by making plausible the expressions for impulse and energy of the material particle in the well known way. The fundamental invariant of the Lorentz transformation is

$$ds^2 = dt^2 - dx^2 - dy^2 - dz^2,$$

or

$$ds = dt(1 - u^2)^{1/2},$$

where

$$u^2 = \left(\frac{dx}{dt}\right)^2 + \left(\frac{dy}{dt}\right)^2 + \left(\frac{dz}{dt}\right)^2 = u_1^2 + u_2^2 + u_3^2.$$

If one divides the components of the contravariant vector (dt, dx, dy, dz) by ds, one obtains the vector

$$\left(\frac{1}{(1 - u^2)^{1/2}}, \frac{u_1}{(1 - u^2)^{1/2}}, \frac{u_2}{(1 - u^2)^{1/2}}, \frac{u_3}{(1 - u^2)^{1/2}}\right).$$

---

* The Eleventh Josiah Willard Gibbs Lecture, delivered at Pittsburgh, December 28, 1934, under the auspices of this Society, at a joint meeting of this Society, the American Physical Society, and Section A of the A.A.A.S.

Let (dt, dx, dy, dz) belong to the world-line of a material particle of mass m. We obtain a vector connected with the motion of the latter by multiplying by m the *four-vector* of velocity that we have just written down. We thus have

$$(\eta^\sigma) = \left(\frac{m}{(1 - u^2)^{1/2}}, \frac{mu_i}{(1 - u^2)^{1/2}}\right),$$

where the index i runs from 1 to 3.

Neglecting the third power of the velocity, we can express the components of this vector by

$$(\eta^\sigma) = (m + \frac{1}{2} mu^2, \ mu_i).$$

The *space* components of (η^σ) express in this approximation the components of the impulse in classical mechanics, while the *time* component, aside from the additive constant m, expresses the kinetic energy of the material point.

If one goes back to the exact expression for (η^σ), it is therefore natural to take

$$\frac{mu_i}{(1 - u^2)^{1/2}}$$

as the impulse and

$$m\left(\frac{1}{(1 - u^2)^{1/2}} - 1\right)$$

as the kinetic energy of the particle. However, how is one to interpret the time component $m/(1-u^2)^{1/2}$ itself, the expression for which after all is the really significant one? Here it is natural to give it directly the meaning of energy, hence to ascribe to the mass-point in a state of rest the *rest-energy* m (with the usual time unit, mc^2).

Of course, this derivation cannot pretend to be a proof since in no way is it shown that this impulse satisfies the impulse-principle and this energy the energy-principle if several particles of the same kind interact with one another; it may be a priori conceivable that in these conservation-principles different expressions of the velocity are involved.

Furthermore, it is not perfectly clear as to what is meant in speaking of the *rest-energy*, as the energy is defined only to within an undetermined additive constant; in connection with this, however, the following is to be remarked. Every system can be looked upon as a material point as long as we consider no processes other than changes in its translation velocity as a whole. It has a clear meaning, however, to consider changes in the rest-energy in case changes are to be considered other than mere changes of translation velocity. The above interpretation asserts, then, that in such a transformation of a material point its inertial mass changes as the rest-energy; this assertion naturally requires a proof.

What we will now show is the following. If the principles of conservation of impulse and energy are to hold for all coordinate systems which are connected with one another by the Lorentz transformations, then impulse and energy are really given by the above expressions and the presumed equivalence of mass and rest-energy also exists.

We start from some simple kinematic consequences of the Lorentz transformation:

$$t = \frac{t' + vx'}{(1 - v^2)^{1/2}}, \qquad x = \frac{x' + vt'}{(1 - v^2)^{1/2}}, \qquad y = y', \qquad z = z',$$

where v is the relative velocity of the coordinate systems K and K'. The same relations hold for the differentials dx, etc. By suitable division one obtains the law of transformation of the velocities:

$$u_1 = \frac{u_1' + v}{1 + u_1' v}, \qquad u_2 = \frac{u_2' (1 - v^2)^{1/2}}{1 + u_1' v}, \qquad u_3 = \frac{u_3' (1 - v^2)^{1/2}}{1 + u_1' v}.$$

From this one obtains

$$u^2 = \frac{u'^2 + 2u_1' v + v^2 - u_2'^2 v^2 - u_3'^2 v^2}{(1 + u_1' v)^2},$$

and

$$\frac{1}{(1 - u^2)^{1/2}} = \frac{1 + u_1' v}{(1 - u'^2)^{1/2}(1 - v^2)^{1/2}},$$

as well as

$$\frac{u_1}{(1-u^2)^{1/2}} = \frac{u_1' + v}{(1-u'^2)^{1/2}(1-v^2)^{1/2}},$$

$$\frac{u_2}{(1-u^2)^{1/2}} = \frac{u_2'}{(1-u'^2)^{1/2}}, \qquad \frac{u_3}{(1-u^2)^{1/2}} = \frac{u_3'}{(1-u'^2)^{1/2}}.$$

We introduce now for the following the concept of the *particle-pair*. By this we understand two material points with equal and opposite velocities referred to K' (and later to be chosen of equal mass). The two particles are designated by the indices $+$ or $-$. Hence $u_+' = u_-'$, $u_{1+}' = -u_{1-}'$, etc. Applying to these our transformation equations after addition, we get

$$(1)\qquad \begin{aligned}
&\frac{1}{(1-u_+^2)^{1/2}} + \frac{1}{(1-u_-^2)^{1/2}} = \frac{2}{(1-u'^2)^{1/2}(1-v^2)^{1/2}},\\
&\frac{u_{1+}}{(1-u_+^2)^{1/2}} + \frac{u_{1-}}{(1-u_-^2)^{1/2}} = \frac{2v}{(1-u'^2)^{1/2}(1-v^2)^{1/2}},\\
&\frac{u_{2+}}{(1-u_+^2)^{1/2}} + \frac{u_{2-}}{(1-u_-^2)^{1/2}} = 0,\\
&\frac{u_{3+}}{(1-u_+^2)^{1/2}} + \frac{u_{3-}}{(1-u_-^2)^{1/2}} = 0.
\end{aligned}$$

The sums on the left sides of these equations depend, therefore, only on the velocity u' of the pair referred to the special system K' and on the relative velocity v of K' with respect to K, not however on the direction in which the particles are moving.

We remark that the equations (1) can be derived more clearly if one considers directly the transformation for the sum of the four-vectors of the velocities of a particle-pair. I have chosen the above representation, however, because the conservation laws indicate the use of this 3-dimensionally inhomogeneous manner of writing.

We now go to the actual considerations. We assume that impulse and energy of a material point are given by expressions of the form

$$I_\nu = mu_\nu F(u), \qquad E = E_0 + n.G(u), \qquad (\nu = 1, 2, 3),$$

where F and G are universal, even functions of the velocity u, which vanish for $u=0$. Then $mG(u)$ is the kinetic energy, E_0 the *rest-energy* of the material point, m the *rest-mass* or, simply, the mass. It is here assumed that impulse and energy of the *mass-point* are independent of direction of motion and of the orientation of the *mass-point* relative to the velocity. It is further assumed that in impulse and energy the *same* mass-constant m occurs, for which, however, we shall find later a partial justification.

We consider now the elastic eccentric collision between two particles of equal mass. One can always choose the coordinate-system K' so that, referred to the latter, the velocities of the masses before the collision are equal in magnitude and opposite in direction; how are the velocities after the collision constituted with reference to K'? If the velocities after the collision were not likewise equal and opposite, the impulse law would be violated. If the common velocity of both masses after the collision were not equal in magnitude to that before the collision, in case the collision is elastic, the energy law would be violated. This holds independently of the particular law of dependence of impulse and energy on the velocity. The collision, therefore, changes only the direction of motion of the two mass-points referred to K'. We can express this briefly as follows. A particle-pair before the collision is transformed by the latter again into a particle-pair of the same velocity u'.

The right side of (1) is therefore not changed by the collision. It follows then, from (1), that, referred to K, we have for the states before and after the collision the equations:

$$
\begin{aligned}
&\frac{1}{(1-u_+^2)^{1/2}}+\frac{1}{(1-u_-^2)^{1/2}}=\frac{1}{(1-\bar{u}_+^2)^{1/2}}+\frac{1}{(1-\bar{u}_-^2)^{1/2}},\\
(2)\qquad&\frac{u_{i+}}{(1-u_+^2)^{1/2}}+\frac{u_{i-}}{(1-u_-^2)^{1/2}}=\frac{\bar{u}_{i+}}{(1-\bar{u}_+^2)^{1/2}}+\frac{\bar{u}_{i-}}{(1-\bar{u}_-^2)^{1/2}}.
\end{aligned}
$$

Barred quantities are those which refer to the state after the collision. These equations, which are valid in general for elastic collisions of equal masses, have the form of conservation equations; it may therefore be taken for granted that no other symmetrical or anti-symmetrical functions of the velocity-components exist which in the present case of the elastic collision of two identically constituted material points give anything analogous. Consequently we shall have to regard

(3) $$\frac{mu_i}{(1-u^2)^{1/2}}$$

as impulse and

(4) $$m\left(\frac{1}{(1-u^2)^{1/2}}-1\right)$$

as the kinetic energy of a particle.*

We now turn to the proof of the assertion that the mass is equal to the *rest-energy*. For the total energy E of a moving particle we shall have to take

(4a) $$E = E_0 + m\left(\frac{1}{(1-u^2)^{1/2}}-1\right),$$

where we imagine that E_0 (rest-energy) and m can be changed in the case of interactions of material points that are not *elastic.*

We now consider the inelastic collision between two particles of equal mass and equal rest-energy, which before the collision again form a *particle-pair* with respect to K' (velocities equal and opposite). We assume here for simplicity that the internal changes which the material points suffer in the collision are equal to each other. From the impulse principle referred to K' it follows as above that the final velocities of the two particles must be equal and opposite ($\bar{u}_+' = -\bar{u}_-'$). The energy law gives then, referred to K' and K, respectively,

$$2E_0 + 2m\left(\frac{1}{(1-u'^2)^{1/2}}-1\right) = 2\bar{E}_0 + 2\bar{m}\left(\frac{1}{(1-\bar{u}'^2)^{1/2}}-1\right),$$

$$2E_0 + m\left(\frac{1}{(1-u_+^2)^{1/2}}-1\right) + m\left(\frac{1}{(1-u_-^2)^{1/2}}-1\right)$$

$$= 2\bar{E}_0 + \bar{m}\left(\frac{1}{(1-\bar{u}_+^2)^{1/2}}-1\right) + \bar{m}\left(\frac{1}{(1-\bar{u}_-^2)^{1/2}}-1\right).$$

As the points form a pair before and after the collision, one can write the last equation on the basis of (1) in the form

* This must naturally vanish for $u=0$; for it is defined as the expenditure of energy necessary to impart the velocity u to the particle initially at rest (without internal change).

$$E_0 - m + \frac{m}{(1-u'^2)^{1/2}(1-v^2)^{1/2}}$$
$$= \bar{E}_0 - \bar{m} + \frac{\bar{m}}{(1-\bar{u}'^2)^{1/2}(1-v^2)^{1/2}}.$$

The first equation we write analogously in the form

$$E_0 - m + \frac{m}{(1-u'^2)^{1/2}} = \bar{E}_0 - \bar{m} + \frac{\bar{m}}{(1-\bar{u}'^2)^{1/2}}. \tag{5}$$

Multiplying the last equation by $1/(1-v^2)^{1/2}$ and subtracting from the previous one we get

$$[(\bar{E}_0 - E_0) - (\bar{m} - m)]\left(\frac{1}{(1-v^2)^{1/2}} - 1\right) = 0,$$

or

$$\bar{E}_0 - E_0 = \bar{m} - m. \tag{6}$$

The rest-energy changes, therefore, in an inelastic collision (additively) like the mass. As the former, from the nature of the concept, is determined only to within an additive constant, one can stipulate that E_0 should vanish together with m. Then we have simply

$$E_0 = m,$$

which states the principle of equivalence of inertial mass and rest-energy.

If we apply the impulse law to the x-component, it follows (for an inelastic collision) that

$$m\frac{u_{+1}}{(1-u_+^2)^{1/2}} + m\frac{u_{-1}}{(1-u_-^2)^{1/2}} = \bar{m}\frac{\bar{u}_{+1}}{(1-\bar{u}_+^2)^{1/2}} + \bar{m}\frac{\bar{u}_{-1}}{(1-\bar{u}_-^2)^{1/2}},$$

or by the application of the second of equations (1), for the state before and after the collision,

$$\frac{m}{(1-u^2)^{1/2}} = \frac{\bar{m}}{(1-\bar{u}^2)^{1/2}}.$$

The same relation follows also from equations (5) and (6) which were obtained from the energy principle. If, from the beginning, we had provided the expression for the impulse with a mass-constant different from that of the energy, these considerations would show that the impulse-mass changes in an inelastic collision like the energy-mass. This is a partial justification for setting both mass-constants equal to each other.

The result of this consideration is therefore as follows. If for collisions of material points the conservation laws are to hold for an arbitrary (Lorentz) coordinate-system, the well known expressions for impulse and energy follow, as well as the validity of the principle of equivalence of mass and rest-energy.

Professor G. D. Birkhoff has called my attention to the fact that in the book which he has written in collaboration with Professor R. E. Langer, *Relativity and Modern Physics*, quite similar considerations are given concerning collisions of particles, as well as concerning impulse and energy. In spite of this, I believe that the present derivations merit a certain amount of interest.

Thus, in the book just mentioned, essential use is made of the concept of *force*, which in the relativity theory has no such direct significance as it has in classical mechanics. This is connected with the fact that, in the latter, the force is to be considered as a given function of the coordinates of all the particles, which is obviously not possible in the relativity theory. Therefore I have avoided introducing the force concept.

Furthermore, I was concerned with avoiding making any assumption concerning the transformation character of impulse and energy with respect to a Lorentz transformation.

25

THE PRINCIPLE OF RELATIVITY AND THE FUNDAMENTAL EQUATIONS OF MECHANICS

M. Planck

This article was translated expressly for this Benchmark volume by R. Bruce Lindsay, Brown University, from "Das Prinzip der Relativität und die Grundgleichungen der Mechanik," Verh. Dtsch. Phys. Ges. **8**:136-141 (1906), *by permission of the publisher.*

The principle of relativity was introduced not long ago by H. A. Lorentz (1) and in more general fashion by A. Einstein (2). It says that of two reference systems (x, y, z, t) and (x', y', z', t') connected by the relations

$$x' = \frac{c}{\sqrt{c^2 - v^2}} \cdot (x - vt), \qquad y' = y, \quad z' = z$$
$$t' = \frac{c}{\sqrt{c^2 - v^2}} \cdot \left(t - \frac{v}{c^2} x\right) \tag{1}$$

where c = velocity of light in free space, neither can be used with greater correctness than the other in the fundamental equations of mechanics and electrodynamics. Hence they can both be considered to be at rest. If the principle should be confirmed, it would bring about such an enormous simplification of all the problems of the electrodynamics of moving bodies that the question of its validity deserves a place in the foreground of every theoretical investigation in this domain. The question might indeed appear to have been already settled by the latest experimental measurements by W. Kaufmann (3), which would suggest that further investigation might well be spared. However, the theory behind these researches of Kaufmann is not too simple, and I do not consider it wholly excluded that in its more elaborate development, the relativity principle will not show itself in agreement with experimental observations. I am not inclined to attribute decisive weight to the consideration that according to the relativity principle, a moving electron is subjected to a special work of deformation since we can, in general, incorporate this work in the kinetic energy of the electron. In any case, the question of an electrodynamic explanation of the inertia of the electron remains an open one. On the other hand, this type of explanation has the growing advantage that we do not need to attribute a spherical shape or indeed any other special form to the electron in order to obtain a definite dependence of the mass on the velocity.

We may go on to say that a physical idea of the simplicity and generality of that contained in the principle of relativity deserves to be tested in more than one way. If it is incorrect, it ought to be shown to lead to absurdity. There can be no better test than by a search for the consequences to which it leads. From this standpoint, perhaps the following investigation will satisfy some needs. We shall undertake here the task of determining the relativistic form of the fundamental equations of mechanics replacing the usual Newtonian equations of motion of a free mass particle, the latter being

$$m\ddot{x} = X, \quad m\ddot{y} = Y, \quad m\ddot{z} = Z \tag{2}$$

According to the principle of relativity, equations 2 can be valid only for the special case in which $\dot{x} = \dot{y} = \dot{z} = 0$, that is, the case in which the particle is at rest. For nonvanishing velocity of the particle, that is,

$$q = \sqrt{\dot{x}^2 + \dot{y}^2 + \dot{z}^2} \tag{3}$$

the Newtonian equations must be altered. For arbitrary values of q, we might indeed simply define the quantities X, Y, Z as the products of the mass and the component accelerations, as is done in many presentations of mechanics. But in this case the forces producing motion would have no independent physical significance, and their simple connection with the potential energy would be lost. According to the principle of relativity, the equations must also hold in the primed coordinates defined by equation 1, that is, we must have

$$m\ddot{x}' = X', \quad m\ddot{y}' = Y', \quad m\ddot{z}' = Z'$$

Hence the relations between X', Y', Z' and X, Y, Z would be in the form of very complicated equations, which we should have to determine from the relations connecting x and x', etc. There would thus be no simple physical meaning for such equations.

To establish the general connection between acceleration and the forces producing motion, it is plausible to begin with a special case in which we know the relation between the force components in both reference systems. Such a case is provided by the action of an electromagnetic field in a vacuum on a particle of mass m charged with quantity of electricity e. Then we get (4) for the electric and magnetic field intensities in both reference systems the relations

$$\begin{aligned} E'_{x'} &= E_x & H'_{x'} &= H_x \\ E'_{y'} &= \frac{c}{\sqrt{c^2 - v^2}}\left(E_y - \frac{v}{c}H_z\right); & H'_{y'} &= \frac{c}{\sqrt{c^2 - v^2}}\left(H_y + \frac{v}{c}E_z\right) \\ E'_{z'} &= \frac{c}{\sqrt{c^2 - v^2}}\left(E_z + \frac{v}{c}H_y\right); & H'_{z'} &= \frac{c}{\sqrt{c^2 - v^2}}\left(H_z - \frac{v}{c}E_y\right) \end{aligned} \tag{4}$$

We suppose that the mass particle is located at the origin of the coordinates of the unprimed system (x, y, z, t) and is given the velocity components $\dot{x}$, $\dot{y}$, $\dot{z}$, relative to this system. We then ask what the equations of motion will be. This question can be answered uniquely by imagining that we locate in the mass particle the origin of a new reference system, which moves with respect to the original reference system with constant velocity components $\dot{x}$, $\dot{y}$, $\dot{z}$. We let the x axis of this system coincide with the direction of the velocity q of the mass particle, whose magnitude is expressed by equation 3. The mass particle is therefore at rest in the new reference system, and in this system the equations of motion in the simple form (2) hold for it. In these equations we may write the motion-producing force as the product of the electrical charge e and the intensity of the electric field. We now transform the equations of motion to a second reference system whose x axis again coincides with the direction of the velocity q, which reference system is now at rest in the system (x, y, z, t). As far as the acceleration components are concerned, the relations (1) serve this purpose. On the other hand, so far as the force components are concerned, the relations (4) serve, if everywhere q is put in place of v. Finally after simple rotation of axes, we go over to the system (x, y, z, t), and after carrying out all these elementary calculations, we obtain the required equations of motion in the form

$$\frac{m\ddot{x}}{\sqrt{1-q^2/c^2}} = eE_x - \frac{e\dot{x}}{c^2}(\dot{x}E_x + \dot{y}E_y + \dot{z}E_z) + \frac{e}{c}(\dot{y}H_z - \dot{z}H_y) \tag{5}$$

with two other similar equations in $\ddot{y}$ and $\ddot{z}$.

We can convince ourselves of the general validity of these equations through the consideration that according to the principle of relativity, the equations must remain correct if throughout we replace the unprimed x, y, z, t by the primed one, keeping e, c, and m unchanged. Because of equations 1 and 4, this is verified for every arbitrary value of v.

We now wish to put the equations of motion into their simplest form. If we multiply them in turn by $\dot{x}$, $\dot{y}$, and $\dot{z}$ and add, we get

$$e(\dot{x}E_x + \dot{y}E_y + \dot{z}E_z) = \frac{m(\dot{x}\ddot{x} + \dot{y}\ddot{y} + \dot{z}\ddot{z})}{(1 - q^2/c^2)^{3/2}}$$

If we set

$$eE_x + \frac{e}{c}(\dot{y}H_z - \dot{z}H_y) = X$$

with the two similar equations for Y and Z, and substitute the equation just above in equations 5, we finally have

$$\frac{d}{dt}\left(\frac{m\dot{x}}{\sqrt{1-q^2/c^2}}\right) = X \tag{6}$$

and the two similar equations for Y and Z.

Equations 6 contain the solution of the problem that has been posed. They form the generalization of the Newtonian equations of motion (2), as demanded by the principle of relativity. If we compare equations 6 with the Lagrangian equations of motion

$$\frac{d}{dt}\left(\frac{\partial H}{\partial x}\right) = X, \text{ etc.} \tag{7}$$

where H is the kinetic potential [*Editor's Note:* This function, which is the difference between the kinetic and potential energies, is written as L in current terminology. It is also called the "Lagrangian function"], we then get

$$H = -mc^2\sqrt{1 - q^2/c^2} + \text{const} \tag{8}$$

We obtain the *vis viva* law if we multiply equations 7 in turn with $\dot{x}dt$, $\dot{y}dt$, and $\dot{z}dt$ and add. There follows

$$d\left(\dot{x}\frac{\partial H}{\partial x} + \dot{y}\frac{\partial H}{\partial x} + \dot{z}\frac{\partial H}{\partial z} - H\right) = Xdx + Ydy + Zdz$$

From this relation we get the expression for the *vis viva* [*Editor's Note:* This is strictly the kinetic energy or one-half the classical *vis viva*. Planck calls it L. In current terminology, it is usually written T] as follows:

$$L = \dot{x}\frac{\partial H}{\partial x} + \dot{y}\frac{\partial H}{\partial y} + \dot{z}\frac{\partial H}{\partial z} - H = \frac{mc^2}{\sqrt{1 - q^2/c^2}} + \text{const}$$

The equations of motion (7) can be represented in the form of Hamilton's principle

$$\int_{t_0}^{t_1} (\delta H + A)dt = 0$$

in which the time t as well as the initial and final points of the path are invariant and A is the virtual work

$$A = X\delta x + Y\delta y + Z\delta z$$

Finally we set up Hamilton's canonical equations of motion. For this purpose we introduce the momentum coordinates ξ, η, ζ, such that

$$\xi = \frac{\partial H}{\partial x} = \frac{m\dot{x}}{\sqrt{1 - q^2/c^2}}$$

with the two other similar equations.

[*Editor's Note:* In current terminology ξ, η, ζ appear as p_x, p_y, and p_z.]

If now we consider the kinetic energy L as a function of ξ, η, ζ and abbreviate by writing

$$\xi^2 + \eta^2 + \zeta^2 = \rho^2$$

we get

$$L = mc^2\sqrt{1 + \rho^2/m^2c^2} + \text{const}$$

The canonical equations then become

$$\frac{d\xi}{dt} = X, \qquad \frac{d\eta}{dt} = Y, \qquad \frac{d\zeta}{dt} = Z$$

$$\frac{dx}{dt} = \frac{\partial L}{\partial \xi}, \qquad \frac{dy}{dt} = \frac{\partial L}{\partial \eta}, \qquad \frac{dz}{dt} = \frac{\partial L}{\partial \zeta}$$

All these equations hold in the reference system (x, y, z, t) as well as for every other reference system (x', y', z', t'), which is connected with the former through equations 1.

REFERENCES

1. H. A. Lorentz, *Versl. Kon. Akad. Wet. Amsterdam,* p. 809, 1904.
2. A. Einstein, *Ann. Phys.,* (4)**17,** 891, 1905.
3. W. Kaufmann, *Sitzungsber. der Preussische Akad. der Wiss.,* p. 949, 1906. See also *Ann. Phys.* (4)**19,** 487, 1906.
4. A. Einstein, *Ann. Phys.* (4)**17,** 909, 1905.

Part VIII
NUCLEAR ENERGY

Editor's Comments on Papers 26, 27, and 28

The domain of nuclear energy is enormous, and we make no pretense of covering it in this volume. It all goes back, however, to the discovery of radioactivity, and in the three papers reprinted in this group, we effectively review some of the early history of the subject, specifically from the standpoint of energy.

Shortly after the news of Röntgen's discovery of x-rays (in 1895) reached Paris, the French physicist Henri Becquerel (1852–1908) noted that in Röntgen's first experiments the x-rays appeared to originate at the fluorescent spot on the glass surface of the discharge tube on which the cathode rays impinged, and he speculated that there might be a connection between fluorescence and the production of x-rays. Becquerel had for some time been carrying out experiments on fluorescence and phosphorescence; he immediately began a new series in which he showed that a crystal of potassium uranyl nitrate previously irradiated by sunlight could fog a photographic plate completely wrapped in black opaque paper. This effect suggested that the crystal in question (a normally fluorescent material) was emitting invisible penetrating radiation in addition to its usual fluorescence. However, shortly thereafter he found that the crystal in question, even when wrapped in black paper so that it had no chance to fluoresce, would still fog a photographic plate wrapped in opaque paper. This suggested that some entirely new phenomenon was taking place, that is, that some form of invisible radiation was being emitted directly by the crystal without the intervention of any stimulating radiation. This was essentially the discovery of what Pierre and Marie Curie later decided to call "radioactivity." The search for other radioactive materials by the Curies led, of course, to the isolation of radium and polonium. Becquerel continued his own researches on the subject; his most important later achievement was to show that electrons form a part of the radiation from radium. He had nothing specifically to say about the energy associated with the radiations from radioactive materials. But the fact, in this connection, of their great penetrating power must have made a deep impression on him as it did on his immediate successors. We reprint here in English translation as Paper 26 Becquerel's famous 1896 paper in which he announced his discovery.

Once radioactivity was established as a fact, it was obvious that the energy associated with the radiation would be investigated. The first attempt at such a research was probably that of Pierre Curie (1859–1906) and A. Laborde, a student of Curie. Pierre Curie was the husband of Marie Sklodoroska Curie (1867–1934) and her collaborator in the discovery of radium and further radioactive researches. The experiment described in the Curie-Laborde paper of 1903, which we reprint in English translation as Paper 27, was performed in Curie's old laboratory in the École Municipale de Physique et Chimie in Paris shortly before he moved into his new laboratory in the Sorbonne and not long before his untimely accidental death in 1906. By means of a sensitive thermocouple, the collaborators measured the temperature difference between two vials, one containing pure barium chloride and the other a mixture of barium chloride and radium chloride. They detected a difference of 1.5°C. More elaborate instrumentation made it possible to measure the actual amount of heat emitted per unit time of radium salts and hence by extrapolation by radium itself. They reached the conclusion that 1 g of radium emits on the average 100 cal/h. They concluded that this could not be due to any ordinary chemical transformation like combustion and pointed to it as indicating the existence of an entirely new energy source on earth.

While the Curies and Becquerel were pursuing their researches on radioactivity in Paris, Ernest Rutherford (1871–1937) was studying radioactive decay at McGill University in Montreal, Canada. Rutherford and his collaborator Howard T. Barnes (1873–1950) showed from their research that the heat emitted by radium is a rather complicated affair connected with what came to be known as "radioactive disintegration," that is, the transformation of one radioactive element into another. In particular, they studied the rate of heat emission from one of the radioactive-decay products, the gas that they called "radium emanation" (now known as an isotope of the rare gas radon). They were able to show that the total amount of heat derived from the emanation from 1 g of radium is about 10,000 cal. Hence they concluded that the amount of heat released per hour by 1 cc of emanation would be in the neighborhood of 10^8 cal. This was another indication of the relatively enormous amount of energy locked up in what later became known (through Rutherford's subsequent work) as the nucleus of the atom. The fundamental 1904 paper of Rutherford and Barnes is reprinted here as Paper 28.

26

ON THE INVISIBLE RADIATION EMITTED BY PHOSPHORESCENT BODIES

H. Becquerel

This article was translated expressly for this Benchmark volume by R. Bruce Lindsay, Brown University, from "Sur les radiations invisibles émises par les corps phosphorescents," C. R. **122**:501–503 (1896), *by permission of the publisher, Gauthier Villars.*

At an earlier meeting I summarized the experiments that I have conducted in order to show evidence of the invisible radiations emitted by certain phosphorescent bodies, radiations that can travel through various objects that are opaque to light.

I have been able to extend these observations, and although I propose to continue and develop the study of these phenomena, the current interest in them prompts me to reveal today the first results I have obtained.

The experiments I am reporting have been performed with radiations emitted by thin crystalline plates of potassium uranyl double sulfate,

$$[SO^4(UO)K + H^2O]$$

The phosphorescence of this substance is very vivid, and the light persists for less than 1/100 s. The character of the luminous radiations emitted by this substance has been studied previously by my father, and I have since had occasion to point out some interesting features presented by these radiations.

One can verify very simply that the radiations emitted by this substance, when it is exposed to sunlight or to diffuse daylight, travel through not only sheets of black paper but also different metals, for example, a plate of aluminum and a thin sheet of copper. I have carried out in particular the following experiment.

A Lumiere photographic plate coated with gelantinized silver bromide enclosed in an opaque frame by a black cloth was covered on one side with a sheet of aluminum. If the film was exposed to full sunlight during a whole day, the plate did not appear to be fogged. But if one placed on the outside of the aluminum sheet a small fragment of the uranium salt, fastened down with a strip of paper, and exposed it to sunlight for many hours, the silhouette of the crystalline fragment could be seen in black on the sensitized place surface after development with ordinary procedures. The silver salt was reduced in front of the phosphorescent fragment. If the sheet of aluminum is a little thicker, the intensity of the effect is less than when traversing two sheets of black paper.

If between the fragment of the uranium salt and the aluminum sheet or the black paper, one interposes a screen made of a thin sheet of copper approximately 0.10 mm in thickness, for example, in the shape of a cross, one observes the silhouette of the cross very clearly, but with a shading indicating that the radiations have passed through the copper sheet. In another experiment, a thin sheet of copper 0.04 mm thick caused a lesser weakening of the original radiation.

Phosphorescence excited either by direct sunlight or by solar radiation reflected by the metallic mirror of a heliostat and then refracted by a prism and quartz lens, has given the same resulting phenomena.

I will particularly stress the following fact that appears to be very important and outside the phenomena that one might have expected to observe. The same crystalline fragments, placed opposite the photographic plates under the same conditions and through the same screen but protected from the excitation of the incident radiation and maintained in darkness, again produced the same photographic impressions.

Here is how I was led to make this observation. Among the preceding experiments, some were to be carried out on Wednesday, February 26, and Thursday, February 27. Since the sun did not shine on those days except in an intermittent manner, I kept the experiments all prepared and put away the frames in the darkness of a drawer in a piece of furniture, leaving the fragment of uranium salt in place. The sun did not show itself during the following days. On March 1 I developed the photographic plate expecting to find very weak images. On the contrary, the silhouette appeared with very high intensity. I immediately thought that the effect was due to the continuation of the action in the darkness, and I carried out the following experiment.

At the bottom of an opaque cardboard box, I placed a photographic plate. Then on the sensitive side of the plate, I put a fragment of uranium salt, the side facing the plate being convex so that it did not touch the gelantinized bromide save only a few points. Then beside it I arranged on the same plate another fragment of the same salt, separated from the surface of the gelatinized bromide by a thin sheet of glass. The operation was carried out in a dark room, the box was closed up, enclosed in another cardboard box and then placed in a drawer. I did the same with a frame closed with a sheet of aluminum, into which I put a photographic plate and on the outside a fragment of uranium salt. The whole was enclosed in an opaque box and put in a drawer. At the end of five hours, I developed the plates, and the silhouette of the crystal fragment appeared in black just as in the preceding experiment and just as if it had been rendered phosphorescent by light. For the fragment lying directly on the gelatine, there was scarcely any difference in the effect between the points of contact and the parts of the fragment that were about a millimeter away from the gelatine. The difference could perhaps be attributed to the different distance from the sources of active radiation. The effect of the fragment placed on the glass plate was very slightly weaker, but the shape of the fragment was distinctly reproduced. Finally, in traversing the aluminum sheet, the effect was considerably weaker but nevertheless clear.

It is important to note that the phenomena do not appear to be attributable to the luminous radiations of phosphorescence since after 1/100 s, these radiations become so weak that they are almost imperceptible.

One hypothesis that comes rather naturally to mind is to suppose that these radiations, with effects analagous to those produced by the radiations studied by Lenard and Röntgen, are invisible radiations emitted by phosphorescence whose time of persistence is infinitely greater than the time of persistence of luminous radiations emitted by these substances. Nevertheless, the present experiments, without being contrary to this hypothesis, do not completely authorize its formulation. The experiments that I am now seeking to perform will, I hope, bring about some clarification of this new kind of phenomenon.

27

ON THE HEAT RELEASED SPONTANEOUSLY BY RADIUM SALTS

P. Curie and A. Laborde

This article was translated expressly for this Benchmark volume by R. Bruce Lindsay, Brown University, from "Sur la Chaleur dégagée Spontanément par les sels de radium," C. R. 136:673-675 (1903), *by permission of the publisher, Gauthier Villars.*

We have confirmed that radium salts release heat in a continuous fashion.

An iron-constantan thermocouple, one of whose junctions is embedded in radioactive barium chloride and the other of which is embedded in pure barium chloride, indicates a difference in temperature between the two bodies.

We have experimented with two small identical vials, placing in the first 1 g of radioactive barium chloride containing approximately one sixth of its mass as radium chloride and putting in the other 1 g of pure barium chloride. The junctions of the thermocouple are placed respectively in the center of each vial in the middle of the material filling it. These vials are isolated in air in the middle of two small identical enclosures, which in turn are themselves situated in a third enclosure that is thermally insulated and in which the temperature is reasonably uniform. Variations of ambient temperature under these conditions affect the two junctions in the same way and do not influence the reading of the couple.

We have thus confirmed a difference in temperature of 1.5° between the radioactive barium chloride and the pure barium chloride, the radioactive salt having the higher temperature. As a control, we have repeated the experiment under the same conditions, with the two vials both containing pure barium chloride. The differences in temperature then observed are only of the order of 1/100 of a degree.

We have tried to evaluated quantitatively the heat released by radium in a given time. To do this, we have first compared this heat with that released by an electric current of known magnitude in a wire of known resistance.

A vial containing the radium is placed in the interior of a block of metal to which it communicates its heat. One of the junctions of the thermocouple is situated in a cavity in the block, the second being placed in a second similar block that contains no radium. When equilibrium is established, the block receives from the radium in a given time interval as much heat as it loses by conduction and radiation to the outside. The couple then indicates a certain difference in temperature between the two blocks.

Having made the above experiment, we substitute for the vial containing the radium a vial containing a fine platinum-iridium wire that can be heated by the passage of a current. We change the magnitude of the current until when equilibrium prevails, the difference in temperature of the two blocks is the same as in the preceding experiment. The heat released by the radium in the preceding experiment is then equal to that released during the same time by the current in the second experiment. The latter quantity is easy to calculate.

We then proceded to evalute the heat emitted by the radium by measuring it directly with a Bunsen calorimeter.

Before doing this experiment, we first made sure that the level of the mercury in the calorimeter stem remained perfectly steady. The vial containing the radium was during this time kept in a tube maintained at temperature 0°C in melting ice. At a given moment, the vial was introduced into the calorimeter, and it was then observed that the mercury was displaced in the stem with a perfectly uniform velocity of the order of 2.5 cm/h, for example. When the vial containing the radium is removed, the mercury stops rising at once.

One g of radioactive barium chloride, with which we made most of our experiments, emits about 14 small calories per hour, but the composition of this product is not precisely known. From its known radioactivity, it should contain about one sixth of its mass as pure radium chloride. We have also made some measurements with a sample of 0.08 g of pure radium chloride. The measurements made by the two methods lead to results of the same order of magnitude without, however, being precisely identical. It has been our intention merely in these first researches to demonstrate without question the existence of the emission of heat under various conditions and to give the order of magnitude of the phenomenon.

One gram of radium emits a quantity of heat of the order of 100 small calories per hour.

One gram atom of radium (225 g) would then emit 22,500 cal of heat per hour, an amount comparable to the heat released by the combustion of 1 g atom of hydrogen in oxygen.

The continuous release of such a quantity of heat cannot be explained by any ordinary chemical transformation. If we seek for the origin of the production of this heat in some internal transformation, this transformation would have to be of a very profound character and would have to result from a modification of the radium atom itself. However, such a transformation, if it exists, takes place with extreme slowness. For the properties of radium do not experience noticeable variation during a period of several years; Demarcay has observed no difference in the spectrum of a given sample of radium chloride in two observations separated by 5 months. If then the preceding hypothesis should prove correct, the energy put in play by atomic transformations would have to be extraordinarily great.

The hypothesis of a continuous atomic transformation is not the only one that is compatible with the observed emission of heat from radium. This emission of heat can also be explained by assuming that radium utilizes a source of energy exterior to itself of an unknown character.

28

Reprinted from *Philos. Mag.* ser. 6, 7:202–219 (1904)

HEATING EFFECT OF THE RADIUM EMANATION*

E. Rutherford and H. T. Barnes†

P. CURIE and Laborde ‡ first observed the rapid rate of heat emission of radium, and deduced that 1 gram of radium emitted heat at the rate of about 100 gram-calories per hour. In a later paper P. Curie § found that the rate of emission of heat depended upon the age of the radium preparation. The heating effect for freshly prepared radium compound was small at first, but gradually increased to a maximum after a month's interval, and remained constant over a further interval of two months.

The present experiments were undertaken with the view of seeing how the heat emission of radium is connected with its radioactivity. It has been shown by Rutherford and Soddy ‖ that the radiation emitted from a radium compound in a state of radioactive equilibrium may be divided into three parts :—

(1) A non-separable radiation consisting entirely of α rays and constituting about 25 per cent. of the total radiation.

(2) The radiation from the emanation occluded in the radium, also consisting entirely of α rays.

(3) The excited radiation produced by the emanation in the mass of the radium, and consisting of α, β, and γ rays.

(2) and (3) together constitute about 75 per cent. of the total radiation.

Some experiments have been recently made to find how much of the activity of radium is supplied directly by the emanation occluded in it. The saturation-current, between parallel plates, due to a radium preparation spread uniformly over a platinum plate, was determined by means of an electrometer. The platinum plate was then heated rapidly to a temperature sufficient to completely drive off the emanation and the saturation-current due to the radium immediately measured. There was a decrease observed corresponding to 18 per cent. of the total. The gradual decay of the excited activity left behind in the radium after the removal of the emanation is shown graphically in fig. 5, curve A (p. 213).

We may thus conclude that the emanation supplies

* A short account of the preliminary results was published in 'Nature' (Oct. 29, p. 622, 1903). Read before the American Physical Society, St. Louis, Dec. 29, 1903.

† Communicated by the Authors.

‡ *Comptes Rendus*, cxxxvi. p. 673 (1903).

§ *Société de Physique*, 1903.

‖ Phil. Mag. April 1903.

18 per cent., the non-separable activity 25 per cent., and the excited activity 57 per cent. of the total activity of radium.

The excited activity produced on bodies has been shown to be due to a deposit of radioactive matter on their surface. The term "excited activity" refers only to the radiations from this active matter. It is convenient to have a definite name for the matter itself. It is suggested that the name "emanation X" be given to it, since the matter which causes excited activity is produced directly from the emanation. This name is given from analogy to the products UrX and and ThX, which are produced directly from uranium and thorium respectively. On this nomenclature, the radium produces the emanation at a constant rate, and this in turn is transformed into the emanation X. The matter of emanation X of radium itself undergoes at least three and probably four successive changes. The nature of these changes and their connexion with the radioactivity will be discussed later.

On heating or dissolving a radium compound in an open vessel, the emanation is released and can be entirely removed by a current of air. The emanation X, which is non-volatile, is left behind with the radium, and it at once commences to lose its activity. In the course of a few hours the activity due to it has practically disappeared. The β and γ rays which are produced only by emanation X disappear from the radium at the same time, and there then remains a non-separable activity of radium consisting entirely of α rays.

At the same time that the emanation X, left behind in the radium, is undergoing change, fresh emanation X is being produced by the separated emanation, and at such a rate that the activity at any time due to the emanation X left in the radium, together with that due to the emanation X formed afresh by the emanation, is equal to the original activity of the emanation X stored up in the radium.

Since fresh emanation is being continually produced by the radium and occluded in it, the activity of the radium after falling to its minimum gradually rises again, and in the course of about a month has nearly reached its original constant value.

The experiments which will now be described were undertaken to see if the heat emission of radium varied in the same way as its activity when the emanation was removed. For this purpose, the heating effect of the radium was first determined. The emanation was then removed from it and collected by condensation in a small glass tube, and the distribution of the heating effect between the emanation and emanation X and the radium was determined, and also the variation with time of the heating effect of both the emanation and the radium from which it was separated.

Description of Apparatus.

As only about 30 milligrams of pure radium bromide were available in the experiments, special methods were devised to measure with accuracy the small heating effects involved. The maximum heat emission from 30 milligrams of radium bromide is not more than 6×10^{-4} gram-calories per second, and it was necessary to measure with certainty a heat emission of at least $\frac{1}{20}$ of this amount.

In the following experiments a simple form of differential air-calorimeter was employed. In fig. 1 is shown a simple

Fig. 1.—Differential Air-Calorimeter.

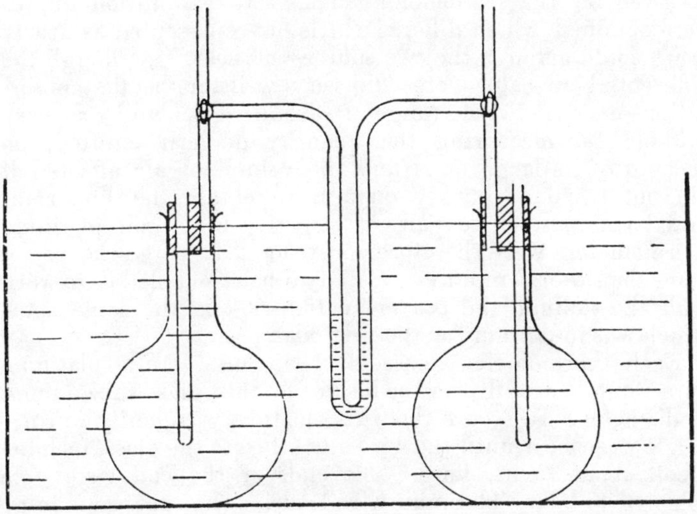

sketch of the apparatus. It consisted of two 1-litre flasks with rubber stoppers through which passed three-way glass cocks, and tubes into which the radium or the emanation-tube could be lowered. The glass cocks connected the air in the flasks either to the outside air or to a manometer-tube which registered the differences in pressure. Some little difficulty was encountered at first in selecting a suitable liquid with sufficient mobility and low vapour-pressure for the manometer-tube. We finally used xylene, which proved in every way satisfactory. The difference in level of the xylene standing in the two arms was observed with a microscope provided with a micrometer eyepiece mounted on a cathetometer-stand. The two flasks were immersed in a constant-temperature water bath, and remained very steady throughout the work. Readings were taken of the position of the xylene in the manometer-tube with the radium or emanation-tube in one flask, and then again when transferred to the other flask, allowing in each case ample time for the attainment of steady conditions. To calibrate the readings of the micrometer-scale directly in gram-calories per hour, two coils of manganin wire were constructed of about 50 ohms each. One was made in a small compact shape of about the same volume as the 30 milligrams of radium bromide introduced, while the other was made of many turns of wire on a frame of approximately the same shape as the emanation-tube. Currents of known strength were sent through the heating-coils, and the corresponding differences of pressure observed on the manometer. Thus two calibration curves were obtained, which differed a little but represented as nearly as we could arrange the two sources of heat. Although the differential air-calorimeter did very well for the first observations over the whole range of several weeks, and was very suitable for measuring the quantity of heat emitted, on account of its large mass and the volume of air affected it did not respond quickly enough to obtain the important initial changes. Two pairs of sensitive inner-coil platinum thermometers were therefore constructed, having a lag of not more than 6 or 7 minutes. The two pairs enabled us to work with the radium and the emanation-tube at the same time, which was important for the first changes.

Each thermometer consisted of 35 cms. of fine platinum wire wound carefully on the inside of a thin glass tube 5 mms. in diameter. To secure the wire, the tube was gently warmed to a low red heat until the wire stuck fast to the glass, forming a coil about 3 cms. long. The ends of the coil were gold soldered to heavy platinum-wire leads. The tube containing the radium, as well as the tube containing the emanation, was selected so as to slide easily into the interior of the coil, the wire thus being in direct contact with the glass envelope containing the source of heat.

Each thermometer was fitted centrally in a larger glass tube, passing through a long narrow water-bath (fig. 2). A stirrer in the water-bath kept the temperature always uniform. The changes in the resistance of the thermometers when the radium or emanation-tube was transferred from one to the other were measured on a Callendar type of compensated resistance-box constructed with great care. This box has already been described by one of us * in another place.

* H. T. Barnes, Phil. Trans. vol. cxcix. p. 185 (1902).

The differences were observed on a scale-and-vernier reading to the $\frac{1}{100}$ of a millimetre. The greatest difference observed

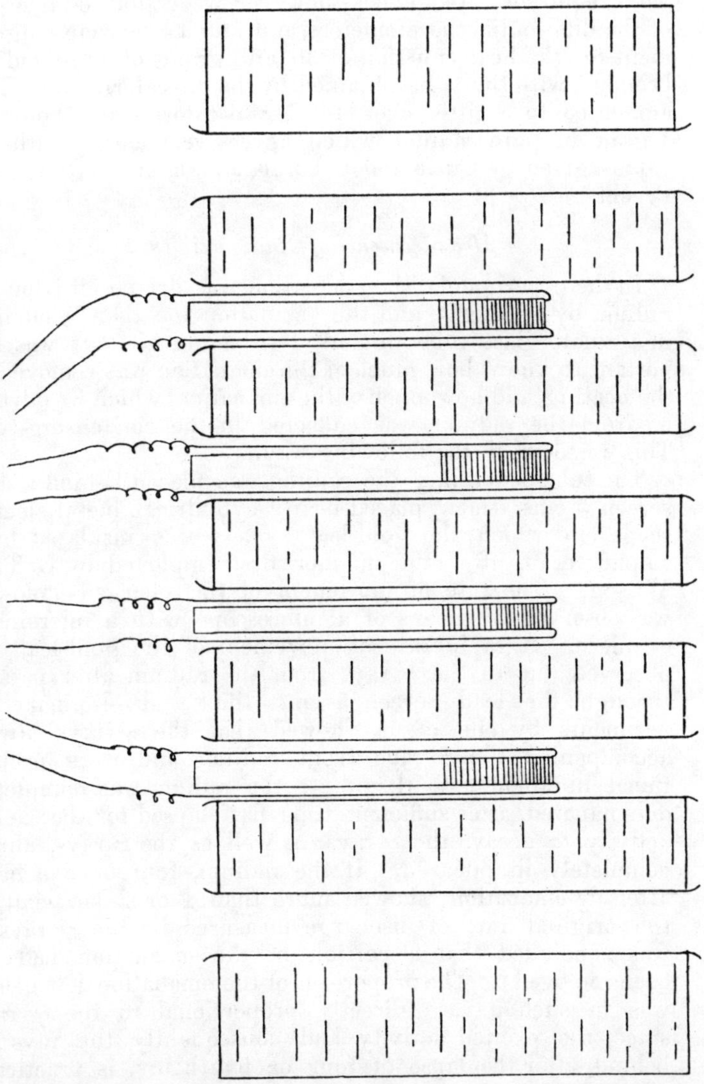

Fig. 2.—Differential Inner-coil Platinum Thermometers in Water-bath.

amounted to somewhat over 7 millimetres, and corresponded to a difference of temperature of about $\frac{3}{10}$° C., and readings could be made with certainty down to one-hundredth of this. We did not aim to obtain quantitative measurements of the heat evolved, as we were directly concerned with the initial portions of the curves, for which relative values were all that were required. As a check, however, we calibrated one pair of the differential thermometers, and found a very close agreement for the heat emission of 30 milligrams of pure radium bromide, with the value obtained by the air-calorimeter. This amounted to a little over 100 gram-calories per hour for 1 gram of pure radium, which agrees very closely with the values given by Curie and Laborde, and later by Runge and Precht*.

Measurements of Radioactivity.

In the experiments, the emanation was driven off from the radium by heating it, and the emanation was then condensed in a small glass tube immersed in liquid air. It was important to know how much of the emanation was removed by the heating, and how much of the emanation, which was driven off from the radium, was collected in the condensing-tube. This was done in the following way.

The tube containing the radium was placed behind a lead screen 5 cms. thick, placed near a cylindrical metal electroscope, in which the gold-leaf system was insulated by a sulphur bead, after the manner first employed by C. T. R. Wilson. The rate of movement of the charged gold-leaf was observed by means of a microscope with a micrometer eyepiece. A rapid rate of movement of the gold-leaf was observed due to the γ rays from the radium after passing through the lead screen 5 cms. thick. Preliminary experiments by one of us, showed that the γ rays always accompanied the β rays from radium and were proportional in amount to them. If the radium was completely de-emanated, after sufficient time has elapsed for the excited activity to decay, the γ rays, as well as the β rays, almost completely disappeared. If the radium, four or five hours after de-emanation, showed more than 1 or 2 per cent. of the original rate of discharge measured by the γ rays, it was concluded that a portion of the emanation had not been removed†. The proportion of the emanation left behind was in such a case directly proportional to the γ rays, since the excited activity and consequently the rays left behind, after the lapse of four or five hours, is practically

* *Sitz. Akad. Wiss. Berlin*, No. 38 (1903).

† In the first heating of the radium bromide to a temperature considerably below that of a red heat, the emanation was completely released. A second heating, after the radium had recovered its activity, only removed 75 per cent. of the occluded emanation.

proportional to the emanation present. If no emanation had been lost, the γ rays from the radium and the emanation-tube together should, at any time, be equal to that from the original radium. If this were not the case, it showed that a portion of the emanation had escaped into the pump, and the amount so lost would be calculated from the difference in the rates of discharge of the electroscope. These measurements could be made with rapidity and accuracy, and served as a guide to the amount of emanation present in a vessel. It must be borne in mind that the γ rays emitted only serve as a measure of the amount of emanation present, when sufficient time has elapsed (3 to 4 hours is enough) for the emanation and its product emanation X to reach a state of approximate radioactive equilibrium. Immediately after the introduction of the emanation into a vessel no β or γ rays are observed, but the intensity of the β rays and γ rays reaches about half their maximum value 45 minutes later.

Description of Experiments.

The heating effect of 30 milligrams of radium bromide, inclosed in a narrow glass tube, was first determined in the air-calorimeter. The radium tube R (fig. 3) was then connected through a phosphorus-pentoxide tube A to a short narrow glass tube T, about 3 cms. long, connected to a mercury pump P. The tube T passed through a small liquid-air vessel, made of ebonite, in order to condense any emanation passing through it.

Liquid air was placed in the vessel, and the tubes partially exhausted. The radium was then heated* with a spirit-lamp to drive off the emanation. The water-vapour given off was absorbed in the tube P and the emanation was condensed in the tube T, and the whole emanation given off was condensed in the tube T by slowly working the pump. The tube T and the radium tube were then sealed off, and the heating effect of each tested as soon as possible afterwards in the air-calorimeter. The results obtained are shown in fig. 4 (p. 210). The heating effect of the radium, when first tested, had fallen considerably, and continued to do so for about three hours, when it reached a minimum corresponding to about 30 per cent. of the original value. At the same time the emanation-tube, when first tested, showed considerable heating effect.

* When the radium was heated to about a red heat a very bright phosphorescence was produced in the radium compound. The luminosity was bright enough to observe in ordinary daylight and persisted for several days. Its spectrum was kindly photographed for us by Dr. Schenck and was found to be continuous.

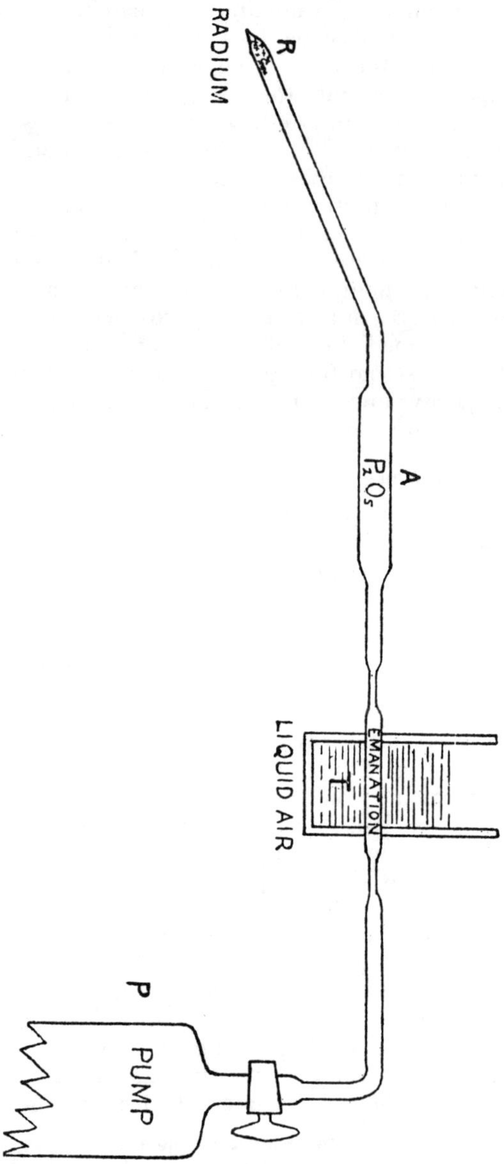

Fig. 3.—Method of Condensing the Emanation from the Radium.

This increased for about three hours, when it reached a maximum. After reaching the minimum, the radium gradually recovered its heating effect, rising to its original value,

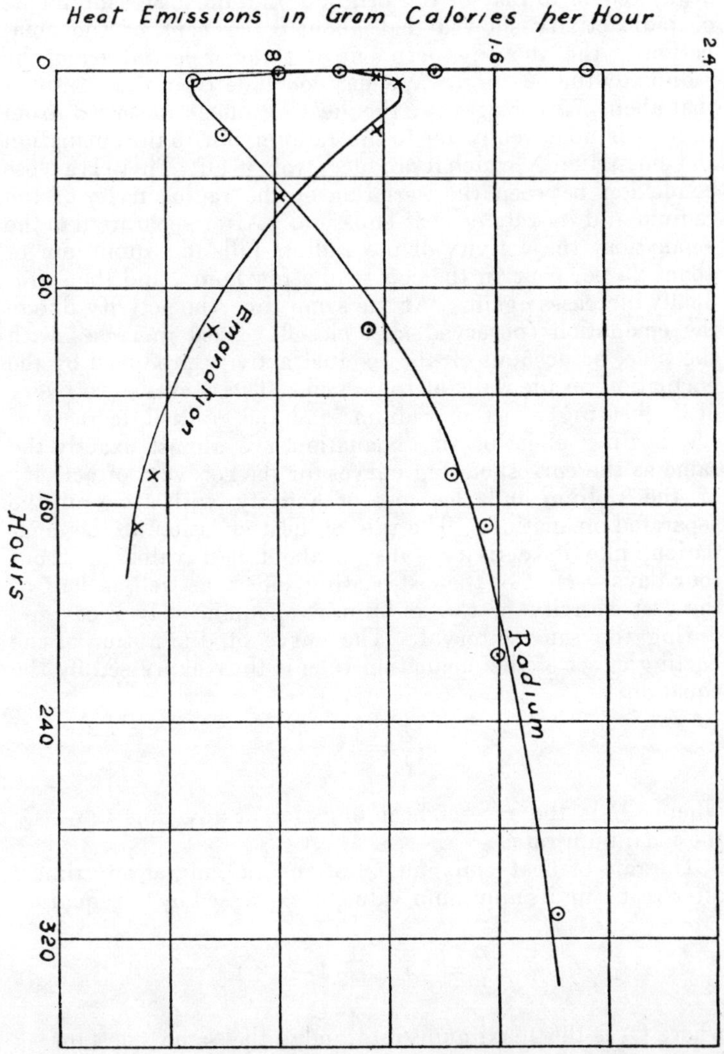

Fig. 4.

after about a month's interval. At the same time the heating effect of the emanation-tube gradually diminished according to an exponential law with the time, falling to half value in about four days.

Within the limit of experimental error the sum total of the heating effect of the radium, together with that of the emanation-tube, was, over the whole course of the experiment, always equal to that of the original radium. Measurements of radioactivity showed that about 6 per cent. of the emanation in the above experiment was not released from the radium by the heating. We may conclude from these results, that about 75 per cent. of the heating effect observed from radium is not directly due to the radium, but to the emanation and emanation X, which it produces from itself. There is a close connexion between the variation of the radioactivity of the radium and its rate of heat emission. After separation of the emanation, the activity of the radium falls to a minimum of about 25 per cent. in the course of a few hours, and then gradually increases again. At the same time, the activity due to the emanation (observed in a closed vessel) increases with the time, on account of the excited activity produced by the emanation on the walls of the vessel. The curves of recovery of the heating effect of radium, and the gradual decrease of the heating effect of the emanation, are almost exactly the same as the corresponding curves for the recovery of activity of the radium and the loss of activity with time of the separated emanation. The rate of heat emission of the emanation, like its activity, falls to about half value in about four days. Half of the lost heating effect, as well as half of the lost activity of the radium, is spontaneously recovered during the same interval. The curve of diminution of the heating effect of the emanation-tube is thus expressed by the equation

$$\frac{Q_t}{Q_0} = e^{-\lambda t},$$

where Q_t is the rate of heat emission at any time t and Q_0 the maximum rate.

The rate of heat emission Q_t of the radium, at any time t after reaching its minimum value, is expressed by the equation

$$\frac{Q_t}{Q_0} = {\cdot}25 + {\cdot}75(1 - e^{-\lambda t}),$$

where Q_0 is the maximum rate and λ the same constant as before. The same numerical constants as well as the same equation are obtained for the recovery curves of activity of de-emanated radium, measured by the α rays (see Rutherford and Soddy, Phil. Mag. April 1903).

These results, as far as they go, are in agreement with the view that the heating effect of radium is, at any time, proportional to its activity measured by the α rays; or, in other words, that the heat emission of radium is an accompaniment of the expulsion of α particles. The heat emission is not proportional to the β or γ rays; for in the experiment the amount of β and γ rays remaining in the radium, after partial de-emanation, was only 6 per cent. of the total, while the heating effect was about 30 per cent. of the total.

These experiments are not, however, sufficient to justify us in concluding that in all cases the heat emission accompanies the expulsion of α particles and is proportional to the number expelled. It will be necessary, in addition, to show that the emanation and emanation X each supplies an amount of heat proportional to its activity, measured by the α rays, and also that the heating effect of each of the succession of changes of emanation X is always proportional to its activity, measured by the α rays.

Decrease of Heating Effect of Radium immediately after De-emanation.

The air-calorimeter, already described, is not suitable for rapid observations of the variation of the heat emission of the de-emanated radium or the emanation immediately after the latter has been removed. For this purpose the differential platinum thermometers, previously described, were used. The emanation was condensed in a small glass tube 3 cms. long, 3 mms. internal diameter. As the object of these experiments was to determine the initial drop of the heating effect of the radium, only a few minutes were occupied in heating the radium and condensing the emanation. In consequence of this, a small amount of the emanation was carried over into the pump and was not condensed in the tube. The radium tube was rapidly removed and placed inside the air-bath, and observations made on its heating effect, as soon as its temperature was steady. The results are shown in fig. 5, curve B. The heating effect of the radium about 10 minutes after removal of the emanation had fallen to about 45 per cent. of the original value. It then decayed more slowly to a minimum corresponding to about 25 per cent. of the original value.

The gradual decay of the heating effect to a minimum is connected with the removal of the emanation and consequent decay of activity of the emanation X left behind. The curve of decrease of the heating effect to a constant minimum should thus be the same as the curve of decrease to zero obtained for the emanation-tube, when the emanation is removed from it. The emanation was allowed to stand four or five hours in the emanation-tube, so that the excited activity on the

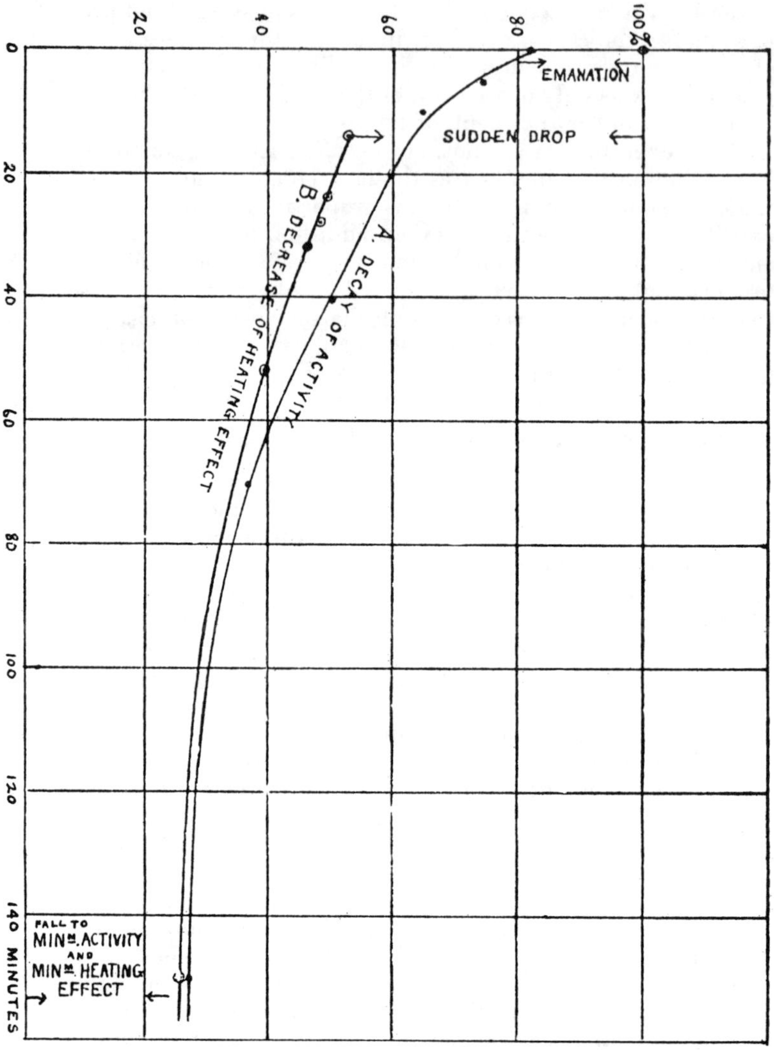

Fig. 5.

walls of the tube had reached a practical maximum. The ends of the tube were then broken, and the emanation rapidly withdrawn. The heating effect of this tube was then determined at regular intervals. The results are shown in fig. 6, curve B. The heating effect dropped rapidly during the first ten minutes after the removal of the emanation, then more

Fig. 6.

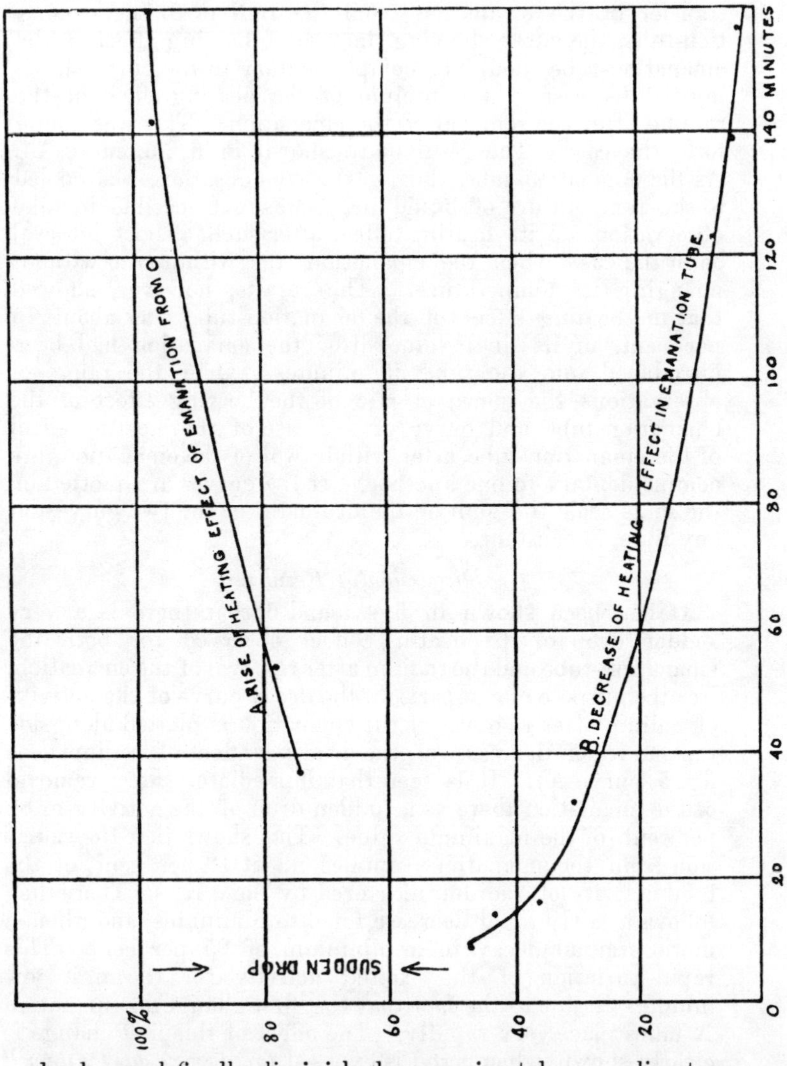

slowly, and finally diminished approximately according to an exponential law with the time, falling to half value about every 30 minutes. The curve of decrease of the heating effect to zero is about the same as the curve of decrease of the heating effect of the radium to a constant minimum.

Increase of Heat Emission after the Introduction of the Emanation into a Tube.

If the total rate of heat emission is not altered by the transfer of the emanation from the radium to the emanation-tube, the curve showing the rise of heating effect of the emanation-tube should be complementary to the curve showing the decrease to a minimum of the heating effect of the radium after the removal of the emanation. This was found to be the case. The results are shown in fig. 6, curve A*. As the emanation-tube, during the condensation, was cooled to the temperature of liquid air, it was not feasible to take observations on its heating effect after such a short interval as in the case when the emanation was withdrawn without changing its temperature. The results, however, showed that the heating effect of the emanation-tube was about 75 per cent. of its final value, after the emanation had been introduced into the tube 35 minutes. Over the range of observations, the curve of rise of the heating effect of the emanation-tube and curve of decrease of the heating effect of the emanation-tube, after withdrawal of the emanation, are complementary to one another. If the curves are plotted on the same scale the sum of the ordinates of the two curves, at any time, is constant.

Discussion of Results.

It has been shown in figs. 5 and 6 that there is a very sudden drop of the heating effect observed for both the emanation-tube and the radium after removal of the emanation. For the purpose of comparison, the decay curve of the activity of radium after removal of the emanation is plotted alongside the curve of decrease of the heating effect of radium (see fig. 5, curve A). It is seen that immediately after removal of the emanation there is a sudden drop of the activity to 82 per cent. of the maximum value. This shows that the radiation from the emanation supplies about 18 per cent. of the total activity of radium, measured by the α rays. There then follows a fairly rapid decrease for 6 to 8 minutes, and then a more gradual decay to a minimum of 25 per cent. This rapid variation of the excited activity for the first few minutes is due to the fact that the first change in emanation X takes place very rapidly. The effect of this first change is clearly shown when a rod is exposed *for a very short interval* in the presence of the radium emanation. On removal of the

* Three sets of curves were obtained, of the increase of the heating effect after introduction of the emanation, and of the decrease of the heating effect after removal of the emanation. The curves in all cases were in good agreement.

rod the activity decreases at first very rapidly, falling to about half value in about 3 minutes. After falling below 20 per cent. of the maximum value the activity varies very little for a space of about 25 minutes, when there follows a gradual decay to zero (see Rutherford and Miss Brook, Phil. Mag. July 1902).

P. Curie and Danne* have shown that the decay of activity of a body exposed for a long interval in the presence of the emanation is given by the equation

$$\frac{I_t}{I_0} = ae^{-\lambda_1 t} - (a-1)e^{-\lambda_2 t},$$

where I_t is the activity at any time t, and I_0 the activity immediately after removal; $\lambda_1 = 1/2420$, $\lambda_2 = 1/1860$ where a second is taken as the unit of time; the numerical constant $a = 4{\cdot}20$. Curie and Danne state that this equation holds accurately over the whole period of decay; there is no doubt, however, that there is a rapid initial drop for about the first ten minutes after removal which is not expressed by this equation. The equation will, however, hold if the initial observations of the activity start from a period about 10 minutes after removal, as the first change is almost completed by that time. The equation will also hold, with the same limitations, for the decay of the activity of the emanation X, left behind in the radium after removal of the emanation.

An analysis of the decay curves of excited activity, produced for different intervals of exposure in the presence of the emanation, shows that there are three well-marked changes occurring in emanation X of radium. In the first change, half the matter is transformed in 3 minutes; in the second, half in 34 minutes; and in the third, half in 28 minutes. A full account of the analysis of these changes and their peculiarities will be given by one of us in a later paper. The first change is accompanied only by α rays, the second change is *not* accompanied by α rays at all, and the third change by α, β, and γ rays.

While the curves of decrease of the rate of heat emission and of the activity of radium are very similar in shape, it is not possible to deduce directly from the result how much of the comparatively sudden drop of the heating effect observed is due to the emanation and how much to the first change in emanation X, on account of the rapid first change in the latter. The activity of the first change of emanation X (half value in 3 minutes) will have been reduced to about 10 per cent. of its original value in 10 minutes. The

* *Comptes Rendus*, cxxxvi. p. 364 (1903).

first certain observation of the heating effect of the radium in the emanation-tube could not be made until about 10 minutes after removal of the emanation. It is essential to allow an interval of at least 5 minutes in order to allow the glass envelope to take up the temperature corresponding to the source of heat inside it.

On account of the rapid initial change in emanation X, we can only conclude from the experiments that about 41 per cent. of the total heating effect of radium is due to the emanation together with the first change in emanation X. The other results all indicate that the heating effects accompany the emission of α rays; for example, the minimum heating effect due to de-emanated radium is 25 per cent. of the total, and the activity is also 25 per cent. of the total activity, and consists entirely of α rays. Both the radiation from the emanation and that from the first change in emanation X consist entirely of α rays; and there can be little doubt that the emanation does give rise to a heating effect of the same order of magnitude as is observed in the other changes, which are accompanied by α rays. Further experiments are, however, in progress to separate, if possible, the heating effect of the emanation from that due to the first change in emanation X.

The decrease of the heating effect due to the excited activity after the first sudden drop is not very rapid at first, and, in this respect, closely resembles the decay of the excited activity. It thus appears probable that the second change in emanation X, which does not give rise to α rays, is also not accompanied by the same emission of heat as in the other changes. Further experiments are in progress to settle this point.

The rate of heat emission of radium, when in a state of radioactive equilibrium, is thus made up as follows :—

Active products.	Nature of rays.	Percentage proportion of activity measured by α rays.		Percentage proportion of total heating effect.
Radium (freed from active products).	α rays.	25		25
↓				
Emanation.	α rays.	18	33	41
↓				
Emanation X (first change).	α rays.	15		
↓				
Second change.	No α rays.	0	42	34
↓				
Third change.	α, β, & γ rays.	42		

Although the experimental results are not yet complete enough to settle definitely whether the heating effect of each of the products is proportional to its activity, measured by the α rays, the results, as far as they have gone, indicate that this is approximately the case. The heating effect is certainly directly connected with the radioactivity of radium, and the time-variation of the heating effect of each of the radioactive products is the same as the time-variation of the activity. This result shows that the heat emission of radium is an accompaniment of the successive changes occurring in radium.

It still remains to be shown how much of the heat emission of radium is due to the kinetic energy of the α particles, and how much to the atomic systems from which they are expelled. In a mass of radium nearly all the α rays, emitted from it, are absorbed in the radium itself. The radium is thus subjected to an intense bombardment by the α particles projected from its own mass. There is no doubt that a proportion of the heating effect of the radium is due to this self bombardment, but probably, also, a part is due to the energy emitted consequent upon the rearrangement of the components of the atoms from which the α particles are expelled. It is not to be expected that the division of heat between the two systems would be the same for each active product, and, in consequence, that the heating effect of each product should be accurately proportional to its activity measured by the α rays.

In the experiments made on the heating effect, the α rays, in all cases, were absorbed in the glass envelope, and thus added their heating effect to that given out by the systems from which they were expelled. A large proportion of the β rays and practically all the γ radiation escaped. It has, however, been shown* that, in all probability, the energy emitted in the form of β and γ rays is only a small fraction of the energy emitted in the form of α rays.

Amount of Heat from the Emanation.

It has been shown that the heating effect from the emanation, together with that of the active product to which it gives rise, is equal to about 75 per cent. of the total heat emission of radium. Since 1 gram of radium emits heat at a rate of 100 gram-calories per hour, the emanation from 1 gram of radium in a state of radioactive equilibrium, a few hours after its removal, radiates heat at the rate of 75 gram-calories per hour. Since the rate of heat emission at any

* Rutherford and Grier, Phil. Mag. Sept. 1902.

time (after it has reached a maximum value) is given by

$$\frac{Q_t}{Q_0} = e^{-\lambda t},$$

the total amount of heat given out from the emanation released from 1 gram of radium is equal to

$$\int_0^{\infty} Q_0 e^{-\lambda t}\, dt = \frac{Q_0}{\lambda}.$$

Since the heating effect of the emanation decays to half value in about 3·73 days $\lambda = {\cdot}0077$ when an hour is taken as the unit of time. The total amount of heat derived from the emanation from 1 gram of radium is thus about 10,000 gram-calories. Now it has been shown (Rutherford, 'Nature,' Aug. 20, p. 366, 1903) that the volume of the emanation released from 1 gram of radium probably lies between 6×10^{-4} and 6×10^{-5} c.c., at standard pressure and temperature. The amount of heat liberated *per hour* from 1 c.c. of the emanation would thus lie between $1{\cdot}25 \times 10^{5}$ and $1{\cdot}25 \times 10^{6}$ gram-calories. This amount of heat from 1 c.c. of the emanation would probably be sufficient to raise to a red heat, if not to melt down, the glass tube containing it.

The emanation behaves as if it were a gas of heavy molecular weight. Assuming for the purpose of calculation that the molecule of the emanation is 100 times as heavy as the molecule of hydrogen, it can readily be deduced that 1 gram of the radium emanation, in its succession of changes, would radiate an amount of energy lying between 2×10^{9} and 2×10^{10} gram-calories. One pound weight of the emanation would initially radiate energy at the rate of 10^{4} to 10^{5} horse-power and, while the heating continued, would emit an amount of energy between 6×10^{4} and 6×10^{5} horse-power-days.

Quite independently of any assumptions, a result of the same order of magnitude can be deduced from the observed heating effect of the emanation, and the fact that the emanation has not, so far, been detected either by its volume or its weight.

There is thus no doubt that matter under special conditions is capable of emitting an amount of energy enormous compared with that released in the most intense chemical reactions. On the disintegration hypothesis (Rutherford and Soddy, Phil. Mag. May 1903) this energy is derived from the energy latent in the radium atoms, and is released in the successive stages of their disintegration.

Part IX

BROWNIAN MOTION AND MOLECULAR REALITY

Editor's Comments on Papers 29 and 30

29 EINSTEIN
Theoretical Comments on Brownian Motion

30 LANGEVIN
On the Theory of Brownian Motion

The origin of the Brownian motion has already been described in Paper 7, which reprints Robert Brown's famous 1828 article. For a long time thereafter, the effect was considered a mystery and indeed a bizarre contradiction of the principles of thermodynamics. It was not until the kinetic theory of gases became well established that a connection between molecular motion in a fluid and the Brownian motion began to be suspected. It was apparently the German mathematician Christian Wiener (1826–1896) of Karlsruhe who first seriously considered this possibility and set it forth in a paper in *Annalen der Physik* (1863). His views were made more specific by the Scottish chemist Willaim Ramsay (1852–1916), who in 1876 definitely attributed the Brownian motion to the impacts of the constituent molecules of the fluid in which the Brownian particles were immersed. Elaborate measurements of the motion were made in 1906 by the Swedish physical chemist The Svedberg (1884–1921), but the most detailed studies of the phenomenon were those of the French physical chemist Jean Perrin (1870–1942), whose book *Brownian Movement and Molecular Reality* (1910), is the classic work on the subject. We reprint here two early theoretical investigations of the Brownian motion, namely, those by Einstein (1907) and Langevin (1908), respectively (Papers 29 and 30).

Einstein's paper is a very short one, but with his usual keen insight, he saw into the heart of the problem of comparing theory with experiment. By a simple application of standard kinetic theory to the motion of Brownian particles, he showed that the average speed of such particles is too large for ordinary experimental measurement. He then assumed that each particle moves through the fluid subject to fluid friction (Stoke's law). On integrating the equation of motion, he found the root-mean-square displacement of such a particle in a single direction in a time interval τ as proportional to τ. This at once provided an equation that the experimentalists could use to determine independently the value of Boltzmann's gas constant *k* and undoubtedly stimulated more elaborate experimental measurements like Perrin's.

In some autobiographical notes (Schillip, 1949), Einstein expressed his strong feeling that the theoretical work he and others had done on the Brownian motion, leading to agreement with experiment, had convinced even skeptics like Wilhelm Ostwald (1853–1952) and Ernst Mach (1838–1916) of the fundamental reality of atoms. This was probably true in the case of Ostwald, though there exists no irrefutable evidence that the phenomenologist Mach was ever converted to a genuine belief in the existence of atoms.

Langevin's method of attack is similar to Einstein's, though he presents somewhat greater detail and puts more emphasis on the use of the concept of energy in his derivation. He of course arrives at the same formula for the root-mean-square displacement of a Brownian particle. This expression naturally contains the Boltzmann constant *k*, which can then be evaluated by the substitution of experimental results. The value obtained turned out to be in good agreement with that from the kinetic theory of gases (Lindsay 1941). This provided encouraging confirmation of what Perrin referred to as "molecular reality."

REFERENCES

Lindsay, R. Bruce, 1941. *Introduction to Physical Statistics.* New York: Wiley. (Reprinted by Dover Publications, 1968, p. 100.)

Perrin, Jean, 1910. *Brownian Movement and Molecular Reality* (trans. Frederick Soddy). London: Taylor and Francis.

Schillip, P. A., ed., 1949. *Albert Einstein: Philosopher-Scientist.* Evanston, Ill.: The Library of Living Philosophers, p. 49.

Wiener, Christian, 1863. "Erklärung der atomistischen Wesens des tropfbarflüssigen Körperzustandes und Bestätigung desselben durch die sogannten Molekular Gewegungen," *Ann. Phys. Chem.*, **118,** 79.

29

THEORETICAL COMMENTS ON BROWNIAN MOTION

A. Einstein

This article was translated expressly for this Benchmark volume by R. Bruce Lindsay, Brown University, from "Theoretische Bemerkungen über die Brownsche Bewegung," Z. Elektrochem. **13**: 41–42 (1907), *by permission of the publisher.*

Stimulated by the report of the investigations of Svedberg on the motion of suspended particles, which has recently appeared in the *Zs. fur Elektrochemie,* I have thought it worthwhile to call attention to some properties of this motion, as demanded by the molecular theory of heat. I hope in what follows to help those physicists who do experiments in this field in the interpretation of their observed results and in the comparison of these results with the theory.

1. The molecular theory of heat permits us to calculate the average value of the instantaneous velocity of a particle at absolute temperature T. The kinetic energy of the motion of the center of mass of a particle is independent of its size and constitution and also independent of the nature of its surroundings, for example, the nature of the fluid in which the particle is suspended. This kinetic energy is equal to that of a monatomic gas molecule. The average velocity $\sqrt{\overline{v^2}}$ of such a particle is therefore determined by the equation

$$\frac{m\overline{v^2}}{2} = \tfrac{3}{2}\,\frac{RT}{N}$$

in which $R = 8.3 \times 10^7$, T is the absolute temperature, and N is the number of actual molecules in a gram molecule (about 4×10^{23}).

[*Editor's Note:* The author is evidently using cgs units.]

We shall compute $\sqrt{\overline{v^2}}$ as well as other quantities connected with particles in a colodial platinum solution such as Svedberg has investigated. For these particle $m = 2.5 \times 10^{-15}$, so that for $T = 292$, we get

$$\sqrt{\overline{v^2}} = \quad \frac{3RT}{mN} = 8.6 \text{ cm/s}$$

2. We now look into the question of whether there is any chance of really observing this enormous velocity for a suspended particle. If we were in complete ignorance of the molecular theory of heat, we would expect the following. Let us assume that through some external impulsive force, we communicate a velocity to a particle suspended in fluid. This velocity will be quickly reduced to zero by the friction ex-

erted by the fluid. We neglect the inertia of the latter and consider that the resistance experienced by the particle moving with velocity v is $6\pi kPv$, where k is the viscosity coefficient of the fluid, and P the radius of the particle

[*Editor's Note:* This is the familiar Stokes law.]

We then obtain the equation

$$m\frac{dv}{dt} = -6\pi kPv$$

For the time θ taken by the particle to have its velocity reduced to $1/10$ of its initial value, we have

$$\theta = \frac{m}{(0.434)\ (6\pi kP)}\,\mathrm{s}$$

In the case of the above-mentioned platinum particles suspended in water, we get[1]

$$\theta = 3.3 \times 10^{-7}\,\mathrm{s}$$

If we now revert to the molecular theory of heat, we have to modify the preceding discussion. We must now also assume that the particle loses nearly all its initial velocity through friction in the very short time θ. At the same time we must also assume that during this time the particle will experience new impulses to motion by a process inverse to friction

[*Editor's Note:* For example, by collisions with the actual molecules of the solution.]

so that it actually retains a velocity that on the average is equal to $\sqrt{\overline{v^2}}$. Since we must assume this motional impulse to be independent in magnitude and direction of the magnitude and direction of the initial motion, we conclude that the speed and direction of motion of the particle must change very quickly in the extraordinary short time θ and indeed in a very irregular fashion.

It is therefore impossible, at least for ultramicroscopic particles, to evaluate $\sqrt{\overline{v^2}}$ from observations.

3. If we limit ourselves to the investigation of the paths of the particles, or more precisely the changes in position in times τ that are essentially greater than θ, then we obtain from the molecular theory of heat

$$\sqrt{\overline{\lambda_x^2}} = \sqrt{\tau}\sqrt{\frac{RT}{N}\cdot\frac{1}{3\pi kP}}$$

1. For "microscopic" particles, θ is significantly greater since other things being equal, θ is proportional to the square of the radius of the particle.

if we designate by λ_x the change in the x coordinate of the particle produced in time τ.

We can define as the average velocity of the particle during the time interval τ the quantity.

$$\frac{\sqrt{\overline{\lambda_x^2}}}{\tau} = \frac{w}{\sqrt{\tau}}$$

in which we have used the abbreviation

$$\frac{RT}{N} \cdot \frac{1}{6\pi kP} = w$$

However, the average velocity is the greater the smaller τ is. As long as τ is large compared to θ, the average velocity does not approach a finite limit as τ decreases.

Since an observer operating with definite equipment in a definite way can never perceive the displacements occurring in arbitrary short time intervals, it follows that a certain average velocity will always appear to him as an instantaneous velocity. It is clear, however, that the velocity so ascertained cannot correspond to any objective property of the motion under consideration, at least if the theory is to agree with facts.

30

ON THE THEORY OF BROWNIAN MOTION

M. P. Langevin

This article was translated expressly for this Benchmark volume by R. Bruce Lindsay, Brown University, from "Sur la theorie du mouvement brownien," C. R. **146**:530–533 (1908), *by permission of the publisher, Gauthier Villars.*

I. The very great theoretical interest presented by the phenomena of the Brownian motion has been indicated by M. Gouy.[1] We owe to this physicist the clear formulation of the hypothesis that views the continuous motion of particles suspended in a fluid a manifestation of molecular thermal agitation as well as its experimental justification at least qualitatively, by demonstrating the permanence of the Brownian motion and its independence of external actions when the latter do not change the temperature of the medium.

A quantitative verification of the theory has been rendered possible by A. Einstein,[2] who has recently provided a formula predicting at the end of a given time τ the mean square $\overline{\Delta_x^2}$ of the displacement Δ_x of a spherical particle in a given direction x as a result of the Brownian motion in a liquid, as a function of the radius a of the particle, of the viscosity μ of the liquid, and the absolute temperature T. This formula is

$$\overline{\Delta_x^2} = \frac{RT}{N} \cdot \frac{1}{3\pi\mu a} \cdot \tau \tag{1}$$

where R is the gas constant for a gram molecule of an ideal gas and N is the number of molecules in a gram molecule, a well-known number, having approximately the value 8×10^{23}.

M. Smoluchowski[3] has tackled the same problem by a method more direct than those employed by Einstein, in the two successive demonstrations he has given of his formula; he has obtained for $\overline{\Delta_x^2}$ an expression of the same form as (1), but differing from it by the factor 64/27.

II. I have been able to confirm at once that a correct application of the Smoluchowski method leads precisely to the Einstein formula and that moreover it is easy to give an infinitely simpler demonstration of a wholly different method.

The point of departure is always the same: The theorem of equipartition of kinetic energy among the various degrees of freedom of a system in thermal equilibrium demands that a particle in suspension in any fluid possesses in the x direction

1. L. G. Gouy, *Journ. de Phys.* (Second Series), **7**, 561, 1888. Cf. also *Comptes Rendus,* **109**, 102, 1889.
2. A. Einstein, *Ann. Phys.* (Fourth Series), **17**, 549, 1905; **19**, 371, 1906.
3. M. Smoluchowski, *Ann. de Physik* (Fourth Series), **21**, 756, 1906.

a mean kinetic energy $RT/2N$ equal to that of a gas molecule of any kind, in a given direction, at the same temperature. If $\xi = dx/dt$ is the velocity at a given instant of the particle in the direction considered, we then have for the average over a large number of identical particles of mass m

$$m\overline{\xi^2} = \frac{RT}{N} \tag{2}$$

A particle like that under consideration, large compared to the mean distance between the molecules of the liquid and moving with respect to them with the velocity ξ is subject to a viscous resistance equal to $6\pi\mu a\xi$ according to Stokes formula. In reality, this value is only an average, and because of the irregularity of the collisions of the surrounding molecules, the action of the fluid on the particle oscillates around the preceding value so that the equation of motion in the x direction is

$$m\frac{d^2x}{dt^2} = -6\pi\mu a \cdot \frac{dx}{dt} + X \tag{3}$$

Concerning the complementary force X, we know that it is indifferently positive and negative and its magnitude is such that it is able only to maintain the motion of the particle that would otherwise come to rest because of the viscous resistance.

If we multiply equation 3 by x, we can write the result

$$\frac{m}{2}\left(\frac{d^2x^2}{dt^2}\right) - m\xi^2 = -3\pi\mu a\frac{dx^2}{dt} + Xx \tag{4}$$

If we consider a large number of identical particles and take the average of the equations 4 for each of them, the average value of the term Xx is clearly zero because of the random nature of the complementary actions X. In placing $z = d\overline{x^2}/dt$, we then have

$$\frac{m}{2}\left(\frac{dz}{dt}\right) + 3\pi\mu az = \frac{RT}{N}$$

The general solution

$$z = \frac{RT}{N} \cdot \frac{1}{3\pi\mu a} + Ce^{-\frac{6\pi\mu a}{m} \cdot t}$$

reduces to the first constant term when a permanent regime is established at the end of a time interval of the order $m/6\pi\mu a$ or about 10^{-8} s for particles that manifest the Brownian motion.

Thus for a permanent regime we have

$$\frac{d\overline{x^2}}{dt} = \frac{RT}{N} \cdot \frac{1}{3\pi\mu a}$$

from which, for an interval of time τ, we obtain

$$\overline{x^2} - \overline{x_0{}^2} = \frac{RT}{N} \cdot \frac{1}{3\pi\mu a} \cdot \tau$$

The displacement Δ_x of a particle is given by

$$x = x_0 + \Delta_x$$

and since the displacements are indifferently positive and negative,

$$\overline{\Delta_x{}^2} = \overline{x^2} - \overline{x_0{}^2} = \frac{RT}{N} \cdot \frac{1}{3\pi\mu a} \cdot \tau$$

which is formula 1.

III. A first attempt at experimental verification has just been made by M. Svedberg[4], whose results differ from those predicted by formula 1 only in the ratio 1:4 and approach indeed more nearly those predicted by Smoluchowski's formula.

[*Editor's Note:* Perrin's later elaborate measurements mentioned in the Editor's Comments on Papers 31 and 32 show decisive agreement with the Einstein-Langevin formula 1.]

The two new demonstrations I have obtained for Einstein's formula, in following for one of them the path suggested by Smoluchowski, appear to depart definitely from the modification suggested by the latter.

Moreover, the fact that Svedberg does not really measure the quantity $\overline{\Delta_x{}^2}$, which appears in the Einstein formula, and the uncertainty with respect to the real diameter of the ultramicroscopic granules that he has observed call for new measurements made preferably on microscopic granules of dimensions that can be known more precisely and for which the application of Stokes formula, which neglects the effects of the inertia of the liquid, is certainly more legitimate.

[*Editor's Note:* Perrin's measurements met the condition laid down here.]

4. T. Svedberg, *Studien zur Lehre von den Kolloiden Losungen,* Upsala, 1907.

Part X

ENERGY AND THE PROPERTIES OF SOLIDS

Editor's Comments on Papers 31, 32, and 33

31 DRUDE
Excerpt from *On the Electron Theory of Metals*

32 DEBYE
Excerpt from *On the Theory of Specific Heats*

33 LANGEVIN
On the Theory of Magnetism

We close this volume with extracts from three significant papers dealing with the role of the concept of energy in the atomic theory of the solid state. These articles are by Paul Drude (1863–1906), Peter Debye (1884–1966), and Paul Langevin (1872–1946), respectively, and represent early twentiety-century efforts to account theoretically for large-scale properties of solids in terms of their atomic constitution and in which the energy associated with the atoms plays a prominent if not indeed the predominant role. The properties in question are the thermal and electrical conductivities, the specific heats, and the magnetic properties of solids.

Of this group of papers we reprint first as Paper 31 the English translation of the first part of Drude's 1900 paper on the electron theory of metals. Here the author applied the standard Maxwell-Boltzmann kinetic theory of gases (without, however, using the Maxwellian distribution of molecular velocities) to the motions of electrified particles in the solid. At first he termed these particles "electrons," but in most of his paper, he referred to them as "nuclei," a rather unfortunate choice of terms since his "nuclei" have no relation to the atomic nuclei of the later Rutherford theory. Drude assumed that his nuclei carry charges that are integral multiples of the universal charge e, which is now associated with the electron as understood in modern atomic theory.

In Drude's theory the conduction of heat is due to the transfer of the kinetic energy of the charged nuclei through the agency of collisions. The standard kinetic theory equations then lead to an expression for the average transfer of heat in terms of the temperature gradient.

From this there follows the thermal conductivity as a function of the average speed of each type of nucleus in the solid, the mean free paths of the nuclei and the number of nuclei per unit volume. The classical Boltzmann gas constant also enters into this expression. Drude considered the kinetic theory expression for the mean free path too questionable and too complicated for his use and hence did not actually attempt to calculate values of thermal conductivity from known universal constants. He expected to be able to compute values of the mean free path by substituting the experimental values of the thermal conductivity into his formula.

Drude treated the electric current in a metal as due to the drift velocity of the charged nuclei superposed on the random kinetic theory velocity by the action of the applied electric field. This drift motion is, of course, impeded by collisions, thus introducing the mean free path again as an important factor in the theory. On the basis of this fundamental assumption, Drude was able to calculate the electrical conductivity in terms of the charge e, the number of nuclei per unit volume, the mean free path, and the temperature. Lack of knowledge of the mean free path again made it impossible to calculate actual values of conductivity. However, Drude's great achievement here was the derivation of the Wiedemann-Franz law for the ratio of the thermal to the electrical conductivity.

In the later portions of his article, which are not reprinted here, Drude applied his theory to the thermoelectric and related effects. Though his results had to be modified later in the century by the use of quantum mechanics, Drude's theory was a pioneer effort of the highest quality and significance. He spent his professional career largely at the Universities of Göttingen, Leipzig, and Giessen. He died shortly after his move to Berlin in 1905. He is most famous for his work in theoretical optics in which he developed extensively Maxwell's electromagnetic theory of light.

Another example of the important role of the concept of energy in atomic physics is provided by the theory of specific heats. Theoretical predictions in this domain go back to the molecular theory of gases of Maxwell and Boltzmann (see Paper 14). Application of this theory to solids yields the simple Dulong and Petit law that the gram molecular heat capacity of all monatomic crystalline solids shall be the same and equal to 6 cal/°C. This result, however, takes no account of the observed dependence of the specific heats of solids on temperature. In 1907 Einstein pointed out that the assumption of equipartition of energy among the various degrees of freedom of the constituent particles leading to the preceding result does not agree with the quantum theory. He proposed that the particles of a solid vibrating with frequency ν should be assigned energy $h\nu$, where h is Planck's constant, and that the average energy should be that given by Planck's formula (see Paper 15). Application of this idea produced a formula in which the specific heat of a solid does indeed decrease as the absolute temperature is lowered, in qualitative agreement with experiment. However, the quantitative agreement left much to be desired. In 1911 Einstein suggested that the assumption of more than a single frequency might be required to rectify the difficulty. In 1912 Debye seized on this idea, and the result was the famous paper reprinted here as Paper 32. Debye evaluated the spectrum of natural frequencies of an elastic solid by the usual acoustical method; but in summing over the energies with the use of the Planck formula, he used for the sake of simplicity an integral instead of a sum. He introduced for each solid a characteristic temperature θ given by the maximum natural frequency and then was able to show that the specific heat of a monatomic solid is a universal function of T/θ, where T is the absolute temperature. He also proved that for sufficiently low temperatures, the specific heat of such a solid is proportional to the third power of T, in general agreement with experiment. This paper set the stage for the further development of the quantum mechanical theory of lattice dynamics of solids (Born and Huang 1954).

Paper 32 is the English translation of the first part of Debye's paper stressing the fundamental method and the agreement of the results with experiment. We omit the later parts with their detailed examination of the basic elastic equations and the setting up of the elastic energy.

Debye was Dutch by birth but spent a large part of his professional career in Switzerland and Germany. He came to the United States in 1940 and served as professor of chemistry in Cornell University until his retirement in 1950.

Paul Langevin was a distinguished French experimental and theoretical physicist whose researches covered a wide domain of physics, including molecular theory of gases, Brownian motion (see Paper 30), x-rays, relativity, ultrasonics, and magnetism. His theoretical studies of the last-named field were particularly important through his attempt to explain paramagnetism and diamagnetism of solids. Langevin's professional career was spent almost entirely in Paris at the College de France and the École Municipale de Physique et Chimie Industrielles; he was for a long time director of this school. Paper 33 is the (complete) English translation of Langevin's celebrated 1905 paper on the theory of magnetism. This paper begins with a brief review of the known facts about the three important manifestations of magnetic phenomena, namely, ferromagnetism, paramagnetism, and diamagnetism, noting the difficulties encountered up to the beginning of the twentieth century in the attempts to account theoretically for the existence of the latter two, which are, of course, the most commonly encountered forms of magnetism. Langevin then calls attention to the development (current in his time) of the electron theory of matter, particularly through the work of J. J. Thomson (see Paper 20) and H. A. Lorentz (1936). He decided to seek the explanation of magnetism in the motion of the electrons constituting a vital part of the atoms making up any solid. He applied his theory to the motions of electrons in closed orbits and showed that in certain cases a nonvanishing magnetic moment results and the material is then paramagnetic. If due to the existence of a high degree of symmetry in the atomic structure, the resultant magnetic moment is zero, the material is polarized diamagnetically by the application of an external magnetic field, and we say the material is intrinsically diamagnetic. Langevin then deduced from his theory the law of Curie that the paramagnetic susceptibility is inversely proportional to the absolute temperature. Langevin did not devote much attention to ferromagnetism, merely emphasizing that it is associated with a more complicated behavior of the constituent charged particles. Nevertheless, his paper was a very important pioneer step in the attempt to provide a theoretical understanding of magnetism and one on which the later quantum mechanical studies could successfully build. The concept of energy entered vitally in all aspects of his work.

REFERENCES

Born, M., and K. Huang, 1954. *Dynamical Theory of Crystal Lattices.* New York: Oxford University Press.

Lorentz, H. A., 1936. "Theory of a System of Charged Particles Moving Through the Aether Without Entraining It," *Collected Papers of H. A. Lorentz,* Vol. II. The Hague: Martinus Nijhoff, pp. 228–255.

31

ON THE ELECTRON THEORY OF METALS

P. Drude

This excerpt was translated expressly for this Benchmark volume by R. Bruce Lindsay, Brown University, from "Zur Elektronentheorie der Metalle," Ann. Phys. (Leipzig) ser. 4, 1:567–613 (1900), *by permission of the publisher.*

INTRODUCTION

W. Weber was the first to suggest that in its fundamental nature, metallic conduction of electricity is not too different from the conduction of electricity in electrolytes, in so far as the electrical current is produced by the transport of small electrified particles. This view was later further developed by Giese.[1] In line with recently developed terminology, I shall call these particles "electrons," or to employ a more convenient expression, "electrical nuclei" or "nuclei" for short.

[*Editor's Note:* It must be remembered that Drude was writing this paper some dozen years before the introduction of the Rutherford-Bohr nuclear atom model. His nuclei must not be confused with contemporary atomic nuclei.]

I prefer to avoid the expressions "corpuscles" or "ions" since both terms imply the view that the electrical particles have a certain small intrinsic mass. I believe it is more appropriate to reserve the term "ion" for the aggregate of electrical nuclei of ponderable mass, which we encounter in electrolysis.

For the time being I leave undecided whether an electron has an intrinsic, ponderable mass. I shall remark only that it is unnecessary to attribute mass to an electron in order to ascribe to it a certain kinetic energy and an inertia with respect to changes of motion. For example, we can do the latter as in the case of the magnetic field deflection of cathode rays and in the optical properties of metals. Since every moving electron represents an electric current that produces a certain number of magnetic lines of force in the surrounding medium (aether), the effort to change the direction and magnitude of the velocity of the electron must give rise to opposing forces that are brought into play by the change in the number of magnetic lines of force, or as we may say for short, the self-induction of the electron. Since the self-induction of a current of given strength is the greater the smaller the area of cross-section into which it is concentrated, the electrically interpreted apparent mass of an electron must depend on its charge and spatial extension. We shall investigate later the way in which the apparent mass of an electron can be approximately calculated.

The viewpoint adopted here has the following advantage over the usual conception in which the electron possesses a real, ponderable mass. From this standpoint metallic conduction differs from electrolytic conduction in that it is only in the

1. W. Giese, Wied. Ann., 37, 576, 1889.

latter that a weighable transport of material takes place through the medium through which the current passes, whereas in conduction in a metal there is no observable transport of material. Moreover, aside from the production of Joule and Thomson heat, the metal is in no way altered by the passage of the current. In order to explain this difference between metals and electrolytes, Giese assumes that a conduction electron in a metal, after going a short distance, transfers its electric charge to a mass particle, which it strikes. This transfer of charge is, however, difficult to understand from a pictorial point of view and leads to complications that cannot be resolved without the introduction of further hypotheses. In any case, it seems preferable to avoid this idea of the release and transfer of charge.

One can accomplish this as follows: One can assume that the electrons in all metals are of two kinds, charged positively and negatively, respectively, but the rest of their properties (charge and mass, etc.) are identical in all metals. Even when the diffusion velocities of the positive and negative electrons are different in an electric circuit formed of metals, no mass displacement will take place like the changes in concentration at the electrodes in electrolysis. It is true that this conception would require that the ratio of charge to mass of the positive and negative electrons are two universal constants. This is inconsistent with the optical properties of metals, if one takes into consideration their galvanomagnetic and thermomagnetic properties. But this difficulty is avoided if the electrons are assumed to be massless.

I have recently considered the optical properties of metals from the standpoint of the electron theory.[2] I have shown there that for the representation of the optical properties of gold and copper, we must necessarily assume the existence of bound electrons, that is, electrons that are located at certain equilibrium positions in the metal or, in other words, are bound to material atoms or molecules. It was this extension of the theory of the optical properties of metals that led me to undertake a revision of the electron theory of electric conductivity, heat conductivity, the Thomson effect, contact potentials, thermoelectricity, and the galvanomagnetic and thermomagnetic properties of metals. I hoped that thereby one can get a grip on the problem of how many electrons in a metal are free and how many are bound. It is only in optical phenomena that the latter play a role. It is through the building of such a foundation that the number of constants needed for the representation of the optical behavior of metals can be limited so that one can use the optical properties to draw conclusions about the properties of the bound electrons and thereby hope not only to achieve a theory that is broad enough to encompass all the properties previously named but also to draw probable conclusions about metallic properties that have so far not been subject to direct observation.

Riecke[3] has recently developed an electron theory of metals, which covers the fields named save for the optical properties. My theory has many points of contact with his work but also many points of difference. In the latter category I might mention that according to my theory, certain numerical relations may be deduced without the introduction of special hypotheses such as those that have already been verified in other domains, e.g., the gas laws, the Loschmidt number, and the assumption of the value of the electron charge, which has now been pretty well established

2. P. Drude, *Phys. Zeitschrift,* **1**, 161, 1900.
3. E. Riecke, *Wiedemanns Ann.,* **66**, 353, 545, 1898.

at any rate in order of magnitude, particularly through the work of J. J. Thomson. The numerical relations in question are those that connect heat conductivity with electric conductivity, the Thomson effect, contact potential differences, thermoelectric power, etc.

Because of the very large domain to be covered, I am not in a position to guarantee that the theory set forth here will not be in contradiction to experience in some points. However, since the theory has been confirmed in so many cases and has not so far led to contradictions, I have decided to forego publication no longer. I am further encouraged to do this since everyone has greeted with pleasure the appearance of Riecke's work, even though that author himself considers his results to be incomplete and has expressed the desire that other investigators should follow up his preliminary efforts in greater detail. I might indeed use these same words of his as the motto of my work. In any case the numerical factors should be more precisely determined since I have not carried out my computations using the Maxwell distribution of velocities. Here I content myself with as thorough a survey as possible.

Part I as presented here will be followed by Part II on the galvanomagnetic and thermomagnetic properties of metals. Part III will handle the optical properties.

1. FUNDAMENTAL ASSUMPTIONS

We begin by assuming arbitrarily many types of freely moving nuclei (electrons) with different charges, $e_1, e_2, e_3, \ldots$ (measured electrostatically). These charges can be different both in sign and magnitude. In order not to throw overboard the well-established position of the fundamental quantum of electricity e, we shall assume that the e_n are all integral multiples of e. It is at least plausible that such multiples of e can exist. If, for example, $e_2 = 2e$, we shall refer to the type as positive double nuclei; if $e_2 = -2e$, we have negative double nuclei, etc. We let the numbers of the various types of nuclei per unit volume of the metal (in cm^3) at absolute temperature T be $N_1, N_2, N_3, \ldots$

We shall accept the representations of the kinetic theory of gases for these moving nuclei. We assume that in spite of their strong mutual attraction, a positive nucleus and a negative nucleus will not coalesce into a neutral aether point but will both remain freely movable. The justification of this assumption is that as soon as the kinetic energy of the nuclei becomes greater than a certain value (compare the analogy of the comets that go round the sun but never return), this will enable them to escape their mutual attraction. The ponderable atoms of the body can also possess kinetic energy, but these atoms are essentially bound to equilibrium sites.

In order not to complicate matters unduly, we shall assume that all nuclei of type 1 have the same speed u_1 at temperature T, though they move irregularly to and fro in all directions in the metal. Let their mean freepaths all be equal to ℓ_1. For type 2, this becomes ℓ_2, and so on for the rest.

We shall write the kinetic energy of a nucleus of type 1 as $\frac{1}{2}m_1u_1^2$. Here m_1 denotes the intrinsic, ponderable mass, if the nucleus has one. If this is not so, then m_1 has another meaning, which will be derived later from the self-induction of the moving nucleus. In that case we call m_1 the apparent mass of the nucleus of type 1.

According to a theorem of Boltzmann[4] for temperature equilibrium, the mean

4. L. Boltzmann, *Sitzungsber. k. Akad. Wiss. zu Wien (II)*, **58**, 517, 1868.

kinetic energy for every moving particle must be the same. Hence we must have

$$\tfrac{1}{2}\, m_1 u_1{}^2 = \tfrac{1}{2}\, m_2 u_2{}^2 = \cdots = \alpha T \tag{1}$$

where T is the absolute temperature and α is a universal constant. We shall not assume an internal kinetic energy of the particles, even if they have a compound structure. A simple nucleus is just a place in the surrounding aether from which $4\pi e$ electric lines of force emerge (or conversely $4\pi e$ lines are absorbed). A double nucleus is a place from which $8\pi e$ lines emerge, etc.

Even when only simple nuclei are in question, all having the same charge and the same apparent mass, differences can show up in different metals since the spatial dimensions of the nuclei, i.e., the divergence of the electric lines of force, can be affected by the ponderable atoms or molecules of the metal. For the sake of generality, we may assume, to begin with, that even in a single homogeneous body, for example, a metallic alloy, several types of the same kind of nucleus can exist and have different apparent masses.

If a particular place in the body contains no free electric charge, it is still not necessary that the equation

$$e_1 N_1 + e_2 N_2 + e_3 N_3 + \cdots + = 0 \tag{2}$$

should,[5] for this equation will be valid only if N is the total number of all nuclei. In our analysis, however, N is only the number of the *moving* nuclei. If equation 2 is not satisfied at a charge-free point, it means that there are some bound nuclei in the body.

In general, we shall assume that bound nuclei exist. The phenomena of cathode rays and canal rays make it plausible to assume that these bound nuclei in metals carry a positive charge. This is also suggested by the fact that in electrolytes the metal ions always appear as positive. Cathode rays correspond to the emission of free, negatively charged nuclei, whereas the canal rays consist of an emission of free and bound positively charged nuclei. The latter carry the ponderable atoms of the metal, to which they are bound, with them, and hence the intrinsic mass corresponding to the atomic weight of the metal.[6] Their kinetic energy therefore corresponds almost entirely to the kinetic energy of the moving mass since because of the transport of the heavy mass, they have a much smaller speed than the free electrons, which have no intrinsic mass. Further quantitative investigations of cathode and canal rays will perhaps make possible the determination of the free and bound electrons at the metal electrode.

2. THE UNIVERSAL CONSTANT α

In the kinetic theory of gases, it is shown that the pressure p exerted at temperature T by a gas having N molecules/cm^3, all moving with speed u and each having mass m, is

$$p = \tfrac{1}{3}\, N m u^2$$

5. Cf. *Wiedemanns Ann.*, **66**, 356, 1898.
6. E. Riecke *(Wiedemanns Ann.*, **66**, 570, 1898) drew this conclusion from an investigation of W. Wien.

If in this we put

$$\tfrac{1}{2} mu^2 = \alpha T$$

we get

$$p = \tfrac{2}{3} \alpha N T \qquad (3)$$

This gas law has also been confirmed for the osmotic pressure produced by the ions in an electrolyte not only formally but also with the same numerical constants.

If now a metal is immersed in an electrolyte, if temperature equilibrium prevails, the free electrons (nuclei) of the metal must possess the same kinetic energy as the ions in the electrolyte. Hence in our equation 1, the constant α can be determined by the use of the gas law.

According to Loschmidt[7] the number N of the molecules at 25°C or 298°K and p = 1 atm pressure lies between the values 0.025×10^{20} and 2.50×10^{20}. For 0°C or 273°K, N therefore lies between 0.027×10^{20} and 2.7×10^{20}. F. Richarz[8] has obtained the same order of magnitude for N. If therefore we set, as an order of magnitude value, $N = 10^{20}$ at T = 273°K and p = 1 atm pressure (= 1.013×10^6 dyn/cm^2), we obtain

$$1.013 \times 10^6 = \tfrac{2}{3} \alpha \cdot 10^{20} \cdot 273$$

or

$$\alpha = 5.6 \times 10^{-17} \qquad (4)$$

in cgs units.

[*Editor's Note:* Examination of the text shows that $\alpha = \tfrac{2}{3} k$, where k is the Boltzmann gas constant. The accepted modern value for this is $k = 1.37 \times 10^{-16}$ erg/°C. This corresponds to $\alpha = 20.5 \times 10^{-17}$ in contrast to Drude's value. This difference is, of course, caused by Drude's choice of the Loschmidt number as 10^{20}. If he had used the modern accepted value of 2.687×10^{19}, he would have obtained the value α corresponding to the modern accepted value for k.]

3. HEAT CONDUCTION FOR THE CASE IN WHICH THE NUMBER OF NUCLEI IS INDEPENDENT OF THE TEMPERATURE

We postulate the general principle that heat conduction can take place only through the collision of electrons, that is, that the ponderable atoms in their motion about their equilibrium positions do not collide with each other and hence cannot transfer energy. If the number of nuclei of any type in a piece of metal is independent of heating, the heat conductivity can be calculated with the use of the equations of the kinetic theory of gases. It is there shown[9] that if any particular

7. F. Richarz, *Wiedemanns Ann.*, **52**, 395, 1894.
8. F. Richarz, *Wiedemanns Ann.*, **52**, 395, 1894.
9. L. Boltzmann, *Gastheorie*, **1**, 12, 1896.

molecule has a property Q and a certain amount of this, say, G, is transferred by each molecule, then in unit time there passes through unit area the amount

$$\Gamma = \frac{u\ell N}{3} \cdot \frac{\partial G}{\partial x} \tag{5}$$

of this particular quantity, transported by the molecules. This assumes that the unit area is normal to the direction in which the property Q is transferred (here taken as the x axis). In the case in hand, each nucleus (according to equation 1) carries with it the amount of heat energy equal to αT, and hence from equation 5 we have for the quantity of heat W transported through unit area per unit time

$$W = \frac{\alpha}{3}(u_1\ell_1 N_1 + u_2\ell_2 N_2 + \cdots)\frac{\partial T}{\partial x} \tag{6}$$

Here W is measured in mechanical energy units, not in calories. The thermal conductivity κ (in mechanical energy units) then becomes[10]

$$\kappa = \frac{\alpha}{3}(u_1\ell_1 N_1 + u_2\ell_2 N_2 + \cdots) \tag{7}$$

Since the number N of nuclei of each type in the whole body is constant, that is, is independent of the temperature, in the process of temperature equilibration (flow of heat down the temperature gradient), the same number of nuclei of any type pass through every cross-section from left to right as from right to left. Thus though there is a net energy transport through the cross-section, since by their collisions the nuclei seek to restore temperature equilibrium, there can be no electric current. In such a body there can be no electrical potential drop due to the temperature gradient.

The mean free path ℓ_1 for nuclei of type 1 is, according to Boltzmann[11] present in

$$\ell_1 = \frac{1}{\pi\left(S_1^2 N_1\sqrt{2} + S_{12}^2 N_2\sqrt{\dfrac{m_1 + m_2}{m_1}}\right)} \tag{8}$$

where S_1 is the radius of the collision sphere for collisions of two nuclei of type 1, and S_{12} is the radius of the collision sphere for collision of a nucleus type 1 with a nucleus of type 2. The radii of the collision spheres can indeed vary with the temperature; they become smaller as the temperature increases because of the larger velocities of the nuclei. However, in any case, formula 8 is applicable to our case only if the nuclei are not yet influenced in their motion by the ponderable atoms of the body. Hence we shall not use formula 8 and shall look on ℓ_1 as a temperature-dependent quantity whose value cannot be determined in advance.

10. The factor ⅓ in front of α in (6) should really be 0.35 according to a more careful calculation. See L. Boltzmann, *Gastheorie,* **1**, 79, 1896.
11. L. Boltzmann, *Gastheorie,* **1**, 70, 1896.

4. ELECTRICAL CONDUCTIVITY

In the time interval between two collisions, an electron is subject to the Lagrangian equation of motion:

$$m_1 \frac{\partial^2 \zeta_1}{\partial t^2} = e_1 X \tag{9}$$

where X is the force exerted by the electric field on unit electric charge (measured electrostatically), and ζ_1 is the displacement of the electron in the x direction. If the time between two collisions is τ_1, we have

$$u_1 \tau_1 = \ell_1 \tag{10}$$

and hence from (9)

$$m_1 \zeta_1 = \tfrac{1}{2} e_1 X \tau_1^2 + a\tau_1 + b \tag{11}$$

[*Editor's Note:* It is assumed that X is constant in time and that a and b are constants of integration.]

Through the action of a force X in an arbitrarily long time interval τ_i, the nucleus moves on the average as if a continuing x component were added to its random velocity u_1 and indeed in the amount

$$u_x = \tfrac{1}{2} e_1 X \frac{\tau_1}{m_1} \tag{12}$$

For according to (11) the nucleus in the time interval τ_1 is displaced by $\tau_1 u_x$ under the action of the force X, and hence in an arbitrarily long time τ by the amount τu_x.

From (10) we can then put into (12)

$$u_x = \tfrac{1}{2} e_1 X \cdot \frac{\ell_1}{m_1 u_1} = \tfrac{1}{2} e_1 X \cdot \frac{\ell_1 u_1}{m_1 u_1^2}$$

If we now look back at equation 1, we have

$$u_x = e_1 X \cdot \frac{\ell_1 u_1}{4xT} \tag{13}$$

We shall denote by ν_1 the quantity multiplying $e_1 X$ in (13). It is the translation velocity which the electron receives if a force of 1 dyn acts on it continuously. We may call ν_1 the *mobility* of the nucleus of type 1. From (13) it follows that it is given by

$$\nu_1 = \frac{\ell_1 u_1}{4\alpha T} \tag{14}$$

Hence in unit time $e_1 X \nu_1 N_1$, nuclei of type 1 pass through unit area. Since every such nucleus carries the charge e_1, through the motion of the nuclei of type 1, the following quantity of electricity is carried per unit time across unit area (the equivalent current density)

$$j = e_1^2 N_1 \nu_1 X \tag{15}$$

For an arbitrary number of nuclear types, the current density becomes

$$j = (e_1^2 N_1 \nu_1 + e_2^2 N_2 \nu_2 + \cdots) X \tag{16}$$

The term in parentheses multiplying X has the significance of the electrical conductivity measured in electrostatic units. If we call this σ, we have

$$\sigma = e_1^2 N_1 \nu_1 + e_2^2 N_2 \nu_2 + \cdots \tag{17}$$

or by use of equation 4

$$\sigma = \frac{1}{4\alpha T}(e_1^2 N_1 \ell_1 u_1 + e_2^2 N_2 \ell_2 u_2 + \cdots) \tag{18}$$

Since, according to experimental observation, σ for pure metals is inversely proportional to T, we shall have to look on the quantities $N_1 \ell_1 u_1$, $N_2 \ell_2 u_2$, etc., as approximately independent of temperature. It can of course happen that only one of these quantities makes an appreciable contribution to σ, and hence only this one needs to be independent of temperature. It follows indeed from the optical properties of metals that if one assumes the existence of only two types of nuclei, for most metals one of these two makes the principal contribution to σ.

From formula 17 it follows that the conductivity of an alloy can be calculated from the conductivities of the components by the use of the usual mixing rule if these components retain their normal number of free nuclei in the alloy. The abnormal increase in resistance of copper when small quantities of phosphorus or arsenic are added therefore seems to mean that in such cases the number of free nuclei N (at any rate of the type making by far the greatest contribution to the conductivity) has decreased in the alloy. If one thinks of the analogous case of an electrolyte, this could be interpreted as a decrease in the dissociation of the body.

5. THE RELATION OF HEAT CONDUCTIVITY TO ELECTRIC CONDUCTIVITY

If we divide equation 7 by equation 17, we get

$$\frac{\kappa}{\sigma} = \tfrac{4}{3}\, \alpha^2 T \left(\frac{u_1 \ell_1 N_1 + u_2 \ell_2 N_2 + \cdots}{e_1^2 u_1 \ell_1 N_1 + e_2^2 u_2 \ell_2 N_2 + \cdots} \right) \tag{19}$$

If only only type of nucleus is active, this becomes

$$\frac{\kappa}{\sigma} = \tfrac{4}{3} \left(\frac{\alpha}{e} \right)^2 T \tag{20}$$

This is the Wiedemann-Franz law for the relation between thermal and electric conductivities. For silver[12] at 18°C, the thermal conductivity measured in gram-calories per square centimeter per degree Celsius is 1.005. Since 1 g-cal is 419×10^5 erg [*Editor's Note:* This is the mechanical equivalent of heat] we have for silver at 18°C in cgs units

$$\kappa = 421 \times 10^5$$

The absolute electric conductivity of silver at 18°C measured in electromagnetic units is 61.4×10^{-5}. To obtain the equivalent value measured in electrostatic units, we must multiply this value by 9×10^{20}, i.e., the square of the velocity of light in free space. Hence in the latter units, we have for silver at 18°C

$$\sigma = 553 \times 10^{15}$$

Since for 18°C, $T = 291$°K, we have from equation 20

$$\frac{\kappa}{\sigma} = \frac{421}{553} \times 10^{-10} = \tfrac{4}{3} \left(\frac{\alpha}{e}\right)^2 (291) = 0.76 \times 10^{-10} \tag{21}$$

or

$$\left(\frac{\alpha}{e}\right)^2 = 1.95 \times 10^{-13}$$

or

$$\frac{\alpha}{e} = 4.42 \times 10^{-7} \tag{22}$$

If now we take for the elementary electric charge e the value found by J. J. Thomson,[13] that is, 6×10^{-10} esu, which agrees approximately with the value obtained from electrolysis, we get for α the value

$$\alpha = 26.5 \times 10^{-17} \tag{23}$$

[*Editor's Note:* From the fact that $\alpha = \tfrac{3}{2} k$, where k is the Boltzmann gas constant (1.37×10^{-16} erg/°C), we should expect $\alpha = 20.6 \times 10^{-17}$, which is not too far out of line with Drude's result.]

This value of α agrees in order of magnitude with that in equation 4, obtained through use of the gas law and the Loschmidt number.

12. W. Jaeger and H. Diesselhorst, *Sitzungsber. k. Akad. Wiss. zu Berlin,* **38,** 1899.
13. J. J. Thomson, *Philos. Mag.* (5) **46,** 528, 1898.

With the help of the value in (23), we can obtain a better value for the Loschmidt number. Thus we get in this way

$$N = 2.1 \times 10^{19} \tag{24}$$

Accordingly, the ratio of the thermal conductivity to the electric conductivity can be calculated as far as order of magnitude is concerned from the gas law and from previously determined values of the Loschmidt number and the elementary electric charge e. Moreover, the second part of the law enunciated in (20), that is, that κ/σ should be proportional to T is in part substantiated by experiment. Jaeger and Diesselhorst have measured κ/σ at 18°C and 100°C for several metals, with results shown in the following table.

The approximate confirmation of the Wiedemann-Franz law leads to this conclusion: Only simple nuclei are present in metals. We make our further calculations on the basis of this result.[14] The exceptions to the law of the constancy of κ/σ at constant temperature cannot be explained by the polymerization of the nuclei since according to (19), κ/σ under that explanation would have to come out smaller than

Table 1. Ratios of thermal conductivity to electrical conductivity for several metals

Metal	$(\kappa/\sigma)_{18} \times 10^{10}$	$(\kappa/\sigma)_{100} \times 10^{10}$	$\dfrac{(\kappa/\sigma)_{100}}{(\kappa/\sigma)_{18}}$
Aluminum	0.706	0.937	1.32
Copper II	0.738	0.957	1.30
Copper III	0.745	0.967	1.30
Silver	0.760	0.978	1.28
Gold	0.807	1.027	1.27
Nickel	0.776	1.006	1.30
Zinc	0.745	0.962	1.29
Cadmium	0.784	1.005	1.28
Lead	0.794	1.035	1.31
Tin	0.816	1.024	1.26
Platinum	0.836	1.124	1.35
Palladium	0.837	1.129	1.35
Iron I	0.890	1.178	1.32
Iron II	0.930	1.287	1.33
Steel	1.105	–	–
Bismuth	1.068	1.195	1.12
Red Brass	0.840	1.060	1.26
Constantan	1.228	1.454	1.18

[*Editor's Note:* The ratio 373:291 = 1.28, which is not far from the values listed, save for bismuth and constantan.]

14. We first meet double nuclei in electrolysis in the form of divalent ions.

the normal value, whereas in bismuth and constantan, etc., it is much larger. To explain the exceptions to the Wiedemann-Franz law, we shall give up the assumption made in Section 3 in the calculation of thermal conductivity, namely, that the number of nuclei is independent of the temperature. Indeed, in the later discussion of Volta's potential law and of thermoelectricity, we shall show that we have to assume for every metal a dependence of the number of nuclei on the temperature (at least for one nuclear type).

From (20), we deduce

$$\frac{(\kappa/\sigma)_{100}}{(\kappa/\sigma)_{18}} = 1.28$$

This number is confirmed approximately for all the metals for which κ/σ has the normal value assumed in (21). The temperature coefficient of κ/σ is, however, smaller for those metals for which the deviation from the Wiedemann-Franz law is particularly large.

[*Editor's Note:* The author continues his article with a discussion of thermal conduction in the case in which the number of nuclei depends on the temperature. He then goes on to develop the theory of the Thomson effect, contact difference of potential, the standard thermoelectric effect, electrolysis, and the apparent mass of the electron. Part I of Drude's work on the electron theory of metals was followed in subsequent issues of the *Annalen der Physik* by studies of the galvanomagnetic and thermomagnetic effects in metals and the optical properties of metals. This early important and pioneer work of Drude was later modified and brought more closely in line with experience by the introduction of quantum statistics in the late 1920s.]

32

Reprinted from pages 650–665 of *The Collected Papers of Peter J. W. Debye.*
New York: Wiley Interscience, 1954, 700pp.

ON THE THEORY OF SPECIFIC HEATS
(Zur Theorie der spezifischen Wärmen*)

P. Debye

Translated from
Annalen der Physik, Vol. 39, 1912, pages 789–839.

The observations on the temperature dependence of specific heats carried out recently in the Nernst Laboratory have proven in a most convincing way that the theorem of equipartition of energy is incorrect for material bodies also. As is well known, Einstein[1] first called attention to the situation and obtained a formula giving the specific heat as a function of temperature using the quantum theory which Planck[2] had developed for radiation. While all measurements show a trend which agrees qualitatively with the Einstein formula, quantitative discrepancies between theory and experiment appear which are greater the lower the temperature. As an improvement, Nernst and Lindemann[3] altered the Einstein formula by introducing a second frequency $\nu/2$ in addition to ν. The introduction of this second frequency leads to a better formula than that of Einstein as far as practical needs are concerned; however, it has been impossible to find any sound basis for this particular value $\nu/2$. It will be seen in the following that there is no fundamental justification for it. Now even if the value $\nu/2$ itself has no theoretical significance, the necessity for the introduction of several frequencies can still be justified, as Einstein[4] has already suggested. The neighbors of a vibrating atom exert so strong an influence on it that its motion can have only a very slight resemblence to simple harmonic motion. The direct application of the Planck formula (with a single frequency) then naturally becomes questionable. If we consider the Fourier analysis of the motion of the atom, then the motion can be characterized by very many frequencies; thus, in agreement with Einstein, the necessity of several frequencies seems plausible.

While this suggestion is correct, the actual details of such a calculation are very laborious; the following method leads much

*Received July 24, 1912.

more directly to a sound formula for the specific heat. It is exactly analogous to Jeans' proof of the Rayleigh radiation formula.

If a solid is thought of as consisting of N atoms which can be treated as point masses, it represents a system of $3N$ degrees of freedom* which can carry out $3N$ different periodic motions with $3N$ different frequencies. The slowest of these vibrations are the readily observable sound vibrations. The most general motion which the body can carry out may be represented as a superposition of the above $3N$ forms of motion and can be described by a sum of these motions with the introduction of $6N$ constants.** Also for heat motion (a special case of general motion) this suggestion must also be valid.

Were the theorem of equipartition of energy correct, the "spectrum" of the solid would not have to be determined. From the outset to each eigenvibration would have to be ascribed the energy kT ($k = 1.47 \times 10^{-16}$ erg/degree, T = the absolute temperature) and the total energy of the solid would be $3NkT$, which, as is well known, corresponds to the Dulong-Petit law. Instead of this, however, Planck's formula states that the average energy at temperature T of a system with the frequency ν has the value

$$\frac{h\nu}{e^{\frac{h\nu}{kT}} - 1}$$

h = quantum of action = 7.10×10^{-27} erg sec.***

k = Boltzmann constant = 1.47×10^{-16} erg/degree.

We must first determine the eigen frequencies of the eigenvibrations of the solid, with each of these eigenvibrations we have to associate an energy corresponding to Planck's formula and finally we have to sum over the whole spectrum. Then the energy of the solid at temperature T is known and by differentiation also its specific heat C.

The first thing to be done is a determination of the "acoustic spectrum" of a solid of some arbitrary shape, for as is known the calculation of U does not depend on this shape. Now for this purpose the solid could be viewed as a point lattice consisting of $3N$ mass points and the $3N$ vibrational frequencies of the system determined from the associated system of $3N$ equations of motion.

*In terms of this definition a linear oscillator is a system with one degree of freedom.

**The number of constants is $6N$ because two data are required for the complete specification of one eigenvibration: amplitude and phase at some initial time.

***The numerical values are those given recently by Paschen-Gerlach, *Ann. Physik 38*, 41 (1912).

Then there would be no choice other than to make special simple hypotheses concerning the interactions of the atoms. There is no question that in so far as the hypotheses about the forces are correct this is completely correct. However, it is not said that a completely exact knowledge of the spectrum is necessary; on the contrary, it can be expected that from the practical point of view a sufficiently exact formula will be obtained in case only the characteristic properties of this spectrum be properly taken into account. We would like to show in the following that this is possible without considering in detail the mutual forces between atoms.

We proceed from the ordinary equations of elasticity. In the derivation of these the solid is treated as a continuum; doing this corresponds to ascribing infinitely many eigenvibrations to an elastic body. Now the latter is certainly false; the solid consisting of N atoms can have only $3N$ different eigenvibrations. Yet in another way a calculation based on these equations of elasticity will be of interest to us. Namely, it will give us a law for the number of spectral lines of the acoustic spectrum per frequency interval $d\nu$. In the following we will consider this law for the density of spectral lines entirely accurate. The discontinuous structure of the solid will be taken into account only in so far as it permits the following conclusion: the spectrum consists in its entirety only of $3N$ and not of infinitely many spectral lines. Accordingly, the spectrum calculated from the equations of elasticity will be cut off at the $3N$th spectral line. With certainty it can be asserted that the spectrum of the actual solid agrees completely with the calculated spectrum at low frequencies. With equal certainty, for high frequencies, at which the wavelength is comparable to the distance between two atoms, the so calculated distribution of spectral lines can be considered only as approximately correct. Yet we will neglect this error; the result supports us and shows that it is probably not of great importance. For, as will be shown in the following, this approximate calculation gives a formula for the specific heats, which agrees in every respect with the experimental facts.*

*A case which can easily be treated rigorously is that of a linear body of atomic structure (compare, for example, Lord Rayleigh, *Theory of Sound*, Vol. I, London, 1877, p. 129). By analogy with this case it can be concluded that the lines are more dense at the high frequency end than our formula indicates.

Since it has often been emphasized that a fundamental difference between the optical series and those otherwise known is to be found in that the first have a finite limit, I will not suppress the (really self-evident) remark that, in reality, besides the spectrum of the black body not a single one exists which does not have a finite limit. Of course, this limit is in no case an accumulation point in the mathematical sense, as must be expected according to the series formula. However, an accumulation of lines (con'd)

In place of the simple frequency of Einstein there appears a complete spectrum, which, according to the previous discussion, is characterized by two properties: (a) by its cut-off; (b) by the number of lines per frequency interval $d\nu$. With these assumptions the cut-off frequency ν_m appears in these two data as the only essential parameter. This remark leads to the theorem:

If the temperature T is expressed as a multiple of a characteristic temperature θ of the substance concerned, the specific heat of all (monatomic) solids is represented by the same curve, in other words, the specific heat of a monatomic solid is a universal function of the ratio T/θ.

A second theorem which is not limited to the case of monatomic solids but claims a completely general validity is obtained if the density of line distribution is taken into account. It is found (compare section X) that in the frequency interval $d\nu$ there is a number of lines proportional to $\nu^2 d\nu$, that is, the same law which according to Jeans holds for black body radiation. This leads to the theorem:

At sufficiently low temperatures the specific heat of all solids is proportional to the third power of the absolute temperature.*

Then the energy content becomes proportional to T^4, analogous to the Stefan-Boltzmann law for black body radiation which holds at all temperatures. If the solid like the black body possessed an "acoustic spectrum" which had no finite limit, the theorem would also be correct at all temperatures. In reality it holds only below a temperature which depends on the solid. We will show later that it is correct within about 1% below temperatures equal to 0.1 of the temperature referred to above as the characteristic temperature.

As will be shown in section V, the cut-off frequency ν_m and with it the characteristic temperature θ can be represented as a

(con'd from preceding page)

always occurs in the neighborhood of this cut-off as far as I can see. Now, since it is possible to give arrangements of electrons whose spectra satisfy the Deslandre law of band spectra, it is perhaps to be expected that the series spectra can also be understood as eigenvibrations of electron systems of a finite number of degrees of freedom.

*After developing the two theorems in the Winter semester in my lectures on thermodynamics, I have reported on them later at the meeting of the Swiss Physical Society at Bern and published them in *Arch. de Genève*, March 1912, p. 256. Somewhat later, Born and von Karman developed independently of these communications a formula for the specificheat which agrees with the two theorems given, as can be verified easily.

function of the elastic constants of the solid. This relation corresponds in its structure with the well known formula of Einstein, which connects his single eigenfrequency with the compressibility. There is a difference in so far as, besides the compressibility, Poisson's ratio plays a roll. The appearane of these two quantities cannot be surprising if it is recalled that the properties of an elastic body are determined by two constants. Generally, it can be said that for crystals, for example, even these two quantities are not sufficient to determine θ but that all elastic constants are required.

For clarity the following is divided into a Part I in which a formula is developed for the specific heat on the basis described above and is compared with experimental results. In this part the formula for the constitution of the "acoustic spectrum" is used without proof. The proof of this is added in a Part II.

Part I

I. Energy Content and Specific Heat of Monatomic Solids

According to the theory of elasticity, a body of volume V possesses an unbounded spectrum. According to Part II, section X, the number of eigenvibrations z below a certain upper limit ν is:

$$z = \nu^3 V F \tag{1}$$

F is a function of the elastic constants and the density ρ which, in terms of the compressibility κ and Poisson's ratio σ, for example, can be expressed in the form:

$$F = \frac{4\pi}{3}\rho^{3/2}\kappa^{3/2}\left[2\left(\frac{2}{3}\right)^{3/2}\left(\frac{1+\sigma}{1-2\sigma}\right)^{3/2} + \frac{1}{3^{3/2}}\left(\frac{1+\sigma}{1-\sigma}\right)^{3/2}\right] \tag{2}$$

For the purposes of these paragraphs it suffices to know that F represents a quantity which can be calculated for any body from its density and its elastic constants.

Now the body whose energy content U will be calculated consists of N atoms and, as a system of $3N$ degrees of freedom, can carry out only $3N$ eigenvibrations. If we assume, as was explained in the introduction, that equation (1) can always be used with sufficient accuracy, the highest frequency ν_m can be calculated from the expression:

$$3N = \nu_m^3 V F \tag{3}$$

or:

$$\nu_m = \left(\frac{3N}{VF}\right)^{1/3} \qquad (3')$$

The density of spectral lines can be calculated immediately by differentiating (1). The number dz which lie in the frequency interval $d\nu$ is:

$$dz = 3\,VF\,\nu^2\,d\nu$$

If, in this equation, VF is replaced by its value following from (3):

$$VF = \frac{3N}{\nu_m^3}$$

then:

$$dz = 9N\frac{\nu^2\,d\nu}{\nu_m^3} \qquad (4)$$

On the other hand, according to the Planck formula, each eigen-vibration with the frequency ν possesses the energy:

$$\frac{h\nu}{e^{\frac{h\nu}{kT}} - 1} \qquad (5)$$

if $h = 7.10 \times 10^{-27}$ erg sec. denotes the fundamental quantum of action, $k = 1.47 \times 10^{-16}$ erg/°C. the Boltzmann constant, and T the absolute temperature. By combining (4) and (5) we obtain for the energy U of the body the expression:

$$U = \frac{9N}{\nu_m^3}\int_0^{\nu_m} \frac{h\nu}{e^{\frac{h\nu}{kT}} - 1}\,\nu^2\,d\nu \qquad (6)$$

where the integral has to be carried out over the entire acoustic spectrum from $\nu = 0$ to $\nu = \nu_m$.

If now a characteristic temperature Θ of the body be defined by the expression:

$$\Theta = \frac{h\nu_m}{k} \qquad (7)$$

and a new dimensionless variable:

$$\xi = \frac{h\nu}{kT} \tag{8}$$

is introduced in the integral, U can be written in the form:

$$U = 9NkT\left(\frac{kT}{h\nu_m}\right)^3 \int_0^{\frac{h\nu_m}{kT}} \frac{\xi^3 d\xi}{e^\xi - 1}$$

or if the definition (7) is taken into account:

$$U = 9NkT\left(\frac{T}{\Theta}\right)^3 \int_0^{\Theta/T} \frac{\xi^3 d\xi}{e^\xi - 1} \tag{9}$$

As is well known (and in addition follows naturally from (9) for high temperatures):

$$U = 3NkT$$

corresponds to the law of Dulong and Petit. The relation expressed by (9) can be put in words as follows:

The energy of a body is obtained if the Dulong Petit value is multiplied by a factor which is a universal function of the ratio T/Θ, *that is, temperature* T *divided by the characteristic temperature* Θ.

If for brevity we put:

$$\frac{\Theta}{T} = x$$

then according to (9) this factor has the value:

$$\frac{3}{x^3}\int_0^x \frac{\xi^3 d\xi}{e^\xi - 1}$$

If N is understood to be the number of atoms per atomic weight, (9) represents the corresponding energy and by differentiation with respect to T the atomic heat at constant volume C_v is obtained, for which we write simply C without an index so long as there is no danger of confusion. Thus from (9):

$$C = 3Nk\left[\frac{12}{x^3}\int_0^x \frac{\xi^3 d\xi}{e^\xi - 1} - \frac{3x}{e^x - 1}\right] \tag{10}$$

if again x is the ratio θ/T.

The quantity $3Nk$ has the value 5.955 cal.; if we denote it by C_∞ because it is reached in the limit $T = \infty$, (10) can be rewritten:

$$\frac{C}{C_\infty} = \frac{12}{x^3}\int_0^x \frac{\xi^3 d\xi}{e^\xi - 1} - \frac{3x}{e^x - 1} \qquad (10')$$

Thus the result stated in the introduction is verified:

The specific heat of a monatomic solid is a universal function of the ratio $T/\theta = 1/x$.

II. Approximate Formula and Numerical Results

(a) Approximation for Small Values of x

If high temperatures at which the ratio θ/T is small compared to 1 are considered first, the integrand in (10') can be replaced by a first approximation:

$$\frac{\xi^3}{e^\xi - 1} = \xi^2$$

Then the integral becomes $x^3/3$. On the other hand, for the limit $x = 0$:

$$\frac{3x}{e^x - 1} = \frac{3x}{x} = 3$$

which leads to C/C_∞:

$$\frac{C}{C_\infty} = \frac{12}{3} - 3 = 1$$

a value which corresponds to the law of Dulong and Petit.

For use in what follows some additional terms will be calculated. With the abbreviation $x = \theta/T$, (9) can be rewritten:

$$\frac{U}{\Theta C_\infty} = \frac{3}{x^4}\int_0^x \frac{\xi^3 d\xi}{e^\xi - 1}$$

Since on the other hand:

$$\frac{d}{dT} = -\frac{\Theta}{T^2}\frac{d}{dx} = -\frac{1}{\Theta}x^2\frac{d}{dx}$$

we obtain for the ratio C/C_∞ the relation:

$$(11) \qquad \frac{C}{C_\infty} = -3x^2 \frac{d}{dx}\left[\frac{1}{x^4}\int_0^x \frac{\xi^3 d\xi}{e^\xi - 1}\right]$$

The development of $1/e^\xi - 1$ in powers of ξ is well known; with the help of the Bernoulli numbers B it can be expressed in the form:*

$$\frac{1}{e^\xi - 1} = \frac{1}{\xi} - \frac{1}{2} + \xi \sum_{n=1}^{n=\infty} (-1)^{n-1} \frac{B_n}{(2n)!} \xi^{2n-2}$$

By means of this expansion and the operations indicated in (11), for C/C_∞ the series:

$$(12) \qquad \frac{C}{C_\infty} = 1 - 3 \sum_{n=1}^{n=\infty} (-1)^{n-1} \frac{B_n}{(2n)!} \frac{2n-1}{2n+3} x^{2n}$$

is obtained. The first Bernoulli numbers have the values:

$$B_1 = \frac{1}{6}, \quad B_2 = \frac{1}{30}, \quad B_3 = \frac{1}{42}, \quad B_4 = \frac{1}{30}, \quad B_5 = \frac{5}{66},$$
$$B_6 = \frac{691}{2730}, \quad B_7 = \frac{7}{6}, \ldots$$

By use of these values (12) can also be written:

$$(12') \frac{C}{C_\infty} = 1 - \frac{x^2}{20} + \frac{x^4}{560} - \frac{x^6}{18144} + \frac{x^8}{633600} - \frac{x^{10}}{23063040} + \ldots$$

In the numerical calculations only the first four terms of this series were used. This results in less than 1% error in the interval $x = 0$ to $x = 2$. If the six terms written out are used, the interval can be extended with the same maximum error to beyond $x = 3$.

(b) Approximation for Large Values of x

In this case it is best to proceed directly from (10'). Again the value attained in the limit of low temperatures, that is, for very large values of x, is calculated first.

The term $3x/e^x - 1$ vanishes exponentially; on the other hand,

*Compare, for example, J. A. Serret, *Cours de calcul differentiel et integral*, Paris 1886, v. 2., p. 213. This series, like that for C/C_∞, converges only for values of the variable which are less than 2π.

the upper limit of the integral can be replaced by ∞ and thus a numerical can be obtained for the integral:

$$\int_0^\infty \frac{\xi^3\, d\xi}{e^\xi - 1} = \int_0^\infty \xi^2 (e^{-\xi} + e^{-2\xi} + e^{-3\xi} + \ldots)\, d\xi$$

$$= 6\left(1 + \frac{1}{2^4} + \frac{1}{3^4} + \frac{1}{4^4} + \ldots\right)$$

The series in the parentheses appears in radiation theory[5]: it has the value. (B_2 is again the second Bernoulli number):

$$\frac{8\pi^4}{4!} B_2 = \frac{\pi^4}{90} = 1{,}0823$$

By substituting in (10') we obtain for low temperatures:

$$\frac{C}{C_\infty} = \frac{4}{5}\pi^4 \frac{1}{x^3} = \frac{4\pi^4}{5}\frac{T^3}{\Theta^3} = 77.938\,\frac{T^3}{\Theta^3} \tag{13}$$

With this the second theorem emphasized in the introduction is verified:

For sufficiently low temperatures the specific heat is proportional to the third power of the absolute temperature.

Besides this the proportionality factor is found; according to (13) is has the value $77.938/\theta^3$. Neither the Einstein nor the Nernst-Lindemann formula agrees with this law; both require that the specific heat vanish exponentially in the neighborhood of absolute zero. In the following (13) will be verified on the basis of the existing observations. It may, perhaps, be noted that (13) does not contradict the Nernst heat theorem, since the specific heat for low temperatures still vanishes sufficiently rapidly to give to the integral:

$$\int_0^T \frac{C}{T}\, dT$$

a finite value.

The proportionality to T^3 is a direct consequence of the circumstance that the number of vibrations per frequency interval $d\nu$ is proportional to the quantity $\nu^2 d\nu$. Now, on the one hand, at low temperatures chiefly slow vibrations come into consideration, while, on the other, the proportionality to $\nu^2 d\nu$ deduced from the elastic equations unquestionably can be applied to low frequencies. Therefore it can be concluded that the proportionality to T^3 follows not only for monatomic but for all bodies at sufficiently low temperatures.*

*Notice that according to (13) there is no temperature below which the radiation energy exceeds the energy of the atoms unless bodies exist with a characteristic temperature higher than 135,000°. This is in contrast to the Einstein and the Nernst-Lindemann formulae, for which such a temperature can always be found.

In order to test the exactness of the law (13) for higher temperatures and at the same time to be able to compute values of the formula (10') in such a region, we will also calculate here the series development of which (13) is the first term. If the integrand in (10') is developed in powers of $e^{-\xi}$:

$$\frac{\xi^3}{e^\xi - 1} = \xi^3 e^{-\xi}(1 + e^{-\xi} + e^{-2\xi} + \ldots) = \sum_{n=1}^{n=\infty} \xi^3 e^{-n\xi}$$

is obtained.

On the other hand there is found easily by integration by parts:

$$\int_0^x \xi^3 e^{-n\xi} d\xi = \int_0^\infty \xi^3 e^{-n\xi} d\xi - \int_x^\infty \xi^3 e^{-n\xi} d\xi$$

$$= \frac{6}{n^4} - x^4 e^{-nx}\left(\frac{1}{nx} + \frac{3}{n^2 x^2} + \frac{6}{n^3 x^3} + \frac{6}{n^4 x^4}\right)$$

so that the integral appearing in (10') can be written:

$$\int_0^x \frac{\xi^3}{e^\xi - 1} d\xi = 6 \sum_{n=1}^{n=\infty} \frac{1}{n^4} - x^4 \sum_{n=1}^{n=\infty} e^{-nx}\left(\frac{1}{nx} + \frac{3}{n^2 x^2} + \frac{6}{n^3 x^3} + \frac{6}{n^4 x^4}\right)$$

For C/C_∞ (10') finally gives:

$$(14)\quad \begin{cases} \dfrac{C}{C_\infty} = \dfrac{72}{x^3} \displaystyle\sum_{n=1}^{n=\infty} \dfrac{1}{n^4} \\ \qquad - 12x \displaystyle\sum_{n=1}^{n=\infty} e^{-nx}\left(\dfrac{1}{nx} + \dfrac{3}{n^2 x^2} + \dfrac{6}{n^3 x^3} + \dfrac{6}{n^4 x^4}\right) - \dfrac{3x}{e^x - 1} \end{cases}$$

If $\Sigma 1/n^4$ is replaced by its value $\pi^4/90$ given earlier, (14) can be rewritten:

$$(14')\quad \begin{cases} \dfrac{C}{C_\infty} = \dfrac{4\pi^4}{5} \dfrac{1}{x^3} - \dfrac{3x}{e^x - 1} \\ \qquad - 12x \displaystyle\sum_{n=1}^{n=\infty} e^{-nx}\left(\dfrac{1}{nx} + \dfrac{3}{n^2 x^2} + \dfrac{6}{n^3 x^3} + \dfrac{6}{n^4 x^4}\right) \end{cases}$$

For the entire interval $x = \infty$ to $x = 2$ an error of less than .1 per cent is made if the first four terms of the above sum are used; for an error less than 1 per cent the first three terms at most suffice. From $x = 12$ on C/C_∞ is given with an error of less than 1 per cent by the expression $(4\pi^4/5)(1/x^3)$ alone; i.e., from $T = \theta/12$ to $T = 0$ the proportionality to T^3 holds with an error of less than 1 per cent.

Finally, by use of (12') and (14'), the values of C/C_∞ were calculated as a function of T/θ. They are recorded in Table I, in which the value of $1/x = T/\theta$ is also given.

Table I

$x = \frac{\theta}{T}$	$\frac{1}{x} = \frac{T}{\theta}$	$\frac{C}{C_\infty}$	$\frac{C}{C_\infty}$ Einstein	Deviation, %	$\frac{C}{C_\infty}$ Nernst-Lindemann	Deviation, %
0.250	4	0.997	0.994	- 0.3	0.997	0
0.333	3	0.994^5	0.989	- 0.5	0.994	0
0.500	2	0.988	0.979	- 0.9	0.988	0
0.667	1.5	0.978	0.959	- 1.9	0.977	- 0.1
1.000	1.0	0.952	0.920	- 3.3	0.951	- 0.1
1.111	0.9	0.941	0.902	- 4.1	0.939	- 0.2
1.250	0.8	0.926	0.878	- 5.2	0.924	- 0.2
1.429	0.7	0.904^5	0.841	- 7.0	0.904	0
1.667	0.6	0.872^5	0.799	- 8.3	0.873	0
2.000	0.5	0.825	0.724	-12.2	0.823	- 0.2
2.500	0.4	0.745	0.610	-18.1	0.744	- 0.1
3.333	0.3	0.607^5	0.428	-29.5	0.615	+ 1.2
4.000	0.25	0.503	0.304	-39.5	0.514	+ 2.2
4.021	0.2488	0.500	---	---	---	---
5.000	0.20	0.369	0.171	-51.1	0.390	+ 5.7
6.667	0.15	0.213	0.0598	-72.0	0.242	+ 13.6
10.00	0.10	0.0758	0.0045	-94.2	0.0875	+ 15.4
13.33	0.075	0.0328	---	---	0.0284	- 13.4
20.00	0.050	0.00974	---	---	0.00227	- 76.3
40.00	0.025	0.00122	---	---	0.00000412	-100
∞	0	0	0	-100	0	-100

In order to make possible a comparison with the two other formulae used up to now: that of Einstein:

$$\frac{C}{C_\infty} = \frac{x^2 e^x}{(e^x - 1)^2} \tag{15}$$

and that of Nernst-Lindemann:

$$\frac{C}{C_\infty} = \frac{1}{2}\left\{\frac{x^2 e^x}{(e^x - 1)^2} + \frac{\left(\frac{x}{2}\right)^2 e^{x/2}}{(e^{x/2} - 1)^2}\right\}$$

the values of these functions are reoorded in the 4th and 6th columns.*

The fifth column contains the deviation of the Einstein function from our function (columns 4 and 3) in per cent of the latter. It is seen that this per cent difference soon takes on relatively high values. The Nernst-Lindemann function behaves much better in this respect; down to T/θ = 0.4 the deviations are almost unnoticeable; they are, nevertheless, present and are negative.** In the table they do not appear because of our limitation to too few decimal places. At T/θ = approximately 0.4, the deviation vanishes. Then the per cent difference becomes larger again and at first is positive; however, it goes down again, becomes zero, and finally quickly attains high negative values. That the difference must be taken to be 100 per cent in the neighborhood of absolute zero is due to the fact that according to the formula of Einstein, or to that of Nernst-Lindemann, the specific heat must vanish exponentially, while according to our formula it vanishes only as T^3.

Finally, in Figure 1 the different curves for C/C_∞ are plotted as a function of $1/x$ = T/θ. The curve marked 1 corresponds to our formula (10'). The Nernst-Lindemann function is represented by curve 2, which is dashed so that it may be better distinguished from 1. The curve 3 represents the Einstein function. According to column 7 the Nernst-Lindemann function should run first below, then above, and again below the curve 1. Within the accuracy of the figure, however, the two curves fall together in the first interval. The Einstein function is smaller than (10') throughout. The ratio for low temperatures is very hard to represent on the scale adopted. However, in the same figure, the lower part of the three curves are plotted on a scale ten times enlarged with abscissae ten times enlarged. The numbers 1', 2', and 3' correspond to the numbers 1, 2, and 3 of the complete curves.

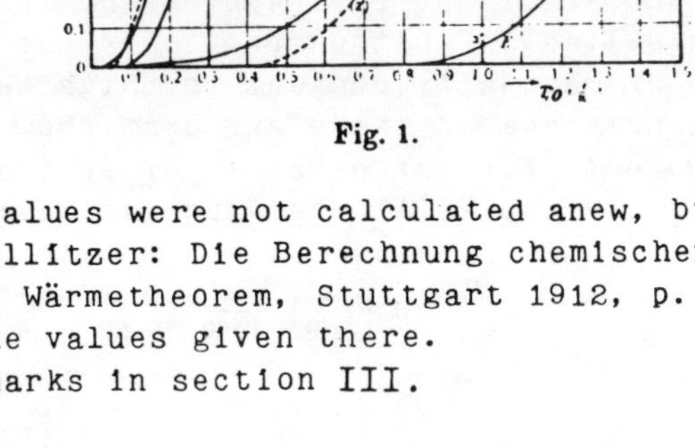

Fig. 1.

*The numerical values were not calculated anew, but taken from the tables of B. Pollitzer: Die Berechnung chemischer Affinitäten nach dem Nernstschen Wärmetheorem, Stuttgart 1912, p. 165 or interpolation between the values given there.

**Compare the remarks in section III.

III. Search for a Basis for the Nernst-Lindemann Form of the Function C/C_∞

The Nernst-Lindemann function has proven to be such a good representation of the observations in a large temperature range that the conclusion is inescapable that it must be connected with the theoretically founded formula in a simple if perhaps only superficial way. We will show in the following in what sense the Nernst-Lindemann function can be interpreted as an approximate representation of our function.

According to our analysis the formula for C/C_∞ is (10'):

$$\frac{C}{C_\infty} = \frac{12}{x^3}\int_0^x \frac{\xi^3 d\xi}{e^\xi - 1} - \frac{3x}{e^x - 1}$$

With the same notation ($x = \Theta/T$) the Nernst-Lindemann formula is:

$$\frac{C}{C_\infty} = \frac{1}{2}\left\{\frac{x^2 e^x}{(e^x - 1)^2} + \frac{\left(\frac{x}{2}\right)^2 e^{x/2}}{(e^{x/2} - 1)^2}\right\}$$

We will now adhere to this introduction of $x/2$ besides x and pose the following question. If it is supposed that:

$$\frac{C}{C_\infty} = a\frac{x^2 e^x}{(e^x - 1)^2} + b\frac{\left(\frac{x}{2}\right)^2 e^{x/2}}{(e^{x/2} - 1)^2} \tag{16}$$

how must the two constants a and b be chosen so that the integral of the difference of (10') and (16) taken over the temperature interval 0 to ∞, vanishes? Graphically, this means, referring to Figure 1, that the area enclosed between curves 1 and 2 should have in its entirety the content zero when sign is taken into account.*

*In the figure it appears as if the requirement is not satisfied. The reason for this lies in that on the scale used the negative differences which are actually present for high temperatures do not show up. The presence of these are best shown on the basis of the series developments for large values of T, that is, small values of x. According to (12'), our function is:

$$\frac{C}{C_\infty} = 1 - \frac{x^2}{20} + \frac{x^4}{560}$$

(Footnote continued)

The integral of (10') over the temperature is nothing more than the energy content, which has already been calculated in (9), divided by $C_\infty = 3Nk$. It has the value:

$$\Theta \frac{3}{x^4}\int_0^x \frac{\xi^3 d\xi}{e^\xi - 1} \tag{17}$$

The corresponding integral of (16), which has a similar meaning, is:

$$\Theta \left\{\frac{a'}{e^x - 1} + \frac{b/2}{e^{x/2} - 1}\right\} \tag{17'}$$

If our condition is to be satisfied, the difference of the two quantities (17) and (17') must vanish after the limits $T = 0$ and $T = \infty$ have been substituted, that is:

$$\left[\frac{3}{x^4}\int_0^x \frac{\xi^3 d\xi}{e^\xi - 1} - \frac{a}{e^x - 1} - \frac{b/2}{e^{x/2} - 1}\right]_{T=0}^{T=\infty} = 0 \tag{18}$$

Now when $T = 0$, $x = \infty$; however, for this value of x the bracket from (18) vanishes. On the other hand, when $T = \infty$, $x = 0$. For small values of x the quantities appearing in (18) can be expanded as follows:

*(continued from preceding page)

The Nernst-Lindemann function is obtained easily by using the development of $1/e^\xi - 1$ given in section II:

$$\frac{C}{C_\infty} = 1 - \frac{5}{96}x^2 + \frac{17}{7680}x^4$$

If the coefficients in our function are made to have the same denominators, our function is:

$$\frac{C}{C_\infty} = 1 - \frac{4.8}{96}x^2 + \frac{13.71}{7680}x^4$$

Since 5 > 4.8, the value of the Nernst-Lindemann function, as asserted, is smaller than the value of our function for small x (large T). On the other hand, since 17 > 13.71, the development indicates that for still smaller temperatures the ordinates become equal, as has been found earlier, when T/Θ is about equal to 0.4.

$$\frac{3}{x^4}\int_0^x \frac{\xi^3 d\xi}{e^\xi - 1} = \frac{3}{x^4}\int_0^x (\xi^2 - \tfrac{1}{2}\xi^3 + ..)d\xi = \frac{1}{x} - \frac{3}{8} + \ldots$$

$$\frac{a}{e^x - 1} = \frac{a}{x} - \frac{a}{2} + \cdots$$

$$\frac{b/2}{e^{x/2} - 1} = \frac{b}{x} - \frac{b}{4} + \cdot\cdot$$

where only terms have to be taken into account which are infinite for $x = 0$ or remain finite. The condition (18) can be written now:

$$(18') \quad \operatorname*{Lim}_{x=0}\left[\left(\frac{1}{x} - \frac{3}{8}\right) - a\left(\frac{1}{x} - \frac{1}{2}\right) - b\left(\frac{1}{x} - \frac{1}{4}\right)\right] = 0$$

that is, a and b must satisfy the two equations:

$$(19) \quad \begin{cases} a + b = 1 \\ \frac{a}{2} + \frac{b}{4} = \frac{3}{8} \end{cases}$$

From this follow:

$$(20) \quad a = \frac{1}{2} \text{ and } b = \frac{1}{2}$$

that is, just those values of a and b which Nernst and Lindemann used in their formula.

Therefore, it can be asserted that in so far as our formula is regarded as correct, the Nernst-Lindemann formula yields an approximation to it in the above sense.

[*Editor's Note:* Material has been omitted at this point.]

33

ON THE THEORY OF MAGNETISM

P. Langevin

This article was translated expressly for this Benchmark volume by R. Bruce Lindsay, Brown University, from "Sur la Theorie du Magnetisme," J. Phys. et le Radium 4:678–693 (1905), *by permission of the publisher.*

I. The magnetic properties of bodies and the many different ways in which they react under the influence of an applied magnetic field appear at first to be extremely complex. Thus ferromagnetic substances acquire an intense magnetization, which is a function not only of the applied magnetic field but also of the history of previous applied fields, thanks to the possibility of remanent magnetization. Other bodies are magnetized weakly in the direction of the magnetizing field and proportionally to the intensity of the applied field, with a proportionality coefficient called the "paramagnetic susceptibility." As shown by M. Curie, the latter varies inversely as the absolute temperature. Furthermore, experience shows that an increase in temperature can make a given substance transform continuously from the ferromagnetic to the paramagnetic state in a fashion similar to the change from the liquid to the gaseous state.

Still other bodies, perhaps the most numerous of all, are polarized very weakly in a direction opposite to that of the magnetizing field. They are called "diamagnetic." It is not possible to observe any transition between this property and the first two just mentioned. Another profound difference is that the diamagnetic susceptibility has been shown (except for the case of bismuth) to be rigorously independent of the temperature and of the physical state of the body. Up to now there has not been any satisfactory representation of this complex of phenomena. Ampere's molecular current hypothesis has been recently revived by Ewing and accounts for certain facts of ferromagnetism and paramagnetism but without any indication of the origin of these currents and the cause maintaining them. In addition, in order to explain diamagnetism, Weber has assumed the existence in the molecules of circuits of zero resistance in which the creation of an external magnetic field produces induced currents of such a nature that the total magnetic flux that traverses the circuits remains zero. By Lenz's law, the direction of these currents corresponds to diamagnetism. In Maxwell's treatise, he has attempted to unite the hypotheses of Ampere and Weber without, however, obtaining a precise solution in default of information about molecular structure.

II. We are in possession today of information of this sort thanks to the theory that treats matter as an aggregate of moving electrons and of centers electrified both positively and negatively, whose displacement produces in the aether of space a magnetic field analogous to that produced by an element of current of magnitude proportional to the charge of the electron and its velocity. The creation of this magnetic field at the moment of the setting into motion of the electron implies an expenditure of energy proportional to the square of the velocity, namely, energy of self-induction of the equivalent current; of the same form as the kinetic energy. An electrically charged center then possesses an inertia due solely to the fact that it is charged, and it appears reasonable today to assume that all inertia is of electromagnetic origin, the inertia of a material molecule being nothing but the sum of the electromagnetic inertias of the electrons that comprise it.

Further, electromagnetic or light radiation is emitted whenever an electron suffers change in velocity or suffers acceleration, and it is natural to seek in the electrons that move in closed orbits in the interior of molecules for the origin of spectral lines. The period of light vibrations, in the neighborhood of 10^{-15} s, should be of the order of the period τ of the revolution of the electrons in their orbits.

We recall how this idea permitted Lorentz to predict certain properties of the Zeeman effect, which experiment completely confirmed.

III. It is natural to liken the electrons moving in their closed orbits to the particle currents of Ampere. It is easy to show that such an orbit of surface area S, traversed with period τ by an electron of charge e, is, from the point of view of an average external magnetic field, equivalent to a magnet directed normally to the plane of the orbit and possessing magnetic moment

$$M = \frac{eS}{\tau} = Si$$

In other words, the orbit is equivalent to a circuit traversed by a current of magnitude e/τ.

It would not be inappropriate to recall how the production of a magnetic field by the motion of an electrified particle with respect to the aether is a necessary consequence of the equations of Hertz. The latter, which have been completely verified by experience, put into mathematical language the existence of Maxwell's displacement current and the phenomena of induction, according to which every variation in a vacuum of one of the two types of field, electric or magnetic, suffices to create the other.

1. The integral of the electric field along a closed circuit is equal to the derivative with respect to the time of the flux of magnetic force traversing the circuit (*induction*).
2. The integral of the magnetic field along a closed circuit is equal to the derivative with respect to the time of the flux of electric force traversing the circuit (*displacement current*).

[*Editor's Note:* The author fails to mention the *negative* sign involved in the right-hand side of the equation implied in item 1.]

The last statement assumes that the surface bounding the circuit over which one calculates the flux is not traversed by any electric charge or any electron.

Let us now consider and electrified particle O of charge e moving along OX with velocity v with respect to the medium (assumed to be completely at rest in the theory of Lorentz), and let us apply the second of the preceding statements above to a circle of radius $AM = l$ in a plane normal to the direction OX (see Figure 1).

It is impossible that the magnetic field, which by reason of symmetry is the same

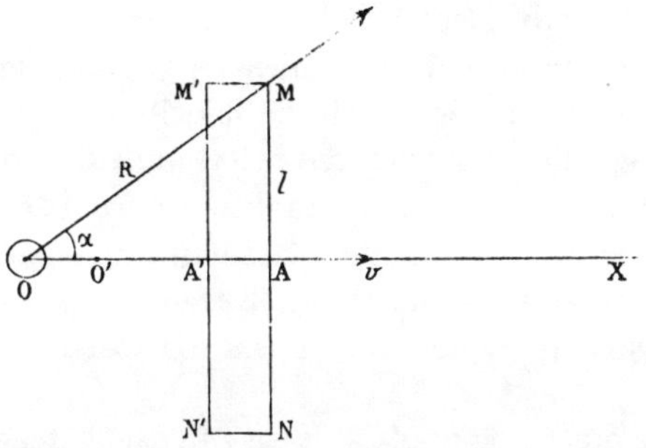

Figure 1. (Reprinted from *J. Phys. et le Radium* 4:680 (1905); copyright ©1905 by M. J. Berger, Directrice des Publications)

at all points of the circumference MN, should vanish on the circumference since the electric flux that traverses it varies by virtue of the motion of the particle. Moreover, the magnetic field H at M is necessarily tangent to the circumference, that is to say, normal to the plane of the figure, since the system as a whole is evidently superposable on its image with respect to this plane, and a magnetic field possessing the symmetry of a rotating cylinder has only one plane of symmetry normal to its direction.[1] The integral of the magnetic field along the circumference of length $2\pi l$ is

$$2\pi l H$$

To calculate the derivative of the flux of electric force, let us seek the variation of this flux when the particle moves in time dt through a distance $OC' = vdt$. The new flux produced by O' across MN is equal to that produced by O across the circumference $M'N'$ equal in size to MN and at a distance from it $AA' = vdt$. But by Gauss's theorem, the difference in the flux produced by O across the two circles is equal to the flux produced across the cylindrical crown $MNM'N'$, which forms with the circles a closed surface containing no charge in its interior. This flux is

$$d\phi = \frac{e}{R^2} \cdot \sin\alpha \cdot 2\pi l v dt$$

$$\frac{d\phi}{dt} = 2\pi l H = \frac{e}{R^2} \cdot \sin\alpha \cdot 2\pi l v$$

$$H = \frac{ev}{R^2} \cdot \sin\alpha$$

This is precisely the magnetic field that would be produced at M by a current element of length δs placed at O, parallel to OX and with moment $i\delta s = ev$.

1. P. Curie, *J. Phys.* (Third Series), **3**, 415, 1884.

The value e/R^2 taken for the electric field at M assumes that this field is not changed as a result of the motion of the charge. This remains approximately correct as long as the velocity v does not approach the velocity of light. The law of induction shows that the magnetic field produced reacts in its turn on the electric field to change its distribution, the lines of force at a distance from O large compared with the radius of the particle tending to accumulate in the plane perpendicular to OX.

The expression obtained for H remains usable even for the electrons present in the molecule since their velocity does not appear to exceed a hundredth part of the velocity of light.

Continuing from the preceding result, it is easy to calculate, as various investigators have already done, the projection H_z in any direction PM of the magnetic field produced at point M by an electrified particle O describing a closed orbit of dimensions small compared to distance R, and following a curve whose projection C on a plane normal to PM is alone of consequence since the component of the velocity parallel to PM produces at M a magnetic field perpendicular to PM and whose projection is zero. Let v be the projected velocity. If p is the normal distance from the line of this velocity to point P, the fact that triangles PQM and ZMH are similar gives (see Figure 2):

$$H_z = H \cdot \frac{R^2}{R \sin \alpha}$$

whence, if we replace H by its value $ev \sin \alpha/R^2$, we get

$$H_z = \frac{epv}{R^3}$$

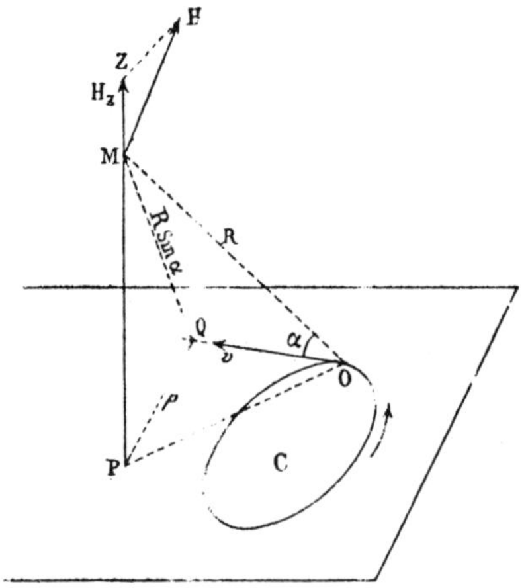

Figure 2. (Reprinted from *J. Phys. et le Radium* **4**:682 (1905); copyright © 1905 by M. J. Berger, Directrice des Publications)

The moment pv of the velocity with respect to PM is equal to twice the area velocity A of the point O in the projected orbit with respect to the point P, taken positively in the trigonometric sense. Then

$$H_z = \frac{2eA}{R^3}$$

This component magnetic field H_z will remain constant in the particular case in which the motion of the electron is due to a central force emanating from a point C interior to the orbit and where the direction MP passes through the point C since the area velocity A will then remain constant, as well as the distance R, assumed large with respect to the dimensions of the orbit.

In all other cases H_z will be variable, and the interesting quantity will be its mean value $\bar{H}_z$ as a function of the time. If the variations of R during the motion of O are of second order with respect to the dimensions of the orbit, implying that R is large compared to p, that is to say that α is always close to a right angle, we can consider R as more or less constant; and if $\bar{A}$ designates the mean value of the area velocity, we have

$$\bar{H}_z = \frac{2e\bar{A}}{R^3}$$

If S represents the surface of the projected orbit and τ is the period of revolution, we have

$$\bar{A} = \frac{S}{\tau}$$

whence

$$\bar{H}_z = \frac{2e}{R^3} \cdot \frac{S}{\tau}$$

The particle current from the point of view of the production of the mean field is then equivalent to an elementary magnet of moment

$$M = \frac{eS}{\tau} = e\bar{A} \qquad (1)$$

It is easy to show that this equivalence holds for points at some distance from the orbit and in any direction whatever, provided that the variations in R are of the order of magnitude of the dimensions of the orbit. Hence the average magnetic field produced by a particle current at a distance is identical to that due to a magnet directed normal to the plane of the orbit and of moment

$$\frac{eS}{\tau} = e\bar{A}$$

A molecule contains similar particle currents due to electrons in motion. The aggregate of equivalent magnets can have a zero resultant magnetic moment. This will be the case if the molecular structure possesses a degree of symmetry sufficiently high to be incompatible with the disymmetry of a magnetic moment, comparable to that of a rotating cylinder. If, on the contrary, the molecule has for its symmetry group a subgroup of that of the rotating cylinder, a resultant nonvanishing moment can exist, and the molecule will be a small magnet.

In every case the creation of an external magnetic field modifies the particle currents, as Weber supposed, in polarizing all the molecules *diamagnetically.* This effect is produced only if the resultant moment is zero: The material is "diamagnetic" in the ordinary sense of the word. If the resultant moment is not zero, then on the initial diamagnetic phenomenon, another phenomenon is superposed, due to the orientation of the molecular magnets by its external field: The substance is then paramagnetic if the mutual actions between the molecular magnets are negligible as is the case of gases or solutions. On the other hand, it is ferromagnetic in the case in which such actions play the essential role. Whenever the paramagnetism appears, it is very large with respect to the initial diamagnetism and masks the latter completely. This explains the absence of continuity between the states of paramagnetism and diamagnetism. Either the substance possesses the dissymmetry of the rotating cylinder or it does not; hence paramagnetism exists or it does not. When it exists, it completely masks the initial diamagnetism.

IV. We can indeed demonstrate that the creation of an external magnetic field which, according to the Hertz-Maxwell theory, is accompanied by the production of an electric field, changes the motion of all the electrons in their orbits in a sense corresponding to diamagnetism. Thereby the magnetic moment of any particle current supposed normal to the direction of the field H is increased by the amount

$$\Delta M = -\frac{He^2S}{4\pi m}$$

if m denotes the mass of the electron.

I shall derive this formula only for the special case of an electron in a circular orbit around a center that attracts according to any law whatever as a function of the distance from the center. The magnetic moment of the magnet equivalent to the current is equal to the product of the charge e of the electron and the area velocity (which remains constant in the absence of any force in addition to the central force) with respect to the center. If ω is the angular velocity of revolution and r is the radius of the orbit, we have

$$M = eA = \frac{e\omega r^2}{2}$$

If the electron is acted on by an external force like the electric force induced by the establishment of a magnetic field H, there will result a change in the area velocity A and consequently a change in M also by virtue of the well-known equation

$$2m\,\frac{dA}{dt} = \text{moment of the external force}$$

If E is the component of the induced electric field along the tangent to the orbit, the moment of the force is Eer. Hence

$$dM = edA = \frac{e^2}{2m} \cdot Erdt$$

The orbital velocity of the electron is ωr so that the element ds of the trajectory covered in time dt is ωrdt, whence

$$dM = \frac{e^2}{2\omega m} \cdot Eds$$

If we suppose that the time of one revolution τ is very short with respect to the duration of the creation of the magnetic field (which is always realized in practice), the change δM during one revolution (ω being supposed constant during this time) becomes

$$\delta M = \frac{e^2}{2\omega m} \int Eds$$

But in accordance with the law of induction just recalled, the integral here, which is the curvilinear integral of the electric field intensity taken along the orbit, is equal to the negative of the derivative of the magnetic flux HS with respect to the time. Thus

$$-\int Eds = \frac{dHS}{dt} = \frac{1}{\tau} \cdot S \cdot \delta H$$

We have here used the relation $\omega = 2\pi/\tau$. The minus sign is, of course, in accordance with Lenz's law. We then get

$$\delta M = -\frac{e^2}{4\pi m} \cdot S \cdot \delta H$$

The total variation ΔM resulting from the establishment of the field H then becomes

$$\Delta M = -\frac{He^2S}{4\pi m}$$

The hypothesis we have made concerning the slowness of the establishment of the field can be avoided if we envisage (that which is in any case necessary in practice) the average effect over a large number of similar orbits oriented in all possible ways.

This diamagnetic modification is independent of the way in which the superposition of the particle current and the external field is realized, whether by creation of the latter by means of a current or by the displacement of the molecule in a preexisting field. The effect is always very weak, for we have

$$\frac{\Delta M}{M} = \frac{H\tau}{4\pi} \cdot \frac{e}{m} < 10^{-9} H$$

since the largest known value of e/m is 1.8×10^7, corresponding to cathode-ray corpuscles. Moreover, τ is of the order of 10^{-15} s. *Hence the most intense fields we know how to produce (always less than 10^5 gauss) do not alter the magnetic moment of the particle currents by more than a ten thousandth part.*

V. The small size of this modification is confirmed by the smallness of the Zeeman effect connected with diamagnetic change. The variation in the magnitude of the particle currents brought about at the moment of establishment of the field corresponds to an increase in the period of the motion of the electrons moving in one direction and a decrease for those moving in the opposite direction, in agreement with the changes associated with the Zeeman effect. The extreme smallness of the latter justifies the preceding statements on the diamagnetic modifications.

Moreover, the general nature of the Zeeman effect confirms the general nature of the diamagnetic effect. Even in the case of iron, which also shows the Zeeman effect, the particle currents are changed in the diamagnetic sense.

It is interesting to show that in the particular case previously discussed of a circular orbit in a central force field $f(r)$, the establishment of the magnetic field does not modify the radius of the orbit but affects only the period of revolution of the electron; at least for magnetic fields weak enough so as to produce only a very small modification. As we have seen this is always realized in practice. In the absense of a magnetic field, the condition of stationary motion is

$$f(r) = m\omega^2 r \tag{2}$$

Let us assume that in the presence of the magnetic field, ω and r are changed by $\Delta\omega$ and Δr, respectively. Taking account of the force $He\omega r$ due to the magnetic field, the new condition for stationary motion can be written

$$f(r + \Delta r) = m(\omega + \Delta\omega)^2 (r + \Delta r) + He\omega r$$

If we keep only terms of the first order, this yields

$$[f'(r) - m\omega^2]\Delta r = 2m\omega r \, \Delta\omega + He\omega r \tag{3}$$

On the other hand, for the diamagnetic modification as discussed previously, we have

$$\Delta A = \Delta\left(\frac{\omega r^2}{2}\right) = -\frac{Her^2}{4\omega} - 4m\omega^2 \Delta r$$
$$= 2m\omega r \, \Delta\omega + He\omega r$$

If we combine this with (2), the result is

$$[f'(r) + 3m\omega^2]\ \Delta r = 0$$

leading to the conclusions

$$\Delta r = 0$$

$$\Delta\omega = -\frac{He}{2m}$$

insofar as the law of force satisfies the specified relation. So we have

$$f'(r) + 3m\omega^2 = 0$$

which, combined with (1), yields

$$\frac{f'(r)}{f(r)} = -\frac{3}{r}$$

or

$$f(r) = \frac{K}{r^3}$$

where K is a constant. Hence if the attractive force that maintains the electron in its orbit varies inversely as the cube of the distance from the center, the diamagnetic modification will not change the dimensions of the orbit but only the period of revolution, in a fashion corresponding to the Zeeman effect.

VI. The particle currents behave as if they were circuits of zero resistance and appropriate self-inductance. For such a circuit of self-inductance, L would be modified by the external field in such a way that the total flux remained constant. Thus

$$\frac{d(Li)}{dt} = -\frac{d(HS)}{dt}$$

whence for the establishment of the field H

$$\Delta(Li) = -HS$$

The circuit under consideration will correspond to a particle current if

$$\Delta M = -\frac{e^2}{4\pi m} \cdot HS = \frac{e^2}{4\pi m} \cdot \Delta(Li)$$

$$L = \frac{4\pi m}{e^2} \cdot \frac{M}{i} = \frac{4\pi^2 r^2 \omega}{e^2}$$

and if the particle orbit is a circle of radius r. The magnetic energy of the circuit will be

$$\frac{Li^2}{2} = \frac{4\pi^2 r^2 m}{2\tau^2} = \frac{mv^2}{2}$$

This says that the particle current will be completely analogous to a circuit of zero resistance if the inertia of the electron producing it is completely electromagnetic and thus if the kinetic energy of the electron is completely magnetic in form.

The diamagnetic modification's being extremely small, the particle currents could be considered as not deformable and of a practically constant magnitude. We shall therefore obtain a complete representation of the facts of paramagnetism and of the exchanges of energy between magnets and currents through the hypothesis that the magnets are constituted of similar circuits to which the laws of ordinary electrodynamics are rigorously applicable.

VII. If the molecules have a zero-resultant moment, no action of the external field will attend the diamagnetic modification, or will there be any change in their motion as a whole. The usual collisions between molecules will take place as before, there will be no change in temperature, and the acquired diamagnetic property alone exists.

In contrast, a change in temperature will not modify this property. All the electrons within the molecule contribute to produce it, and the motion of these electrons does not depend on the temperature, as is shown by the invariability of spectral lines. *We thus obtain the law found experimentally by M. Curie on the lack of dependence of the diamagnetic constant on the temperature.*

If using the experimentally determined diamagnetic constants, we seek to calculate the dimensions of the orbital surfaces like S, we obtain a result in complete agreement with previous estimates of molecular dimensions and in agreement with the hypothesis that the orbits in question are truly intramolecular. Thus the smallness of the diamagnetic effects is definitely linked with the smallness of molecular dimensions.

VIII. If the molecules have a resultant nonvanishing moment, they do at first suffer the diamagnetic modification. Then the external magnetic field tends to turn the magnetic axes parallel to its direction. There results a change in the motion of the ensemble of molecules from their thermal agitation, corresponding to an unequal distribution of kinetic energy among the molecules, those whose magnetic axes are parallel to the field having a greater kinetic energy than the others. This is incompatible with thermal equilibrium and, if a gas is in question, by the action of molecular collisions a rearrangement takes place after which the temperature becomes uniform and the molecular magnets will by preference be directed along the external field. The property of paramagnetism will then appear. If from this rearrangement there results an increase dM in the resultant magnetic moment, the producer of the magnetic field does work on the ensemble of molecules. In the case of a gas, this is represented by an increase in the temperature of the medium and is given by

$$dW = HdM$$

To maintain the gas at constant temperature, it is necessary to take away from it the equivalent amount of heat

$$dQ = HdM = H\left(\frac{\partial M}{\partial H} dH + \frac{\partial M}{\partial T} dT\right)$$

Since dQ/T is an exact differential, we obtain the relation

$$M = f\left(\frac{H}{T}\right) \tag{4}$$

In the special case in which the magnetization is proportional to the field, we have

$$M = \frac{kH}{T}$$

This is Curie's law that the paramagnetic susceptibility is inversely proportional to the absolute temperature.

The following comparison will help make the preceding theory comprehensible. Imagine a mass of gas enclosed in a closed vessel and subject to gravity. The molecules distribute themselves in such a fashion that the density of the gas remains the same at all points in the vessel. In the absence of an external magnetic field, the molecules of a magnetic gas like oxygen have their axes distributed uniformly in all directions.

Due to the gravitational field, the molecules acquire an acceleration directed toward the base of the vessel, and in the absence of collisions, each molecule will have a larger velocity at the base of the vessel than at the top. But this inequality in velocity is incompatible with thermal equilibrium, and a redistribution must take place, thanks to the collisions. As a consequence of this, we arrive at the distribution corresponding to the variation of barometric pressure with height. The center of gravity is lowered, and to maintain the gas at its initial temperature, it is necessary to remove from it a quantity of heat equivalent to the product of the weight of the gas and the decrease in the center of gravity. By thermodynamic reasoning analogous to that given previously, we can deduce that the lowering of the center of gravity is inversely proportional to the absolute temperature.

Thermodynamic reasoning does not permit us to calculate the function $f(T)$ in expression 3 for the magnetic moment. To proceed further, it is necessary to bring in kinetic considerations and to determine, when thermal equilibrium is reestablished, the distribution of magnetized molecules in the various directions, or in the analogous case of gravity the, variation of density with height corresponding to the formula for the variation of barometric pressure with height.

In a mass of gas at uniform temperature, after the redistribution the molecules distribute themselves in such a way that they are most numerous where the potential energy is least, that is to say, in places nearest the base in the case of gravity. The law of distribution is that given by Boltzmann[2] as the generalization of the law

2. L. Boltzmann, *Theorie des gaz*, first part, trans. Gallotti, p. 127.

of variation of the barometric pressure with height. [*Editor's Note:* For English translation see Ludwig Boltzmann's *Lectures on Gas Theory* (translated by Stephen G. Brush) Univ. of California Press, Berkeley, 1964 p. 141ff.] The ratio of the densities of the gas at two points between which the potential energy of a molecule varies by W is

$$e^{W/rT}$$

where e is the base of Naperian logarithms, T is the absolute temperature, and r the gas constant per molecule for an ideal gas.

[*Editor's Note:* The gas constant per molecule in modern notation is denoted by k and called the "Boltzmann constant."]

This constant is such according to kinetic theory that rT is two-thirds the average kinetic energy for each of the three degrees of freedom of translational motion.

But we know that a molecule like that of oxygen possesses two extra degrees of freedom of rotation, as indicated by the value of C/c [*Editor's Note:* This is the ratio of the specific heats]. To each of these degrees of freedom corresponds an average kinetic energy equal to $rT/2$, or equal to rT for the whole molecule.

The distribution among the different orientations will be determined by the static equilibrium that is established under the influence of the magnetic potential energy—$MH \cos \alpha$—and the energy rT of the thermal agitation, the molecules being preferentially oriented in the direction of lowest potential energy, that is to say, with their magnetic axes lined up in the direction of the field. If we distribute the magnetic axes among the different possible directions, the density per unit solid angle varies from one direction to the other proportionally to

$$\frac{e^{MH \cos \alpha}}{rT}$$

All directions are thus equally probable if M or H are zero.

The number of molecules whose magnetic axes are directed in the solid angle $d\omega$ is consequently

$$dn = Ke^{MH/rT \cdot \cos \alpha} d\omega$$

Let us take for the element $d\omega$ a zone with opening $d\alpha$ around the direction of the field. Thus

$$d\omega = 2\pi \sin \alpha \cdot d\alpha$$

where α varies from zero to π.

If N is the total number of molecules, K is determined. Thus if we place

$$\frac{MH}{rT} = a$$

we have

$$N = 2\pi K \int_0^{\pi} e^{a \cos \alpha} \sin \alpha \, d\alpha$$

$$= 2\pi K \int_{-1}^{+1} e^{ax} \, dx$$

(with $x = \cos \alpha$, $dx = -\sin \alpha \cdot d\alpha$)

and

$$N = \frac{2\pi K}{a} (e^{a} - e^{-a})$$

or

$$K = \frac{Na}{4\pi \sinh a}$$

The total magnetic moment of the N molecules is clearly directed parallel to the field and is equal to the sum of the projections on this direction of the component moments. If the N molecules are in unit volume of the medium, the resultant moment corresponds to intensity of magnetization I, where

$$I = \int M \cos \alpha \, dn = \int_{-1}^{+1} 2\pi M K \alpha e^{ax} \, dx$$

But

$$\int_{-1}^{+1} x e^{ax} dx = 2\left(\frac{\cosh a}{a} - \frac{\sinh a}{a^2}\right)$$

where

$$I = MN\left(\frac{\cosh a}{\sinh a} - \frac{1}{a}\right)$$

For a given number of molecules and a given mass of gas, I is then a function of a alone, that is, of H/T, a result agreeing with that furnished by thermodynamics.

IX. If mutual actions between molecular magnets are possible and if the substance in question is ferromagnetic, the heat released is equivalent only to the work done in a closed cycle, and we then obtain for the heat of hysteresis

$$Q = \int H dM$$

The case of ferromagnetic materials is naturally much more complex than the preceding cases of diamagnetic and paramagnetic materials because of the interaction of the molecular magnets. We recall that the theory of the nature of liquid and solid media in which forces of cohesion play a role is infinitely more difficult and less advanced than that of a gas in which cohesion does not enter in any significant way. The ideas I am presenting here are consequently in some measure the magnetic analogy of the kinetic theory of gases. To proceed further, it will be necessary to study the interactions of magnetized particles. Perhaps the simplest case here is that of crystalline media, on the magnetic properties of which the works of M. Weiss have already shed some light and promise even more useful indications. The same may be said of the results of M. Maurain on electrolytic deposits of iron, in which the directive actions appear to be exerted at considerable distances.

AUTHOR CITATION INDEX

SUBJECT INDEX

About the Editor

ROBERT LINDSAY is the Brownell-Jarvis Professor of Natural Philosophy and Physics at Trinity College in Hartford, Connecticut. He received the Sc.B. degree in physics from Brown University in 1947, following service in World War II as a weather forecaster in the United States Army Air Force. In 1951 he received the Ph.D. degree in physics from Rice University. Prior to coming to Trinity College in 1956, Dr. Lindsay was employed as a physicist in the Thermodynamics Section of the National Bureau of Standards and held an appointment on the physics faculty of Southern Methodist University. At Trinity College, Dr. Lindsay was promoted to professor of physics in 1965 and named to the Brownell-Jarvis chair in 1978.

Dr. Lindsay's principal research has been concerned with the magnetic properties of iron group chalcogenides and rare earth-platinum group ternary hydrides. He has authored or co-authored nine research articles published in physics and chemistry journals. In addition he has contributed sections on physics topics to several handbooks and encyclopedias including the *American Institute of Physics Handbook* (McGraw-Hill, 1974).

Dr. Lindsay is a member of the American Physical Society, the Research Society of America, the Society of Sigma Xi, and the American Association of Physics Teachers.